Student Solutions Manual
for
McKeague's Beginning and Intermediate Algebra
A Combined Text/Workbook

Ross Rueger
College of the Sequoias

THOMSON

BROOKS/COLE

Australia • Canada • Mexico • Singapore • Spain • United Kingdom • United States

Printed in the United States of America
1 2 3 4 5 6 7   07 06 05 04 03

Printer: Victor Graphics, Inc.

ISBN: 0-534-39881-2

For more information about our products, contact us at:
Thomson Learning Academic Resource Center
1-800-423-0563

For permission to use material from this text,
contact us by:
Phone: 1-800-730-2214
Fax: 1-800-731-2215
Web: http://www.thomsonrights.com

Asia
Thomson Learning
5 Shenton Way #01-01
UIC Building
Singapore 068808

Australia/New Zealand
Thomson Learning
102 Dodds Street
Southbank, Victoria 3006
Australia

Canada
Nelson
1120 Birchmount Road
Toronto, Ontario M1K 5G4
Canada

Europe/Middle East/South Africa
Thomson Learning
High Holborn House
50/51 Bedford Row
London WC1R 4LR
United Kingdom

Latin America
Thomson Learning
Seneca, 53
Colonia Polanco
11560 Mexico D.F.
Mexico

Spain/Portugal
Paraninfo
Calle/Magallanes, 25
28015 Madrid, Spain

Contents

Preface

This *Student Solutions Manual* contains complete solutions to all odd-numbered exercises (and all chapter test exercises) of *Beginning and Intermediate Algebra: A Combined Course* by Charles P. McKeague. I have attempted to format solutions for readability and accuracy, and apologize to you for any errors that you may encounter. If you have any comments, suggestions, error corrections, or alternative solutions please feel free to drop me a note or send an email (address below).

Please use this manual with some degree of caution. Be sure that you have attempted a solution, and re-attempted it, before you look it up in this manual. Mathematics can only be learned by *doing*, and not by observing! As you use this manual, do not just read the solution but work it along with the manual, using my solution to check your work. If you use this manual in that fashion then it should be helpful to you in your studying.

I would like to thank a number of people for their assistance in preparing this manual. Thanks go to Rachael Sturgeon and Molly Nance at Brooks/Cole Publishing for their valuable assistance and support. Special thanks go to Matt Bourez of College of the Sequoias for his meticulous error-checking of my solutions, and prompt return of my manuscript under tight deadlines.

I wish to express my appreciation to Pat McKeague for asking me to be involved with this textbook. This book provides a complete course in beginning and intermediate algebra, and you will find the text very easy to read and understand. Good luck!

Ross Rueger
College of the Sequoias
matmanross@aol.com

December, 2002

Chapter 1
The Basics

1.1 Notation and Symbols

1. The equivalent expression is $x + 5 = 14$.

3. The equivalent expression is $5y < 30$.

5. The equivalent expression is $3y \leq y + 6$.

7. The equivalent expression is $\dfrac{x}{3} = x + 2$.

9. Expanding the expression: $3^2 = 3 \cdot 3 = 9$

11. Expanding the expression: $7^2 = 7 \cdot 7 = 49$

13. Expanding the expression: $2^3 = 2 \cdot 2 \cdot 2 = 8$

15. Expanding the expression: $4^3 = 4 \cdot 4 \cdot 4 = 64$

17. Expanding the expression: $2^4 = 2 \cdot 2 \cdot 2 \cdot 2 = 16$

19. Expanding the expression: $10^2 = 10 \cdot 10 = 100$

21. Expanding the expression: $11^2 = 11 \cdot 11 = 121$

23. Using the order of operations: $2 \cdot 3 + 5 = 6 + 5 = 11$

25. Using the order of operations: $2(3 + 5) = 2(8) = 16$

27. Using the order of operations: $5 + 2 \cdot 6 = 5 + 12 = 17$

29. Using the order of operations: $(5 + 2) \cdot 6 = 7 \cdot 6 = 42$

31. Using the order of operations: $5 \cdot 4 + 5 \cdot 2 = 20 + 10 = 30$

33. Using the order of operations: $5(4 + 2) = 5(6) = 30$

35. Using the order of operations: $8 + 2(5 + 3) = 8 + 2(8) = 8 + 16 = 24$

37. Using the order of operations: $(8 + 2)(5 + 3) = (10)(8) = 80$

39. Using the order of operations: $20 + 2(8 - 5) + 1 = 20 + 2(3) + 1 = 20 + 6 + 1 = 27$

41. Using the order of operations: $5 + 2(3 \cdot 4 - 1) + 8 = 5 + 2(12 - 1) + 8 = 5 + 2(11) + 8 = 5 + 22 + 8 = 35$

43. Using the order of operations: $8 + 10 \div 2 = 8 + 5 = 13$

45. Using the order of operations: $4 + 8 \div 4 - 2 = 4 + 2 - 2 = 4$

47. Using the order of operations: $3 + 12 \div 3 + 6 \cdot 5 = 3 + 4 + 30 = 37$

49. Using the order of operations: $3 \cdot 8 + 10 \div 2 + 4 \cdot 2 = 24 + 5 + 8 = 37$

51. Using the order of operations: $(5 + 3)(5 - 3) = (8)(2) = 16$

53. Using the order of operations: $5^2 - 3^2 = 5 \cdot 5 - 3 \cdot 3 = 25 - 9 = 16$

55. Using the order of operations: $(4 + 5)^2 = 9^2 = 9 \cdot 9 = 81$

57. Using the order of operations: $4^2 + 5^2 = 4 \cdot 4 + 5 \cdot 5 = 16 + 25 = 41$

59. Using the order of operations: $3 \cdot 10^2 + 4 \cdot 10 + 5 = 300 + 40 + 5 = 345$

61. Using the order of operations: $2 \cdot 10^3 + 3 \cdot 10^2 + 4 \cdot 10 + 5 = 2000 + 300 + 40 + 5 = 2345$

63. Using the order of operations: $10 - 2(4 \cdot 5 - 16) = 10 - 2(20 - 16) = 10 - 2(4) = 10 - 8 = 2$

65. Using the order of operations: $4[7 + 3(2 \cdot 9 - 8)] = 4[7 + 3(18 - 8)] = 4[7 + 3(10)] = 4(7 + 30) = 4(37) = 148$

67. Using the order of operations: $5(7 - 3) + 8(6 - 4) = 5(4) + 8(2) = 20 + 16 = 36$

69. Using the order of operations: $3(4 \cdot 5 - 12) + 6(7 \cdot 6 - 40) = 3(20 - 12) + 6(42 - 40) = 3(8) + 6(2) = 24 + 12 = 36$

71. Using the order of operations: $3^4 + 4^2 \div 2^3 - 5^2 = 81 + 16 \div 8 - 25 = 81 + 2 - 25 = 58$

73. Using the order of operations: $5^2 + 3^4 \div 9^2 + 6^2 = 25 + 81 \div 81 + 36 = 25 + 1 + 36 = 62$

75. There are $5 \cdot 2 = 10$ cookies in the package.

77. The total number of calories is $210 \cdot 2 = 420$ calories.

79. There are $7 \cdot 32 = 224$ chips in the bag.

81. For a person eating 3,000 calories per day, the recommended amount of fat would be $80 + 15 = 95$ grams.

83. **a.**   The amount of caffeine is: $6(100) = 600$ mg

   b.   The amount of caffeine is: $2(45) + 3(47) = 90 + 141 = 231$ mg

85. Completing the table:

Activity	Calories burned in 1 hour
Bicycling	374
Bowling	265
Handball	680
Jogging	680
Skiing	544

87. The next number is 5.

89. The next number is 10.

91. The next number is $5^2 = 25$.

93. Since $2 + 2 = 4$ and $2 + 4 = 6$, the next number is $4 + 6 = 10$.

1.2 Real Numbers

1. Labeling the point:

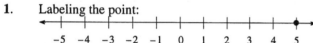

3. Labeling the point:

5. Labeling the point:

7. Labeling the point:

9. Building the fraction: $\frac{3}{4} = \frac{3}{4} \cdot \frac{6}{6} = \frac{18}{24}$

11. Building the fraction: $\frac{1}{2} = \frac{1}{2} \cdot \frac{12}{12} = \frac{12}{24}$

13. Building the fraction: $\frac{5}{8} = \frac{5}{8} \cdot \frac{3}{3} = \frac{15}{24}$

15. Building the fraction: $\frac{3}{5} = \frac{3}{5} \cdot \frac{12}{12} = \frac{36}{60}$

17. Building the fraction: $\frac{11}{30} = \frac{11}{30} \cdot \frac{2}{2} = \frac{22}{60}$

19. The opposite of 10 is -10, the reciprocal is $\frac{1}{10}$, and the absolute value is $|10| = 10$.

21. The opposite of $\frac{3}{4}$ is $-\frac{3}{4}$, the reciprocal is $\frac{4}{3}$, and the absolute value is $\left|\frac{3}{4}\right| = \frac{3}{4}$.

23. The opposite of $\frac{11}{2}$ is $-\frac{11}{2}$, the reciprocal is $\frac{2}{11}$, and the absolute value is $\left|\frac{11}{2}\right| = \frac{11}{2}$.

25. The opposite of -3 is 3, the reciprocal is $-\frac{1}{3}$, and the absolute value is $|-3| = 3$.

27. The opposite of $-\frac{2}{5}$ is $\frac{2}{5}$, the reciprocal is $-\frac{5}{2}$, and the absolute value is $\left|-\frac{2}{5}\right| = \frac{2}{5}$.

29. The opposite of x is $-x$, the reciprocal is $\dfrac{1}{x}$, and the absolute value is $|x|$.

31. The correct symbol is <: $-5 < -3$ **33.**   The correct symbol is >: $-3 > -7$

35. Since $|-4| = 4$ and $-|-4| = -4$, the correct symbol is >: $|-4| > -|-4|$

37. Since $-|-7| = -7$, the correct symbol is >: $7 > -|-7|$

39. The correct symbol is <: $-\frac{3}{4} < -\frac{1}{4}$ **41.**   The correct symbol is <: $-\frac{3}{2} < -\frac{3}{4}$

43. Simplifying the expression: $|8 - 2| = |6| = 6$

45. Simplifying the expression: $\left|5 \cdot 2^3 - 2 \cdot 3^2\right| = |5 \cdot 8 - 2 \cdot 9| = |40 - 18| = |22| = 22$

47. Simplifying the expression: $|7 - 2| - |4 - 2| = |5| - |2| = 5 - 2 = 3$

49. Simplifying the expression: $10 - |7 - 2(5 - 3)| = 10 - |7 - 2(2)| = 10 - |7 - 4| = 10 - |3| = 10 - 3 = 7$

51. Simplifying the expression:
$$\begin{aligned}
15 - |8 - 2(3 \cdot 4 - 9)| - 10 &= 15 - |8 - 2(12 - 9)| - 10 \\
&= 15 - |8 - 2(3)| - 10 \\
&= 15 - |8 - 6| - 10 \\
&= 15 - |2| - 10 \\
&= 15 - 2 - 10 \\
&= 3
\end{aligned}$$

53. Multiplying the fractions: $\frac{2}{3} \cdot \frac{4}{5} = \frac{8}{15}$

55. Multiplying the fractions: $\frac{1}{2}(3) = \frac{1}{2} \cdot \frac{3}{1} = \frac{3}{2}$

57. Multiplying the fractions: $\frac{1}{4}(5) = \frac{1}{4} \cdot \frac{5}{1} = \frac{5}{4}$

59. Multiplying the fractions: $\frac{4}{3} \cdot \frac{3}{4} = \frac{12}{12} = 1$

61. Multiplying the fractions: $6\left(\frac{1}{6}\right) = \frac{6}{1} \cdot \frac{1}{6} = \frac{6}{6} = 1$

63. Multiplying the fractions: $3 \cdot \frac{1}{3} = \frac{3}{1} \cdot \frac{1}{3} = \frac{3}{3} = 1$

65. Expanding the exponent: $\left(\frac{3}{4}\right)^2 = \frac{3}{4} \cdot \frac{3}{4} = \frac{9}{16}$

67. Expanding the exponent: $\left(\frac{2}{3}\right)^3 = \frac{2}{3} \cdot \frac{2}{3} \cdot \frac{2}{3} = \frac{8}{27}$

69. Expanding the exponent: $\left(\frac{1}{10}\right)^4 = \frac{1}{10} \cdot \frac{1}{10} \cdot \frac{1}{10} \cdot \frac{1}{10} = \frac{1}{10,000}$

71. The next number is $\frac{1}{9}$.

73. The next number is $\frac{1}{5^2} = \frac{1}{25}$.

75. The perimeter is $4(1 \text{ in.}) = 4 \text{ in.}$, and the area is $(1 \text{ in.})^2 = 1 \text{ in.}^2$.

77. The perimeter is $2(1.5 \text{ in.}) + 2(0.75 \text{ in.}) = 3.0 \text{ in.} + 1.5 \text{ in.} = 4.5 \text{ in.}$, and the area is $(1.5 \text{ in.})(0.75 \text{ in.}) = 1.125 \text{ in.}^2$.

79. The perimeter is $2.75 \text{ cm} + 4 \text{ cm} + 3.5 \text{ cm} = 10.25 \text{ cm}$, and the area is $\frac{1}{2}(4 \text{ cm})(2.5 \text{ cm}) = 5.0 \text{ cm}^2$.

81. A loss of 8 yards corresponds to –8 on a number line. The total yards gained corresponds to –2 yards.

83. The temperature can be represented as –64°. The new (warmer) temperature corresponds to –54°.

85. The wind chill temperature is –15°.

87. Completing the table:

Month	Temperature (°F)
January	–36
February	–30
March	–14
April	–2
May	19
June	22
July	35
August	30
September	19
October	15
November	–11
December	–26

89. His position corresponds to –100 feet. His new (deeper) position corresponds to –105 feet.

91. The area is given by: $8\frac{1}{2} \cdot 11 = \frac{17}{2} \cdot \frac{11}{1} = \frac{187}{2} = 93.5 \text{ in.}^2$

The perimeter is given by: $2\left(8\frac{1}{2}\right) + 2(11) = 17 + 22 = 39 \text{ in.}$

93. The calories consumed would be: $2(544) + 299 = 1,387$ calories

95. The calories consumed by the 180 lb person would be $3(653) = 1,959$ calories, while the calories consumed by the 120 lb person would be $3(435) = 1,305$ calories. Thus the 180 lb person consumed $1959 - 1306 = 654$ more calories.

1.3 Addition of Real Numbers

1. Adding all positive and negative combinations of 3 and 5:
$3 + 5 = 8$
$3 + (-5) = -2$
$-3 + 5 = 2$
$(-3) + (-5) = -8$

3. Adding all positive and negative combinations of 15 and 20:
$15 + 20 = 35$
$15 + (-20) = -5$
$-15 + 20 = 5$
$(-15) + (-20) = -35$

5. Adding the numbers: $6 + (-3) = 3$

7. Adding the numbers: $13 + (-20) = -7$

9. Adding the numbers: $18 + (-32) = -14$

11. Adding the numbers: $-6 + 3 = -3$

13. Adding the numbers: $-30 + 5 = -25$

15. Adding the numbers: $-6 + (-6) = -12$

17. Adding the numbers: $-9 + (-10) = -19$

19. Adding the numbers: $-10 + (-15) = -25$

21. Performing the additions: $5 + (-6) + (-7) = 5 + (-13) = -8$

23. Performing the additions: $-7 + 8 + (-5) = -12 + 8 = -4$

25. Performing the additions: $5 + [6 + (-2)] + (-3) = 5 + 4 + (-3) = 9 + (-3) = 6$

27. Performing the additions: $[6 + (-2)] + [3 + (-1)] = 4 + 2 = 6$

29. Performing the additions: $20 + (-6) + [3 + (-9)] = 20 + (-6) + (-6) = 20 + (-12) = 8$

31. Performing the additions: $-3 + (-2) + [5 + (-4)] = -3 + (-2) + 1 = -5 + 1 = -4$

33. Performing the additions: $(-9 + 2) + [5 + (-8)] + (-4) = -7 + (-3) + (-4) = -14$

35. Performing the additions: $[-6 + (-4)] + [7 + (-5)] + (-9) = -10 + 2 + (-9) = -19 + 2 = -17$

37. Performing the additions: $(-6 + 9) + (-5) + (-4 + 3) + 7 = 3 + (-5) + (-1) + 7 = 10 + (-6) = 4$

39. Using order of operations: $-5 + 2(-3 + 7) = -5 + 2(4) = -5 + 8 = 3$

41. Using order of operations: $9 + 3(-8 + 10) = 9 + 3(2) = 9 + 6 = 15$

43. Using order of operations: $-10 + 2(-6 + 8) + (-2) = -10 + 2(2) + (-2) = -10 + 4 + (-2) = -12 + 4 = -8$

45. Using order of operations: $2(-4 + 7) + 3(-6 + 8) = 2(3) + 3(2) = 6 + 6 = 12$

47. The pattern is to add 5, so the next two terms are $18 + 5 = 23$ and $23 + 5 = 28$.

49. The pattern is to add 5, so the next two terms are $25 + 5 = 30$ and $30 + 5 = 35$.

51. The pattern is to add –5, so the next two terms are $5 + (-5) = 0$ and $0 + (-5) = -5$.

53. The pattern is to add –6, so the next two terms are $-6 + (-6) = -12$ and $-12 + (-6) = -18$.

55. The pattern is to add –4, so the next two terms are $0 + (-4) = -4$ and $-4 + (-4) = -8$.

57. Yes, since each successive odd number is 2 added to the previous one.

59. The expression is: $5 + 9 = 14$

61. The expression is: $[-7 + (-5)] + 4 = -12 + 4 = -8$

63. The expression is: $[-2 + (-3)] + 10 = -5 + 10 = 5$

65. The number is 3, since $-8 + 3 = -5$.

67. The number is –3, since $-6 + (-3) = -9$.

69. The expression is $-12° + 4° = -8°$.

71. The expression is $\$10 + (-\$6) + (-\$8) = \$10 + (-\$14) = -\$4$.

73. The new balance is $-\$30 + \$40 = \$10$.

75. **a.** Completing the table:

Professions	Salaries
Engineering	$42,862
Computer Science	$40,920
Math / Statistics	$40,523
Chemistry	$36,036
Business Administration	$34,831
Accounting	$33,702
Sales / Marketing	$33,252
Teaching	$25,735

b. The average family income is: $\$36,036 + \$42,862 = \$78,898$

1.4 Subtraction of Real Numbers

1. Subtracting the numbers: $5 - 8 = 5 + (-8) = -3$

3. Subtracting the numbers: $3 - 9 = 3 + (-9) = -6$

5. Subtracting the numbers: $5 - 5 = 5 + (-5) = 0$

7. Subtracting the numbers: $-8 - 2 = -8 + (-2) = -10$

9. Subtracting the numbers: $-4 - 12 = -4 + (-12) = -16$

11. Subtracting the numbers: $-6 - 6 = -6 + (-6) = -12$

13. Subtracting the numbers: $-8 - (-1) = -8 + 1 = -7$

15. Subtracting the numbers: $15 - (-20) = 15 + 20 = 35$

17. Subtracting the numbers: $-4 - (-4) = -4 + 4 = 0$

19. Using order of operations: $3 - 2 - 5 = 3 + (-2) + (-5) = 3 + (-7) = -4$

21. Using order of operations: $9 - 2 - 3 = 9 + (-2) + (-3) = 9 + (-5) = 4$

23. Using order of operations: $-6 - 8 - 10 = -6 + (-8) + (-10) = -24$

25. Using order of operations: $-22 + 4 - 10 = -22 + 4 + (-10) = -32 + 4 = -28$

27. Using order of operations: $10 - (-20) - 5 = 10 + 20 + (-5) = 30 + (-5) = 25$

29. Using order of operations: $8 - (2 - 3) - 5 = 8 - (-1) - 5 = 8 + 1 + (-5) = 9 + (-5) = 4$

31. Using order of operations: $7 - (3 - 9) - 6 = 7 - (-6) - 6 = 7 + 6 + (-6) = 13 + (-6) = 7$

33. Using order of operations: $5 - (-8 - 6) - 2 = 5 - (-14) - 2 = 5 + 14 + (-2) = 19 + (-2) = 17$

35. Using order of operations: $-(5 - 7) - (2 - 8) = -(-2) - (-6) = 2 + 6 = 8$

37. Using order of operations: $-(3 - 10) - (6 - 3) = -(-7) - 3 = 7 + (-3) = 4$

39. Using order of operations: $16 - \left[(4 - 5) - 1\right] = 16 - (-1 - 1) = 16 - (-2) = 16 + 2 = 18$

41. Using order of operations: $5 - \left[(2 - 3) - 4\right] = 5 - (-1 - 4) = 5 - (-5) = 5 + 5 = 10$

43. Using order of operations:

$$21 - \left[-(3 - 4) - 2\right] - 5 = 21 - \left[-(-1) - 2\right] - 5 = 21 - (1 - 2) - 5 = 21 - (-1) - 5 = 21 + 1 + (-5) = 22 + (-5) = 17$$

45. Using order of operations: $2 \bullet 8 - 3 \bullet 5 = 16 - 15 = 16 + (-15) = 1$

47. Using order of operations: $3 \bullet 5 - 2 \bullet 7 = 15 - 14 = 15 + (-14) = 1$

49. Using order of operations: $5 \bullet 9 - 2 \bullet 3 - 6 \bullet 2 = 45 - 6 - 12 = 45 + (-6) + (-12) = 45 + (-18) = 27$

51. Using order of operations: $3 \bullet 8 - 2 \bullet 4 - 6 \bullet 7 = 24 - 8 - 42 = 24 + (-8) + (-42) = 24 + (-50) = -26$

53. Using order of operations: $2 \bullet 3^2 - 5 \bullet 2^2 = 2 \bullet 9 - 5 \bullet 4 = 18 - 20 = 18 + (-20) = -2$

55. Using order of operations: $4 \bullet 3^3 - 5 \bullet 2^3 = 4 \bullet 27 - 5 \bullet 8 = 108 - 40 = 108 + (-40) = 68$

57. Writing the expression: $-7 - 4 = -7 + (-4) = -11$

59. Writing the expression: $12 - (-8) = 12 + 8 = 20$

61. Writing the expression: $-5 - (-7) = -5 + 7 = 2$

63. Writing the expression: $\left[4 + (-5)\right] - 17 = -1 - 17 = -1 + (-17) = -18$

65. Writing the expression: $8 - 5 = 8 + (-5) = 3$

67. Writing the expression: $-8 - 5 = -8 + (-5) = -13$

69. Writing the expression: $8 - (-5) = 8 + 5 = 13$

71. The number is 10, since $8 - 10 = 8 + (-10) = -2$.

73. The number is -2, since $8 - (-2) = 8 + 2 = 10$.

75. The expression is $\$1,500 - \$730 = \$770$.

77. The expression is $-\$35 + \$15 - \$20 = -\$35 + (-\$20) + \$15 = -\$55 + \$15 = -\$40$.

79. The expression is $\$98 - \$65 - \$53 = \$98 + (-\$65) + (-\$53) = \$98 + (-\$118) = -\$20$.

81. The sequence of values is $4500, $3950, $3400, $2850, and $2300. This is an arithmetic sequence, since $-\$550$ is added to each value to obtain the new value.

83. The difference is $1000 \text{ feet} - 231 \text{ feet} = 769 \text{ feet}$.

85. He is 439 feet from the starting line.

87. 2 seconds have gone by.

89. The angles add to $90°$, so $x = 90° - 55° = 35°$.

91. The angles add to $180°$, so $x = 180° - 120° = 60°$.

93. **a.** Completing the table:

Year	Garbage (Millions of tons)
1960	88
1970	121
1980	152
1990	205
1997	217

   b. Subtracting: $205 - 121 = 84$. There was 84 million tons more garbage in 1990 than in 1970.

95. **a.** Completing the table:

Year	Cost (cents / minute)
1998	33
1999	28
2000	25
2001	23
2002	22
2003	20

   b. The difference is cost is $33 - 28 = 5$ cents/minute.

1.5 Properties of Real Numbers

1. commutative property (of addition)

3. multiplicative inverse property

5. commutative property (of addition)

7. distributive property

9. commutative and associative properties (of addition)

11. commutative and associative properties (of addition)

13. commutative property (of addition)

15. commutative and associative properties (of multiplication)

17. commutative property (of multiplication)

19. additive inverse property

21. The expression should read $3(x + 2) = 3x + 6$.

23. The expression should read $9(a + b) = 9a + 9b$.

25. The expression should read $3(0) = 0$.

27. The expression should read $3 + (-3) = 0$.

29. The expression should read $10(1) = 10$.

31. Simplifying the expression: $4 + (2 + x) = (4 + 2) + x = 6 + x$

33. Simplifying the expression: $(x + 2) + 7 = x + (2 + 7) = x + 9$

35. Simplifying the expression: $3(5x) = (3 \cdot 5)x = 15x$

37. Simplifying the expression: $9(6y) = (9 \cdot 6)y = 54y$

39. Simplifying the expression: $\frac{1}{2}(3a) = \left(\frac{1}{2} \cdot 3\right)a = \frac{3}{2}a$

41. Simplifying the expression: $\frac{1}{3}(3x) = \left(\frac{1}{3} \cdot 3\right)x = 1x = x$

43. Simplifying the expression: $\frac{1}{2}(2y) = \left(\frac{1}{2} \cdot 2\right)y = 1y = y$

45. Simplifying the expression: $\frac{3}{4}\left(\frac{4}{3}x\right) = \left(\frac{3}{4} \cdot \frac{4}{3}\right)x = 1x = x$

47. Simplifying the expression: $\frac{6}{5}\left(\frac{5}{6}a\right) = \left(\frac{6}{5} \cdot \frac{5}{6}\right)a = 1a = a$

49. Applying the distributive property: $8(x + 2) = 8 \cdot x + 8 \cdot 2 = 8x + 16$

51. Applying the distributive property: $8(x - 2) = 8 \cdot x - 8 \cdot 2 = 8x - 16$

53. Applying the distributive property: $4(y + 1) = 4 \cdot y + 4 \cdot 1 = 4y + 4$

55. Applying the distributive property: $3(6x + 5) = 3 \cdot 6x + 3 \cdot 5 = 18x + 15$

57. Applying the distributive property: $2(3a + 7) = 2 \cdot 3a + 2 \cdot 7 = 6a + 14$

59. Applying the distributive property: $9(6y - 8) = 9 \cdot 6y - 9 \cdot 8 = 54y - 72$

61. Applying the distributive property: $\frac{1}{2}(3x - 6) = \frac{1}{2} \cdot 3x - \frac{1}{2} \cdot 6 = \frac{3}{2}x - 3$

63. Applying the distributive property: $\frac{1}{3}(3x+6) = \frac{1}{3} \cdot 3x + \frac{1}{3} \cdot 6 = x+2$

65. Applying the distributive property: $3(x+y) = 3x+3y$

67. Applying the distributive property: $8(a-b) = 8a-8b$

69. Applying the distributive property: $6(2x+3y) = 6 \cdot 2x + 6 \cdot 3y = 12x+18y$

71. Applying the distributive property: $4(3a-2b) = 4 \cdot 3a - 4 \cdot 2b = 12a-8b$

73. Applying the distributive property: $\frac{1}{2}(6x+4y) = \frac{1}{2} \cdot 6x + \frac{1}{2} \cdot 4y = 3x+2y$

75. Applying the distributive property: $4(a+4)+9 = 4a+16+9 = 4a+25$

77. Applying the distributive property: $2(3x+5)+2 = 6x+10+2 = 6x+12$

79. Applying the distributive property: $7(2x+4)+10 = 14x+28+10 = 14x+38$

81. No. The man cannot reverse the order of putting on his socks and putting on his shoes.

83. No. The skydiver must jump out of the plane before pulling the rip cord.

85. Division is not a commutative operation. For example, $8 \div 4 = 2$ while $4 \div 8 = \frac{1}{2}$.

87. Computing the yearly take-home pay:
$$12(2400-480) = 12(1920) = \$23,040 \qquad 12 \cdot 2400 - 12 \cdot 480 = 28,800 - 5,760 = \$23,040$$

89. Rewriting the formula: $P = 2w+2l = 2(w+l)$

1.6 Multiplication of Real Numbers

1. Finding the product: $7(-6) = -42$

3. Finding the product: $-8(2) = -16$

5. Finding the product: $-3(-1) = 3$

7. Finding the product: $-11(-11) = 121$

9. Using order of operations: $-3(2)(-1) = 6$

11. Using order of operations: $-3(-4)(-5) = -60$

13. Using order of operations: $-2(-4)(-3)(-1) = 24$

15. Using order of operations: $(-7)^2 = (-7)(-7) = 49$

17. Using order of operations: $(-3)^3 = (-3)(-3)(-3) = -27$

19. Using order of operations: $-2(2-5) = -2(-3) = 6$

21. Using order of operations: $-5(8-10) = -5(-2) = 10$

23. Using order of operations: $(4-7)(6-9) = (-3)(-3) = 9$

25. Using order of operations: $(-3-2)(-5-4) = (-5)(-9) = 45$

27. Using order of operations: $-3(-6)+4(-1) = 18+(-4) = 14$

29. Using order of operations: $2(3)-3(-4)+4(-5) = 6+12+(-20) = 18+(-20) = -2$

31. Using order of operations: $4(-3)^2 + 5(-6)^2 = 4(9)+5(36) = 36+180 = 216$

33. Using order of operations: $7(-2)^3 - 2(-3)^3 = 7(-8)-2(-27) = -56+54 = -2$

35. Using order of operations: $6-4(8-2) = 6-4(6) = 6-24 = 6+(-24) = -18$

37. Using order of operations: $9-4(3-8) = 9-4(-5) = 9+20 = 29$

39. Using order of operations: $-4(3-8)-6(2-5) = -4(-5)-6(-3) = 20+18 = 38$

41. Using order of operations: $7-2[-6-4(-3)] = 7-2(-6+12) = 7-2(6) = 7-12 = 7+(-12) = -5$

43. Using order of operations:
$$7-3[2(-4-4)-3(-1-1)] = 7-3[2(-8)-3(-2)] = 7-3(-16+6) = 7-3(-10) = 7+30 = 37$$

45. Using order of operations:
$$8-6[-2(-3-1)+4(-2-3)] = 8-6[-2(-4)+4(-5)] = 8-6[8+(-20)] = 8-6(-12) = 8+72 = 80$$

47. Multiplying the fractions: $-\frac{2}{3} \cdot \frac{5}{7} = -\frac{2 \cdot 5}{3 \cdot 7} = -\frac{10}{21}$

49. Multiplying the fractions: $-8\left(\frac{1}{2}\right) = -\frac{8}{1} \cdot \frac{1}{2} = -\frac{8}{2} = -4$

51. Multiplying the fractions: $-\frac{3}{4}\left(-\frac{4}{3}\right) = -\frac{3}{4} \cdot \left(-\frac{4}{3}\right) = \frac{12}{12} = 1$

53. Multiplying the fractions: $\left(-\frac{3}{4}\right)^2 = \left(-\frac{3}{4}\right)\left(-\frac{3}{4}\right) = \frac{9}{16}$

55. Multiplying the expressions: $-2(4x) = (-2 \cdot 4)x = -8x$

57. Multiplying the expressions: $-7(-6x) = [-7 \cdot (-6)]x = 42x$

59. Multiplying the expressions: $-\frac{1}{3}(-3x) = \left[-\frac{1}{3} \cdot (-3)\right]x = 1x = x$

61. Simplifying the expression: $-4(a+2) = -4a + (-4)(2) = -4a-8$

63.   Simplifying the expression: $-\frac{1}{2}(3x-6)=-\frac{3}{2}x-\frac{1}{2}(-6)=-\frac{3}{2}x+3$

65.   Simplifying the expression: $-3(2x-5)-7=-6x+15-7=-6x+8$

67.   Simplifying the expression: $-5(3x+4)-10=-15x-20-10=-15x-30$

69.   Writing the expression: $3(-10)+5=-30+5=-25$ **71.**   Writing the expression: $2(-4x)=-8x$

73.   Writing the expression: $-9\cdot 2-8=-18+(-8)=-26$

75.   The pattern is to multiply by 2, so the next number is $4\cdot 2=8$.

77.   The pattern is to multiply by –2, so the next number is $40\cdot(-2)=-80$.

79.   The pattern is to multiply by $\frac{1}{2}$, so the next number is $\frac{1}{4}\cdot\frac{1}{2}=\frac{1}{8}$.

81.   The pattern is to multiply by –2, so the next number is $12\cdot(-2)=-24$.

83.   The amount lost is: $20(\$3)=\$60$

85.   The temperature is: $25°-4(6°)=25°-24°=1°$

87.   The sequence of values is \$500, \$1000, \$2000, \$4000, \$8000, and \$16000. Yes, this is a geometric sequence, since each value is 2 times the preceding value.

89.   The net change in calories is: $2(630)-3(265)=1260-795=465$ calories

1.7 Division of Real Numbers

1.   Finding the quotient: $\frac{8}{-4}=-2$

3.   Finding the quotient: $\frac{-48}{16}=-3$

5.   Finding the quotient: $\frac{-7}{21}=-\frac{1}{3}$

7.   Finding the quotient: $\frac{-39}{-13}=3$

9.   Finding the quotient: $\frac{-6}{-42}=\frac{1}{7}$

11.   Finding the quotient: $\frac{0}{-32}=0$

13.   Performing the operations: $-3+12=9$

15.   Performing the operations: $-3-12=-3+(-12)=-15$

17.   Performing the operations: $-3(12)=-36$

19.   Performing the operations: $-3\div 12=\frac{-3}{12}=-\frac{1}{4}$

21.   Dividing and reducing: $\frac{4}{5}\div\frac{3}{4}=\frac{4}{5}\cdot\frac{4}{3}=\frac{16}{15}$

23.   Dividing and reducing: $-\frac{5}{6}\div\left(-\frac{5}{8}\right)=-\frac{5}{6}\cdot\left(-\frac{8}{5}\right)=\frac{40}{30}=\frac{4}{3}$

25.   Dividing and reducing: $\frac{10}{13}\div\left(-\frac{5}{4}\right)=\frac{10}{13}\cdot\left(-\frac{4}{5}\right)=-\frac{40}{65}=-\frac{8}{13}$

27.   Dividing and reducing: $-\frac{5}{6}\div\frac{5}{6}=-\frac{5}{6}\cdot\frac{6}{5}=-\frac{30}{30}=-1$

29.   Dividing and reducing: $-\frac{3}{4}\div\left(-\frac{3}{4}\right)=-\frac{3}{4}\cdot\left(-\frac{4}{3}\right)=\frac{12}{12}=1$

31.   Using order of operations: $\frac{3(-2)}{-10}=\frac{-6}{-10}=\frac{3}{5}$

33.   Using order of operations: $\frac{-5(-5)}{-15}=\frac{25}{-15}=-\frac{5}{3}$

35.   Using order of operations: $\frac{-8(-7)}{-28}=\frac{56}{-28}=-2$

37.   Using order of operations: $\frac{27}{4-13}=\frac{27}{-9}=-3$

39.   Using order of operations: $\frac{20-6}{5-5}=\frac{14}{0}=\text{undefined}$

41.   Using order of operations: $\frac{-3+9}{2\cdot 5-10}=\frac{6}{10-10}=\frac{6}{0}=\text{undefined}$

43.   Using order of operations: $\frac{15(-5)-25}{2(-10)}=\frac{-75-25}{-20}=\frac{-100}{-20}=5$

45.   Using order of operations: $\frac{27-2(-4)}{-3(5)}=\frac{27+8}{-15}=\frac{35}{-15}=-\frac{7}{3}$

47.   Using order of operations: $\frac{12-6(-2)}{12(-2)}=\frac{12+12}{-24}=\frac{24}{-24}=-1$

49.   Using order of operations: $\frac{5^2-2^2}{-5+2}=\frac{25-4}{-3}=\frac{21}{-3}=-7$

51. Using order of operations: $\dfrac{8^2 - 2^2}{8^2 + 2^2} = \dfrac{64 - 4}{64 + 4} = \dfrac{60}{68} = \dfrac{15}{17}$

53. Using order of operations: $\dfrac{(5+3)^2}{-5^2 - 3^2} = \dfrac{8^2}{-25 - 9} = \dfrac{64}{-34} = -\dfrac{32}{17}$

55. Using order of operations: $\dfrac{(8-4)^2}{8^2 - 4^2} = \dfrac{4^2}{64 - 16} = \dfrac{16}{48} = \dfrac{1}{3}$

57. Using order of operations: $\dfrac{-4 \cdot 3^2 - 5 \cdot 2^2}{-8(7)} = \dfrac{-4 \cdot 9 - 5 \cdot 4}{-56} = \dfrac{-36 - 20}{-56} = \dfrac{-56}{-56} = 1$

59. Using order of operations: $\dfrac{3 \cdot 10^2 + 4 \cdot 10 + 5}{345} = \dfrac{300 + 40 + 5}{345} = \dfrac{345}{345} = 1$

61. Using order of operations: $\dfrac{7 - [(2-3) - 4]}{-1 - 2 - 3} = \dfrac{7 - (-1 - 4)}{-6} = \dfrac{7 - (-5)}{-6} = \dfrac{7 + 5}{-6} = \dfrac{12}{-6} = -2$

63. Using order of operations: $\dfrac{6(-4) - 2(5-8)}{-6 - 3 - 5} = \dfrac{-24 - 2(-3)}{-14} = \dfrac{-24 + 6}{-14} = \dfrac{-18}{-14} = \dfrac{9}{7}$

65. Using order of operations: $\dfrac{3(-5 - 3) + 4(7 - 9)}{5(-2) + 3(-4)} = \dfrac{3(-8) + 4(-2)}{-10 + (-12)} = \dfrac{-24 + (-8)}{-22} = \dfrac{-32}{-22} = \dfrac{16}{11}$

67. Using order of operations: $\dfrac{|3 - 9|}{3 - 9} = \dfrac{|-6|}{-6} = \dfrac{6}{-6} = -1$ **69.** The quotient is $\dfrac{-12}{-4} = 3$.

71. The number is -10, since $\dfrac{-10}{-5} = 2$.   **73.** The number is -3, since $\dfrac{27}{-3} = -9$.

75. The expression is: $\dfrac{-20}{4} - 3 = -5 - 3 = -8$

77. Each person would lose: $\dfrac{13600 - 15000}{4} = \dfrac{-1400}{4} = -350 = \$350$ loss

79. The change per hour is: $\dfrac{61° - 75°}{4} = \dfrac{-14°}{4} = -3.5°$ per hour

81. For the year 2000, the profit was: $\$11,500 - \$9,500 = \$2,000$

83. In the year 2002, the largest increase in costs was: $\$8,250 - \$6,750 = \$1,500$

1.8 Subsets of the Real Numbers

1. The set is $A \cup B = \{0, 1, 2, 3, 4, 5, 6\}$.   **3.** The set is $\{4, 5\}$.

5. The set is $B \cap C = \{1, 3, 5\}$.   **7.** The set is $\{0, 1, 2, 3, 4, 5, 6\}$.

9. The set is $\{0, 2\}$.   **11.** The set is $A \cap B = \{2, 4\}$.

13. The whole numbers are: 0, 1   **15.** The rational numbers are: -3, -2.5, 0, 1, $\dfrac{3}{2}$

17. The real numbers are: -3, -2.5, 0, 1, $\dfrac{3}{2}$, $\sqrt{15}$ **19.** The integers are: -10, -8, -2, 9

21. The irrational numbers are: π **23.** true

25. false   **27.** false

29. true

31. This number is composite: $48 = 6 \cdot 8 = (2 \cdot 3) \cdot (2 \cdot 2 \cdot 2) = 2^4 \cdot 3$

33. This number is prime.   **35.** This number is composite: $1023 = 3 \cdot 341 = 3 \cdot 11 \cdot 31$

37. Factoring the number: $144 = 12 \cdot 12 = (3 \cdot 4) \cdot (3 \cdot 4) = (3 \cdot 2 \cdot 2) \cdot (3 \cdot 2 \cdot 2) = 2^4 \cdot 3^2$

39. Factoring the number: $38 = 2 \cdot 19$

41. Factoring the number: $105 = 5 \cdot 21 = 5 \cdot (3 \cdot 7) = 3 \cdot 5 \cdot 7$

43. Factoring the number: $180 = 10 \cdot 18 = (2 \cdot 5) \cdot (3 \cdot 6) = (2 \cdot 5) \cdot (3 \cdot 2 \cdot 3) = 2^2 \cdot 3^2 \cdot 5$

45. Factoring the number: $385 = 5 \cdot 77 = 5 \cdot (7 \cdot 11) = 5 \cdot 7 \cdot 11$

47. Factoring the number: $121 = 11 \cdot 11 = 11^2$

49. Factoring the number: $420 = 10 \cdot 42 = (2 \cdot 5) \cdot (7 \cdot 6) = (2 \cdot 5) \cdot (7 \cdot 2 \cdot 3) = 2^2 \cdot 3 \cdot 5 \cdot 7$

51. Factoring the number: $620 = 10 \cdot 62 = (2 \cdot 5) \cdot (2 \cdot 31) = 2^2 \cdot 5 \cdot 31$

53. Reducing the fraction: $\frac{105}{165} = \frac{3 \cdot 5 \cdot 7}{3 \cdot 5 \cdot 11} = \frac{7}{11}$ **55.** Reducing the fraction: $\frac{525}{735} = \frac{3 \cdot 5 \cdot 5 \cdot 7}{3 \cdot 5 \cdot 7 \cdot 7} = \frac{5}{7}$

57. Reducing the fraction: $\frac{385}{455} = \frac{5 \cdot 7 \cdot 11}{5 \cdot 7 \cdot 13} = \frac{11}{13}$ **59.** Reducing the fraction: $\frac{322}{345} = \frac{2 \cdot 7 \cdot 23}{3 \cdot 5 \cdot 23} = \frac{2 \cdot 7}{3 \cdot 5} = \frac{14}{15}$

61. Reducing the fraction: $\frac{205}{369} = \frac{5 \cdot 41}{3 \cdot 3 \cdot 41} = \frac{5}{3 \cdot 3} = \frac{5}{9}$ **63.** Reducing the fraction: $\frac{215}{344} = \frac{5 \cdot 43}{2 \cdot 2 \cdot 2 \cdot 43} = \frac{5}{2 \cdot 2 \cdot 2} = \frac{5}{8}$

65. Factoring into prime numbers: $6^3 = (2 \cdot 3)^3 = 2^3 \cdot 3^3$

67. Factoring into prime numbers: $9^4 \cdot 16^2 = (3 \cdot 3)^4 \cdot (2 \cdot 2 \cdot 2 \cdot 2)^2 = 3^4 \cdot 3^4 \cdot 2^2 \cdot 2^2 \cdot 2^2 \cdot 2^2 = 2^8 \cdot 3^8$

69. Simplifying and factoring: $3 \cdot 8 + 3 \cdot 7 + 3 \cdot 5 = 24 + 21 + 15 = 60 = 6 \cdot 10 = (2 \cdot 3) \cdot (2 \cdot 5) = 2^2 \cdot 3 \cdot 5$

71. They are not a subset of the irrational numbers.

73. 8, 21, and 34 are Fibonacci numbers that are composite numbers.

1.9 Addition and Subtraction with Fractions

1. Combining the fractions: $\frac{3}{6} + \frac{1}{6} = \frac{4}{6} = \frac{2}{3}$ **3.** Combining the fractions: $\frac{3}{8} - \frac{5}{8} = -\frac{2}{8} = -\frac{1}{4}$

5. Combining the fractions: $-\frac{1}{4} + \frac{3}{4} = \frac{2}{4} = \frac{1}{2}$ **7.** Combining the fractions: $\frac{x}{3} - \frac{1}{3} = \frac{x-1}{3}$

9. Combining the fractions: $\frac{1}{4} + \frac{2}{4} + \frac{3}{4} = \frac{6}{4} = \frac{3}{2}$

11. Combining the fractions: $\frac{x+7}{2} - \frac{1}{2} = \frac{x+7-1}{2} = \frac{x+6}{2}$

13. Combining the fractions: $\frac{1}{10} - \frac{3}{10} - \frac{4}{10} = -\frac{6}{10} = -\frac{3}{5}$ **15.** Combining the fractions: $\frac{1}{a} + \frac{4}{a} + \frac{5}{a} = \frac{10}{a}$

17. Combining the fractions: $\frac{1}{8} + \frac{3}{4} = \frac{1}{8} + \frac{3 \cdot 2}{4 \cdot 2} = \frac{1}{8} + \frac{6}{8} = \frac{7}{8}$

19. Combining the fractions: $\frac{3}{10} - \frac{1}{5} = \frac{3}{10} - \frac{1 \cdot 2}{5 \cdot 2} = \frac{3}{10} - \frac{2}{10} = \frac{1}{10}$

21. Combining the fractions: $\frac{4}{9} + \frac{1}{3} = \frac{4}{9} + \frac{1 \cdot 3}{3 \cdot 3} = \frac{4}{9} + \frac{3}{9} = \frac{7}{9}$

23. Combining the fractions: $2 + \frac{1}{3} = \frac{2 \cdot 3}{1 \cdot 3} + \frac{1}{3} = \frac{6}{3} + \frac{1}{3} = \frac{7}{3}$

25. Combining the fractions: $-\frac{3}{4} + 1 = -\frac{3}{4} + \frac{1 \cdot 4}{1 \cdot 4} = -\frac{3}{4} + \frac{4}{4} = \frac{1}{4}$

27. Combining the fractions: $\frac{1}{2} + \frac{2}{3} = \frac{1 \cdot 3}{2 \cdot 3} + \frac{2 \cdot 2}{3 \cdot 2} = \frac{3}{6} + \frac{4}{6} = \frac{7}{6}$

29. Combining the fractions: $\frac{5}{12} - \left(-\frac{3}{8}\right) = \frac{5}{12} + \frac{3}{8} = \frac{5 \cdot 2}{12 \cdot 2} + \frac{3 \cdot 3}{8 \cdot 3} = \frac{10}{24} + \frac{9}{24} = \frac{19}{24}$

31. Combining the fractions: $-\frac{1}{20} + \frac{8}{30} = -\frac{1 \cdot 3}{20 \cdot 3} + \frac{8 \cdot 2}{30 \cdot 2} = -\frac{3}{60} + \frac{16}{60} = \frac{13}{60}$

33. First factor the denominators to find the LCM:

$30 = 2 \cdot 3 \cdot 5$

$42 = 2 \cdot 3 \cdot 7$

$LCM = 2 \cdot 3 \cdot 5 \cdot 7 = 210$

Combining the fractions: $\frac{17}{30} + \frac{11}{42} = \frac{17 \cdot 7}{30 \cdot 7} + \frac{11 \cdot 5}{42 \cdot 5} = \frac{119}{210} + \frac{55}{210} = \frac{174}{210} = \frac{2 \cdot 3 \cdot 29}{2 \cdot 3 \cdot 5 \cdot 7} = \frac{29}{5 \cdot 7} = \frac{29}{35}$

35. First factor the denominators to find the LCM:
$$84 = 2 \cdot 2 \cdot 3 \cdot 7$$
$$90 = 2 \cdot 3 \cdot 3 \cdot 5$$
$$LCM = 2 \cdot 2 \cdot 3 \cdot 3 \cdot 5 \cdot 7 = 1260$$
Combining the fractions: $\dfrac{25}{84} + \dfrac{41}{90} = \dfrac{25 \cdot 15}{84 \cdot 15} + \dfrac{41 \cdot 14}{90 \cdot 14} = \dfrac{375}{1260} + \dfrac{574}{1260} = \dfrac{949}{1260}$

37. First factor the denominators to find the LCM:
$$126 = 2 \cdot 3 \cdot 3 \cdot 7$$
$$180 = 2 \cdot 2 \cdot 3 \cdot 3 \cdot 5$$
$$LCM = 2 \cdot 2 \cdot 3 \cdot 3 \cdot 5 \cdot 7 = 1260$$
Combining the fractions: $\dfrac{13}{126} - \dfrac{13}{180} = \dfrac{13 \cdot 10}{126 \cdot 10} - \dfrac{13 \cdot 7}{180 \cdot 7} = \dfrac{130}{1260} - \dfrac{91}{1260} = \dfrac{39}{1260} = \dfrac{3 \cdot 13}{2 \cdot 2 \cdot 3 \cdot 3 \cdot 5 \cdot 7} = \dfrac{13}{2 \cdot 2 \cdot 3 \cdot 5 \cdot 7} = \dfrac{13}{420}$

39. Combining the fractions: $\dfrac{3}{4} + \dfrac{1}{8} + \dfrac{5}{6} = \dfrac{3 \cdot 6}{4 \cdot 6} + \dfrac{1 \cdot 3}{8 \cdot 3} + \dfrac{5 \cdot 4}{6 \cdot 4} = \dfrac{18}{24} + \dfrac{3}{24} + \dfrac{20}{24} = \dfrac{41}{24}$

41. Combining the fractions: $\dfrac{1}{2} + \dfrac{1}{3} + \dfrac{1}{4} + \dfrac{1}{6} = \dfrac{1 \cdot 6}{2 \cdot 6} + \dfrac{1 \cdot 4}{3 \cdot 4} + \dfrac{1 \cdot 3}{4 \cdot 3} + \dfrac{1 \cdot 2}{6 \cdot 2} = \dfrac{6}{12} + \dfrac{4}{12} + \dfrac{3}{12} + \dfrac{2}{12} = \dfrac{15}{12} = \dfrac{5}{4}$

43. The sum is given by: $\dfrac{3}{7} + 2 + \dfrac{1}{9} = \dfrac{3 \cdot 9}{7 \cdot 9} + \dfrac{2 \cdot 63}{1 \cdot 63} + \dfrac{1 \cdot 7}{9 \cdot 7} = \dfrac{27}{63} + \dfrac{126}{63} + \dfrac{7}{63} = \dfrac{160}{63}$

45. The difference is given by: $\dfrac{7}{8} - \dfrac{1}{4} = \dfrac{7}{8} - \dfrac{1 \cdot 2}{4 \cdot 2} = \dfrac{7}{8} - \dfrac{2}{8} = \dfrac{5}{8}$

47. The pattern is to add $-\dfrac{1}{3}$, so the fourth term is: $-\dfrac{1}{3} + \left(-\dfrac{1}{3}\right) = -\dfrac{2}{3}$

49. The pattern is to add $\dfrac{2}{3}$, so the fourth term is: $\dfrac{5}{3} + \dfrac{2}{3} = \dfrac{7}{3}$

51. The pattern is to multiply by $\dfrac{1}{5}$, so the fourth term is: $\dfrac{1}{25} \cdot \dfrac{1}{5} = \dfrac{1}{125}$

53. The perimeter is: $\dfrac{3}{8} + \dfrac{3}{8} + \dfrac{3}{8} + \dfrac{3}{8} = \dfrac{12}{8} = \dfrac{3}{2} = 1\dfrac{1}{2}$ feet

55. The perimeter is: $\dfrac{4}{5} + \dfrac{3}{10} + \dfrac{4}{5} + \dfrac{3}{10} = \dfrac{8}{10} + \dfrac{3}{10} + \dfrac{8}{10} + \dfrac{3}{10} = \dfrac{22}{10} = \dfrac{11}{5} = 2\dfrac{1}{5}$ centimeters (cm)

Chapter 1 Review

1. The expression is: $-7 + (-10) = -17$

3. The expression is: $(-3 + 12) + 5 = 9 + 5 = 14$

5. The expression is: $9 - (-3) = 9 + 3 = 12$

7. The expression is: $(-3)(-7) - 6 = 21 - 6 = 15$

9. The expression is: $2[-8(3x)] = 2(-24x) = -48x$

11. The expression is: $\dfrac{-40}{8} - 7 = -5 - 7 = -5 + (-7) = -12$

13. Labeling the point:

15. Labeling the point (note that $\dfrac{24}{8} = 3$):

17. Labeling the point:

19. The absolute value is: $|12| = 12$

21. The absolute value is: $\left|-\dfrac{4}{5}\right| = \dfrac{4}{5}$

23. Simplifying: $|-1.8| = 1.8$

25. The opposite is –6, and the reciprocal is $\dfrac{1}{6}$.

27. The opposite is 9, and the reciprocal is $-\dfrac{1}{9}$.

29. Multiplying the fractions: $\left(\dfrac{2}{5}\right)\left(\dfrac{3}{7}\right) = \dfrac{6}{35}$

31. Multiplying the fractions: $\left(-\dfrac{4}{5}\right)\left(\dfrac{25}{16}\right) = -\dfrac{100}{80} = -\dfrac{5}{4}$

33. Adding: $-18 + (-20) = -38$

35. Adding: $(-5) + (-10) + (-7) = -22$

37. Adding: $(-21) + 40 + (-23) + 5 = -44 + 45 = 1$

39. Subtracting: $14 - (-8) = 14 + 8 = 22$

41. Subtracting: $4 - 9 - 15 = 4 + (-9) + (-15) = 4 + (-24) = -20$

43.   Simplifying: $5-(-10-2)-3 = 5-(-12)-3 = 5+12+(-3) = 17+(-3) = 14$
45.   Simplifying: $20-\left[-(10-3)-8\right] = 20-(-7-8)-7 = 20-(-15)-7 = 20+15+(-7) = 35+(-7) = 28$
47.   Multiplying: $4(-3) = -12$ **49.**   Multiplying: $(-1)(-3)(-1)(-4) = 12$
51.   Finding the quotient: $\dfrac{-9}{36} = -\dfrac{1}{4}$
53.   Simplifying using order of operations: $4\cdot 5+3 = 20+3 = 23$
55.   Simplifying using order of operations: $2^3 - 4\cdot 3^2 + 5^2 = 8-4\cdot 9+25 = 8-36+25 = 33-36 = -3$
57.   Simplifying using order of operations: $20+8\div 4+2\cdot 5 = 20+2+10 = 32$
59.   Simplifying using order of operations: $-4(-5)+10 = 20+10 = 30$
61.   Simplifying using order of operations:
   $3(4-7)^2 - 5(3-8)^2 = 3(-3)^2 - 5(-5)^2 = 3\cdot 9-5\cdot 25 = 27-125 = 27+(-125) = -98$
63.   Simplifying using order of operations: $\dfrac{4(-3)}{-6} = \dfrac{-12}{-6} = 2$
65.   Simplifying using order of operations: $\dfrac{15-10}{6-6} = \dfrac{5}{0} = $ undefined
67.   Simplifying using order of operations: $\dfrac{2(-7)+(-11)(-4)}{7-(-3)} = \dfrac{-14+44}{7+3} = \dfrac{30}{10} = 3$
69.   multiplicative identity property   **71.**   additive inverse property
73.   additive identity property   **75.**   distributive property
77.   Simplifying the expression: $4(7a) = (4\cdot 7)a = 28a$
79.   Simplifying the expression: $\frac{4}{5}\left(\frac{5}{4}y\right) = \left(\frac{4}{5}\cdot\frac{5}{4}\right)y = 1y = y$
81.   Applying the distributive property: $3(2a-4) = 3\cdot 2a-3\cdot 4 = 6a-12$
83.   Applying the distributive property: $-\frac{1}{2}(3x-6) = -\frac{1}{2}\cdot 3x-\left(-\frac{1}{2}\right)\cdot 6 = -\frac{3}{2}x+3$
85.   The whole numbers are: 0, 5   **87.**   The integers are: 0, 5, –3
89.   Factoring the number: $840 = 84\cdot 10 = (21\cdot 4)\cdot(2\cdot 5) = (3\cdot 7\cdot 2\cdot 2)\cdot(2\cdot 5) = 2^3\cdot 3\cdot 5\cdot 7$
91.   First factor the denominators to find the LCM:
   $70 = 2\cdot 5\cdot 7$
   $84 = 2\cdot 2\cdot 3\cdot 7$
   $LCM = 2\cdot 2\cdot 3\cdot 5\cdot 7 = 420$
   Combining the fractions: $\frac{9}{70}+\frac{11}{84} = \frac{9\cdot 6}{70\cdot 6}+\frac{11\cdot 5}{84\cdot 5} = \frac{54}{420}+\frac{55}{420} = \frac{109}{420}$
93.   The pattern is to multiply by –3, so the next number is: $-270\cdot(-3) = 810$
95.   The pattern is to add 2, so the next number is: $10+2 = 12$
97.   The pattern is to multiply by $-\frac{1}{2}$, so the next number is: $-\frac{1}{8}\cdot\left(-\frac{1}{2}\right) = \frac{1}{16}$

Chapter 1 Test

1.   Translating into symbols: $x+3 = 8$ **2.**   Translating into symbols: $5y = 15$
3.   Simplifying using order of operations: $5^2 + 3(9-7)+3^2 = 5^2 + 3(2)+3^2 = 25+6+9 = 40$
4.   Simplifying using order of operations: $10-6\div 3+2^3 = 10-6\div 3+8 = 10-2+8 = 18-2 = 16$
5.   The opposite of –4 is 4, the reciprocal is $-\frac{1}{4}$, and the absolute value is $|-4| = 4$.
6.   The opposite of $\frac{3}{4}$ is $-\frac{3}{4}$, the reciprocal is $\frac{4}{3}$, and the absolute value is $\left|\frac{3}{4}\right| = \frac{3}{4}$.
7.   Adding: $3+(-7) = -4$
8.   Adding: $\left|-9+(-6)\right|+\left|-3+5\right| = \left|-15\right|+\left|2\right| = 15+2 = 17$
9.   Subtracting: $-4-8 = -4+(-8) = -12$
10.   Subtracting: $9-(7-2)-4 = 9-5-4 = 9+(-5)+(-4) = 9+(-9) = 0$
11.   c (associative property of addition)   **12.**   e (distributive property)
13.   d (associative property of multiplication)   **14.**   a (commutative property of addition)

15. Multiplying: $-3(7) = -21$

16. Multiplying: $-4(8)(-2) = 64$

17. Multiplying: $8\left(-\frac{1}{4}\right) = \frac{8}{1} \cdot \left(-\frac{1}{4}\right) = -\frac{8}{4} = -2$

18. Multiplying: $\left(-\frac{2}{3}\right)^3 = \left(-\frac{2}{3}\right) \cdot \left(-\frac{2}{3}\right) \cdot \left(-\frac{2}{3}\right) = -\frac{8}{27}$

19. Simplifying using order of operations: $-3(-4) - 8 = 12 - 8 = 4$

20. Simplifying using order of operations: $5(-6)^2 - 3(-2)^3 = 5(36) - 3(-8) = 180 + 24 = 204$

21. Simplifying using order of operations: $7 - 3(2 - 8) = 7 - 3(-6) = 7 + 18 = 25$

22 Simplifying using order of operations:
$$4 - 2[-3(-1 + 5) + 4(-3)] = 4 - 2[-3(4) + 4(-3)] = 4 - 2(-12 - 12) = 4 - 2(-24) = 4 + 48 = 52$$

23. Simplifying using order of operations: $\dfrac{4(-5) - 2(7)}{-10 - 7} = \dfrac{-20 - 14}{-17} = \dfrac{-34}{-17} = 2$

24. Simplifying using order of operations: $\dfrac{2(-3 - 1) + 4(-5 + 2)}{-3(2) - 4} = \dfrac{2(-4) + 4(-3)}{-6 - 4} = \dfrac{-8 - 12}{-10} = \dfrac{-20}{-10} = 2$

25. Simplifying: $3 + (5 + 2x) = (3 + 5) + 2x = 8 + 2x$

26. Simplifying: $-2(-5x) = [-2(-5)]x = 10x$

27. Multiplying: $2(3x + 5) = 2 \cdot 3x + 2 \cdot 5 = 6x + 10$

28. Multiplying: $-\frac{1}{2}(4x - 2) = -\frac{1}{2}(4x) - \left(-\frac{1}{2}\right)(2) = -2x + 1$

29. The integers are 1 and –8.

30. The rational numbers are 1, 1.5, $\frac{3}{4}$, and –8.

31. The irrational numbers are $\sqrt{2}$.

32. The real numbers are 1, 1.5, $\sqrt{2}$, $\frac{3}{4}$, and –8.

33. Factoring the number: $592 = 4 \cdot 148 = (2 \cdot 2) \cdot (4 \cdot 37) = (2 \cdot 2) \cdot (2 \cdot 2 \cdot 37) = 2^4 \cdot 37$

34. Factoring the number: $1340 = 10 \cdot 134 = (2 \cdot 5) \cdot (2 \cdot 67) = 2^2 \cdot 5 \cdot 67$

35. First factor the denominators to find the LCM:
$$15 = 3 \cdot 5$$
$$42 = 2 \cdot 3 \cdot 7$$
$$\text{LCM} = 2 \cdot 3 \cdot 5 \cdot 7 = 210$$
Combining the fractions: $\frac{5}{15} + \frac{11}{42} = \frac{5 \cdot 14}{15 \cdot 14} + \frac{11 \cdot 5}{42 \cdot 5} = \frac{70}{210} + \frac{55}{210} = \frac{125}{210} = \frac{5 \cdot 25}{5 \cdot 42} = \frac{25}{42}$

36. Combining the fractions: $\dfrac{5}{x} + \dfrac{3}{x} = \dfrac{5 + 3}{x} = \dfrac{8}{x}$

37. The expression is: $8 + (-3) = 5$

38. The expression is: $-24 - 2 = -24 + (-2) = -26$

39. The expression is: $-5(-4) = 20$

40. The expression is: $\dfrac{-24}{-2} = 12$

41. The pattern is to add 5, so the next term is $7 + 5 = 12$.

42. The pattern is to multiply by $-\frac{1}{2}$, so the next term is $-1\left(-\frac{1}{2}\right) = \frac{1}{2}$.

Chapter 2
Linear Equations and Inequalities

2.1 Simplifying Expressions

1. Simplifying the expression: $3x - 6x = (3 - 6)x = -3x$
3. Simplifying the expression: $-2a + a = (-2 + 1)a = -a$
5. Simplifying the expression: $7x + 3x + 2x = (7 + 3 + 2)x = 12x$
7. Simplifying the expression: $3a - 2a + 5a = (3 - 2 + 5)a = 6a$
9. Simplifying the expression: $4x - 3 + 2x = 4x + 2x - 3 = 6x - 3$
11. Simplifying the expression: $3a + 4a + 5 = 7a + 5$
13. Simplifying the expression: $2x - 3 + 3x - 2 = 2x + 3x - 3 - 2 = 5x - 5$
15. Simplifying the expression: $3a - 1 + a + 3 = 3a + a - 1 + 3 = 4a + 2$
17. Simplifying the expression: $-4x + 8 - 5x - 10 = -4x - 5x + 8 - 10 = -9x - 2$
19. Simplifying the expression: $7a + 3 + 2a + 3a = 7a + 2a + 3a + 3 = 12a + 3$
21. Simplifying the expression: $5(2x - 1) + 4 = 10x - 5 + 4 = 10x - 1$
23. Simplifying the expression: $7(3y + 2) - 8 = 21y + 14 - 8 = 21y + 6$
25. Simplifying the expression: $-3(2x - 1) + 5 = -6x + 3 + 5 = -6x + 8$
27. Simplifying the expression: $5 - 2(a + 1) = 5 - 2a - 2 = -2a - 2 + 5 = -2a + 3$
29. Simplifying the expression: $6 - 4(x - 5) = 6 - 4x + 20 = -4x + 20 + 6 = -4x + 26$
31. Simplifying the expression: $-9 - 4(2 - y) + 1 = -9 - 8 + 4y + 1 = 4y + 1 - 9 - 8 = 4y - 16$
33. Simplifying the expression: $-6 + 2(2 - 3x) + 1 = -6 + 4 - 6x + 1 = -6x - 6 + 4 + 1 = -6x - 1$
35. Simplifying the expression: $(4x - 7) - (2x + 5) = 4x - 7 - 2x - 5 = 4x - 2x - 7 - 5 = 2x - 12$
37. Simplifying the expression: $8(2a + 4) - (6a - 1) = 16a + 32 - 6a + 1 = 16a - 6a + 32 + 1 = 10a + 33$
39. Simplifying the expression: $3(x - 2) + (x - 3) = 3x - 6 + x - 3 = 3x + x - 6 - 3 = 4x - 9$
41. Simplifying the expression: $4(2y - 8) - (y + 7) = 8y - 32 - y - 7 = 8y - y - 32 - 7 = 7y - 39$
43. Simplifying the expression: $-9(2x + 1) - (x + 5) = -18x - 9 - x - 5 = -18x - x - 9 - 5 = -19x - 14$
45. Evaluating when $x = 2$: $3x - 1 = 3(2) - 1 = 6 - 1 = 5$
47. Evaluating when $x = 2$: $-2x - 5 = -2(2) - 5 = -4 - 5 = -9$
49. Evaluating when $x = 2$: $x^2 - 8x + 16 = (2)^2 - 8(2) + 16 = 4 - 16 + 16 = 4$
51. Evaluating when $x = 2$: $(x - 4)^2 = (2 - 4)^2 = (-2)^2 = 4$
53. Evaluating when $x = -5$: $7x - 4 - x - 3 = 7(-5) - 4 - (-5) - 3 = -35 - 4 + 5 - 3 = -42 + 5 = -37$
    Now simplifying the expression: $7x - 4 - x - 3 = 7x - x - 4 - 3 = 6x - 7$
    Evaluating when $x = -5$: $6x - 7 = 6(-5) - 7 = -30 - 7 = -37$
    Note that the two values are the same.

55. Evaluating when $x = -5$: $5(2x+1)+4 = 5[2(-5)+1]+4 = 5(-10+1)+4 = 5(-9)+4 = -45+4 = -41$
Now simplifying the expression: $5(2x+1)+4 = 10x+5+4 = 10x+9$
Evaluating when $x = -5$: $10x+9 = 10(-5)+9 = -50+9 = -41$
Note that the two values are the same.

57. Evaluating when $x = -3$ and $y = 5$: $x^2 - 2xy + y^2 = (-3)^2 - 2(-3)(5) + (5)^2 = 9 + 30 + 25 = 64$

59. Evaluating when $x = -3$ and $y = 5$: $(x-y)^2 = (-3-5)^2 = (-8)^2 = 64$

61. Evaluating when $x = -3$ and $y = 5$: $x^2 + 6xy + 9y^2 = (-3)^2 + 6(-3)(5) + 9(5)^2 = 9 - 90 + 225 = 144$

63. Evaluating when $x = -3$ and $y = 5$: $(x+3y)^2 = [-3+3(5)]^2 = (-3+15)^2 = (12)^2 = 144$

65. Evaluating when $x = \frac{1}{2}$: $12x - 3 = 12\left(\frac{1}{2}\right) - 3 = 6 - 3 = 3$

67. Evaluating when $x = \frac{1}{4}$: $12x - 3 = 12\left(\frac{1}{4}\right) - 3 = 3 - 3 = 0$

69. Evaluating when $x = \frac{3}{2}$: $12x - 3 = 12\left(\frac{3}{2}\right) - 3 = 18 - 3 = 15$

71. Evaluating when $x = \frac{3}{4}$: $12x - 3 = 12\left(\frac{3}{4}\right) - 3 = 9 - 3 = 6$

73. **a.** Substituting the values for n:

n	1	2	3	4
$3n$	3	6	9	12

b. Substituting the values for n:

n	1	2	3	4
n^3	1	8	27	64

75. Substituting $n = 1, 2, 3, 4$:
$n = 1$: $3(1) - 2 = 3 - 2 = 1$
$n = 2$: $3(2) - 2 = 6 - 2 = 4$
$n = 3$: $3(3) - 2 = 9 - 2 = 7$
$n = 4$: $3(4) - 2 = 12 - 2 = 10$
The sequence is 1, 4, 7, 10, ..., which is an arithmetic sequence.

77. Substituting $n = 1, 2, 3, 4$:
$n = 1$: $(1)^2 - 2(1) + 1 = 1 - 2 + 1 = 0$
$n = 2$: $(2)^2 - 2(2) + 1 = 4 - 4 + 1 = 1$
$n = 3$: $(3)^2 - 2(3) + 1 = 9 - 6 + 1 = 4$
$n = 4$: $(4)^2 - 2(4) + 1 = 16 - 8 + 1 = 9$
The sequence is 0, 1, 4, 9, ..., which is a sequence of squares.

79. Translating into an algebraic expression: $x + 5$. Evaluating when $x = -2$: $x + 5 = (-2) + 5 = 3$

81. Translating into an algebraic expression: $x - 5$. Evaluating when $x = -2$: $x - 5 = (-2) - 5 = -7$

83. Translating into an algebraic expression: $2(x+10)$. Evaluating when $x = -2$: $2(x+10) = 2(-2+10) = 2(8) = 16$

85. Translating into an algebraic expression: $\dfrac{10}{x}$. Evaluating when $x = -2$: $\dfrac{10}{x} = \dfrac{10}{-2} = -5$

87. Translating into an algebraic expression, and simplifying: $[3x + (-2)] - 5 = 3x - 2 - 5 = 3x - 7$
Evaluating when $x = -2$: $3x - 7 = 3(-2) - 7 = -6 - 7 = -13$

89. **a.** Substituting $x = 8,000$: $-0.0035(8000) + 70 = 42°F$
b. Substituting $x = 12,000$: $-0.0035(12000) + 70 = 28°F$
c. Substituting $x = 24,000$: $-0.0035(24000) + 70 = -14°F$

91. **a.** Substituting $t = 10$: $35 + 0.25(10) = \$37.50$ **b.** Substituting $t = 20$: $35 + 0.25(20) = \$40.00$
c. Substituting $t = 30$: $35 + 0.25(30) = \$42.50$

93. Simplifying the expression: $G - 0.21G - 0.08G = 0.71G$. Substituting $G = \$1,250$: $0.71(\$1,250) = \$887.50$

95. Subtracting: $-3 - \frac{1}{2} = \frac{-3}{1} - \frac{1}{2} = \frac{-3 \cdot 2}{1 \cdot 2} - \frac{1}{2} = \frac{-6}{2} - \frac{1}{2} = -\frac{7}{2}$

97. Adding: $\frac{4}{5} + \frac{1}{10} + \frac{3}{8} = \frac{4 \cdot 8}{5 \cdot 8} + \frac{1 \cdot 4}{10 \cdot 4} + \frac{3 \cdot 5}{8 \cdot 5} = \frac{32}{40} + \frac{4}{40} + \frac{15}{40} = \frac{51}{40}$

2.2 Addition Property of Equality

1. Solving the equation:
$$x - 3 = 8$$
$$x - 3 + 3 = 8 + 3$$
$$x = 11$$

3. Solving the equation:
$$x + 2 = 6$$
$$x + 2 + (-2) = 6 + (-2)$$
$$x = 4$$

5. Solving the equation:
$$a + \tfrac{1}{2} = -\tfrac{1}{4}$$
$$a + \tfrac{1}{2} + \left(-\tfrac{1}{2}\right) = -\tfrac{1}{4} + \left(-\tfrac{1}{2}\right)$$
$$a = -\tfrac{1}{4} + \left(-\tfrac{2}{4}\right)$$
$$a = -\tfrac{3}{4}$$

7. Solving the equation:
$$x + 2.3 = -3.5$$
$$x + 2.3 + (-2.3) = -3.5 + (-2.3)$$
$$x = -5.8$$

9. Solving the equation:
$$y + 11 = -6$$
$$y + 11 + (-11) = -6 + (-11)$$
$$y = -17$$

11. Solving the equation:
$$x - \tfrac{5}{8} = -\tfrac{3}{4}$$
$$x - \tfrac{5}{8} + \tfrac{5}{8} = -\tfrac{3}{4} + \tfrac{5}{8}$$
$$x = -\tfrac{6}{8} + \tfrac{5}{8}$$
$$x = -\tfrac{1}{8}$$

13. Solving the equation:
$$m - 6 = -10$$
$$m - 6 + 6 = -10 + 6$$
$$m = -4$$

15. Solving the equation:
$$6.9 + x = 3.3$$
$$-6.9 + 6.9 + x = -6.9 + 3.3$$
$$x = -3.6$$

17. Solving the equation:
$$5 = a + 4$$
$$5 + (-4) = a + 4 + (-4)$$
$$a = 1$$

18. Solving the equation:
$$12 = a - 3$$
$$12 + 3 = a - 3 + 3$$
$$a = 15$$

19. Solving the equation:
$$-\tfrac{5}{9} = x - \tfrac{2}{5}$$
$$-\tfrac{5}{9} + \tfrac{2}{5} = x - \tfrac{2}{5} + \tfrac{2}{5}$$
$$-\tfrac{25}{45} + \tfrac{18}{45} = x$$
$$x = -\tfrac{7}{45}$$

21. Solving the equation:
$$-\tfrac{5}{9} = x - \tfrac{2}{5}$$
$$-\tfrac{5}{9} + \tfrac{2}{5} = x - \tfrac{2}{5} + \tfrac{2}{5}$$
$$-\tfrac{25}{45} + \tfrac{18}{45} = x$$
$$x = -\tfrac{7}{45}$$

23. Solving the equation:
$$8a - \tfrac{1}{2} - 7a = \tfrac{3}{4} + \tfrac{1}{8}$$
$$a - \tfrac{1}{2} = \tfrac{6}{8} + \tfrac{1}{8}$$
$$a - \tfrac{1}{2} = \tfrac{7}{8}$$
$$a - \tfrac{1}{2} + \tfrac{1}{2} = \tfrac{7}{8} + \tfrac{1}{2}$$
$$a = \tfrac{7}{8} + \tfrac{4}{8}$$
$$a = \tfrac{11}{8}$$

25. Solving the equation:
$$-3 - 4x + 5x = 18$$
$$-3 + x = 18$$
$$3 - 3 + x = 3 + 18$$
$$x = 21$$

27. Solving the equation:
$$-11x + 2 + 10x + 2x = 9$$
$$x + 2 = 9$$
$$x + 2 + (-2) = 9 + (-2)$$
$$x = 7$$

29. Solving the equation:
$$-2.5 + 4.8 = 8x - 1.2 - 7x$$
$$2.3 = x - 1.2$$
$$2.3 + 1.2 = x - 1.2 + 1.2$$
$$x = 3.5$$

31. Solving the equation:
$$2y - 10 + 3y - 4y = 18 - 6$$
$$y - 10 = 12$$
$$y - 10 + 10 = 12 + 10$$
$$y = 22$$

33. Solving the equation:
$$2(x + 3) - x = 4$$
$$2x + 6 - x = 4$$
$$x + 6 = 4$$
$$x + 6 + (-6) = 4 + (-6)$$
$$x = -2$$

35. Solving the equation:
$$-3(x - 4) + 4x = 3 - 7$$
$$-3x + 12 + 4x = -4$$
$$x + 12 = -4$$
$$x + 12 + (-12) = -4 + (-12)$$
$$x = -16$$

37. Solving the equation:
$$5(2a + 1) - 9a = 8 - 6$$
$$10a + 5 - 9a = 2$$
$$a + 5 = 2$$
$$a + 5 + (-5) = 2 + (-5)$$
$$a = -3$$

39. Solving the equation:
$$-(x + 3) + 2x - 1 = 6$$
$$-x - 3 + 2x - 1 = 6$$
$$x - 4 = 6$$
$$x - 4 + 4 = 6 + 4$$
$$x = 10$$

41. Solving the equation:
$$4y - 3(y - 6) + 2 = 8$$
$$4y - 3y + 18 + 2 = 8$$
$$y + 20 = 8$$
$$y + 20 + (-20) = 8 + (-20)$$
$$y = -12$$

43. Solving the equation:
$$-3(2m - 9) + 7(m - 4) = 12 - 9$$
$$-6m + 27 + 7m - 28 = 3$$
$$m - 1 = 3$$
$$m - 1 + 1 = 3 + 1$$
$$m = 4$$

45. Solving the equation:
$$4x = 3x + 2$$
$$4x + (-3x) = 3x + (-3x) + 2$$
$$x = 2$$

47. Solving the equation:
$$8a = 7a - 5$$
$$8a + (-7a) = 7a + (-7a) - 5$$
$$a = -5$$

49. Solving the equation:
$$2x = 3x + 1$$
$$(-2x) + 2x = (-2x) + 3x + 1$$
$$0 = x + 1$$
$$0 + (-1) = x + 1 + (-1)$$
$$x = -1$$

51. Solving the equation:
$$3y + 4 = 2y + 1$$
$$3y + (-2y) + 4 = 2y + (-2y) + 1$$
$$y + 4 = 1$$
$$y + 4 + (-4) = 1 + (-4)$$
$$y = -3$$

53. Solving the equation:
$$2m - 3 = m + 5$$
$$2m + (-m) - 3 = m + (-m) + 5$$
$$m - 3 = 5$$
$$m - 3 + 3 = 5 + 3$$
$$m = 8$$

55. Solving the equation:
$$4x - 7 = 5x + 1$$
$$4x + (-4x) - 7 = 5x + (-4x) + 1$$
$$-7 = x + 1$$
$$-7 + (-1) = x + 1 + (-1)$$
$$x = -8$$

57. Solving the equation:
$$5x - \frac{2}{3} = 4x + \frac{4}{3}$$
$$5x + (-4x) - \frac{2}{3} = 4x + (-4x) + \frac{4}{3}$$
$$x - \frac{2}{3} = \frac{4}{3}$$
$$x - \frac{2}{3} + \frac{2}{3} = \frac{4}{3} + \frac{2}{3}$$
$$x = \frac{6}{3} = 2$$

59. Solving the equation:
$$8a - 7.1 = 7a + 3.9$$
$$8a + (-7a) - 7.1 = 7a + (-7a) + 3.9$$
$$a - 7.1 = 3.9$$
$$a - 7.1 + 7.1 = 3.9 + 7.1$$
$$a = 11$$

61. **a.**   Solving for R:

$$T + R + A = 100$$
$$88 + R + 6 = 100$$
$$94 + R = 100$$
$$R = 6\%$$

b.   Solving for R:

$$T + R + A = 100$$
$$0 + R + 95 = 100$$
$$95 + R = 100$$
$$R = 5\%$$

c.   Solving for A:

$$T + R + A = 100$$
$$0 + 98 + A = 100$$
$$98 + A = 100$$
$$A = 2\%$$

d.   Solving for R:

$$T + R + A = 100$$
$$0 + R + 25 = 100$$
$$25 + R = 100$$
$$R = 75\%$$

63.   Solving for x:

$$x + 55 + 55 = 180$$
$$x + 110 = 180$$
$$x = 70°$$

65.   Applying the associative property: $3(6x) = (3 \cdot 6)x = 18x$

67.   Applying the associative property: $\frac{1}{5}(5x) = \left(\frac{1}{5} \cdot 5\right)x = 1x = x$

69.   Applying the associative property: $8\left(\frac{1}{8}y\right) = \left(8 \cdot \frac{1}{8}\right)y = 1y = y$

71.   Applying the associative property: $-2\left(-\frac{1}{2}x\right) = \left[-2 \cdot \left(-\frac{1}{2}\right)\right]x = 1x = x$

73.   Applying the associative property: $-\frac{4}{3}\left(-\frac{3}{4}a\right) = \left[-\frac{4}{3} \cdot \left(-\frac{3}{4}\right)\right]a = 1a = a$

2.3 Multiplication Property of Equality

1.   Solving the equation:

$$5x = 10$$
$$\frac{1}{5}(5x) = \frac{1}{5}(10)$$
$$x = 2$$

3.   Solving the equation:

$$7a = 28$$
$$\frac{1}{7}(7a) = \frac{1}{7}(28)$$
$$a = 4$$

5.   Solving the equation:

$$-8x = 4$$
$$-\frac{1}{8}(-8x) = -\frac{1}{8}(4)$$
$$x = -\frac{1}{2}$$

7.   Solving the equation:

$$8m = -16$$
$$\frac{1}{8}(8m) = \frac{1}{8}(-16)$$
$$m = -2$$

9.   Solving the equation:

$$-3x = -9$$
$$-\frac{1}{3}(-3x) = -\frac{1}{3}(-9)$$
$$x = 3$$

11.   Solving the equation:

$$-7y = -28$$
$$-\frac{1}{7}(-7y) = -\frac{1}{7}(-28)$$
$$y = 4$$

13.   Solving the equation:

$$2x = 0$$
$$\frac{1}{2}(2x) = \frac{1}{2}(0)$$
$$x = 0$$

15.   Solving the equation:

$$-5x = 0$$
$$-\frac{1}{5}(-5x) = -\frac{1}{5}(0)$$
$$x = 0$$

17.   Solving the equation:

$$\frac{x}{3} = 2$$
$$3\left(\frac{x}{3}\right) = 3(2)$$
$$x = 6$$

19.   Solving the equation:

$$-\frac{m}{5} = 10$$
$$-5\left(-\frac{m}{5}\right) = -5(10)$$
$$m = -50$$

21. Solving the equation:
$$-\frac{x}{2} = -\frac{3}{4}$$
$$-2\left(-\frac{x}{2}\right) = -2\left(-\frac{3}{4}\right)$$
$$x = \frac{3}{2}$$

23. Solving the equation:
$$\frac{2}{3}a = 8$$
$$\frac{3}{2}\left(\frac{2}{3}a\right) = \frac{3}{2}(8)$$
$$a = 12$$

25. Solving the equation:
$$-\frac{3}{5}x = \frac{9}{5}$$
$$-\frac{5}{3}\left(-\frac{3}{5}x\right) = -\frac{5}{3}\left(\frac{9}{5}\right)$$
$$x = -3$$

27. Solving the equation:
$$-\frac{5}{8}y = -20$$
$$-\frac{8}{5}\left(-\frac{5}{8}y\right) = -\frac{8}{5}(-20)$$
$$y = 32$$

29. Simplifying and then solving the equation:
$$-4x - 2x + 3x = 24$$
$$-3x = 24$$
$$-\frac{1}{3}(-3x) = -\frac{1}{3}(24)$$
$$x = -8$$

31. Simplifying and then solving the equation:
$$4x + 8x - 2x = 15 - 10$$
$$10x = 5$$
$$\frac{1}{10}(10x) = \frac{1}{10}(5)$$
$$x = \frac{1}{2}$$

33. Simplifying and then solving the equation:
$$-3 - 5 = 3x + 5x - 10x$$
$$-8 = -2x$$
$$-\frac{1}{2}(-8) = -\frac{1}{2}(-2x)$$
$$x = 4$$

35. Using Method 2 to eliminate fractions:
$$18 - 13 = \frac{1}{2}a + \frac{3}{4}a - \frac{5}{8}a$$
$$8(5) = 8\left(\frac{1}{2}a + \frac{3}{4}a - \frac{5}{8}a\right)$$
$$40 = 4a + 6a - 5a$$
$$40 = 5a$$
$$\frac{1}{5}(40) = \frac{1}{5}(5a)$$
$$a = 8$$

37. Solving by multiplying both sides of the equation by –1:
$$-x = 4$$
$$-1(-x) = -1(4)$$
$$x = -4$$

39. Solving by multiplying both sides of the equation by –1:
$$-x = -4$$
$$-1(-x) = -1(-4)$$
$$x = 4$$

41. Solving by multiplying both sides of the equation by –1:
$$15 = -a$$
$$-1(15) = -1(-a)$$
$$a = -15$$

43. Solving by multiplying both sides of the equation by –1:
$$-y = \frac{1}{2}$$
$$-1(-y) = -1\left(\frac{1}{2}\right)$$
$$y = -\frac{1}{2}$$

45. Solving the equation:
$$3x - 2 = 7$$
$$3x - 2 + 2 = 7 + 2$$
$$3x = 9$$
$$\frac{1}{3}(3x) = \frac{1}{3}(9)$$
$$x = 3$$

47. Solving the equation:
$$2a + 1 = 3$$
$$2a + 1 + (-1) = 3 + (-1)$$
$$2a = 2$$
$$\frac{1}{2}(2a) = \frac{1}{2}(2)$$
$$a = 1$$

49. Using Method 2 to eliminate fractions:

$$\tfrac{1}{8} + \tfrac{1}{2}x = \tfrac{1}{4}$$
$$8\left(\tfrac{1}{8} + \tfrac{1}{2}x\right) = 8\left(\tfrac{1}{4}\right)$$
$$1 + 4x = 2$$
$$(-1) + 1 + 4x = (-1) + 2$$
$$4x = 1$$
$$\tfrac{1}{4}(4x) = \tfrac{1}{4}(1)$$
$$x = \tfrac{1}{4}$$

51. Solving the equation:

$$6x = 2x - 12$$
$$6x + (-2x) = 2x + (-2x) - 12$$
$$4x = -12$$
$$\tfrac{1}{4}(4x) = \tfrac{1}{4}(-12)$$
$$x = -3$$

53. Solving the equation:

$$2y = -4y + 18$$
$$2y + 4y = -4y + 4y + 18$$
$$6y = 18$$
$$\tfrac{1}{6}(6y) = \tfrac{1}{6}(18)$$
$$y = 3$$

55. Solving the equation:

$$-7x = -3x - 8$$
$$-7x + 3x = -3x + 3x - 8$$
$$-4x = -8$$
$$-\tfrac{1}{4}(-4x) = -\tfrac{1}{4}(-8)$$
$$x = 2$$

57. Solving the equation:

$$8x + 4 = 2x - 5$$
$$8x + (-2x) + 4 = 2x + (-2x) - 5$$
$$6x + 4 = -5$$
$$6x + 4 + (-4) = -5 + (-4)$$
$$6x = -9$$
$$\tfrac{1}{6}(6x) = \tfrac{1}{6}(-9)$$
$$x = -\tfrac{3}{2}$$

59. Using Method 2 to eliminate fractions:

$$x + \tfrac{1}{2} = \tfrac{1}{4}x - \tfrac{5}{8}$$
$$8\left(x + \tfrac{1}{2}\right) = 8\left(\tfrac{1}{4}x - \tfrac{5}{8}\right)$$
$$8x + 4 = 2x - 5$$
$$8x + (-2x) + 4 = 2x + (-2x) - 5$$
$$6x + 4 = -5$$
$$6x + 4 + (-4) = -5 + (-4)$$
$$6x = -9$$
$$\tfrac{1}{6}(6x) = \tfrac{1}{6}(-9)$$
$$x = -\tfrac{3}{2}$$

61. Solving the equation:

$$6m - 3 = m + 2$$
$$6m + (-m) - 3 = m + (-m) + 2$$
$$5m - 3 = 2$$
$$5m - 3 + 3 = 2 + 3$$
$$5m = 5$$
$$\tfrac{1}{5}(5m) = \tfrac{1}{5}(5)$$
$$m = 1$$

63. Using Method 2 to eliminate fractions:

$$\tfrac{1}{2}m - \tfrac{1}{4} = \tfrac{1}{12}m + \tfrac{1}{6}$$
$$12\left(\tfrac{1}{2}m - \tfrac{1}{4}\right) = 12\left(\tfrac{1}{12}m + \tfrac{1}{6}\right)$$
$$6m - 3 = m + 2$$
$$6m + (-m) - 3 = m + (-m) + 2$$
$$5m - 3 = 2$$
$$5m - 3 + 3 = 2 + 3$$
$$5m = 5$$
$$\tfrac{1}{5}(5m) = \tfrac{1}{5}(5)$$
$$m = 1$$

65. Solving the equation:

$$9y + 2 = 6y - 4$$
$$9y + (-6y) + 2 = 6y + (-6y) - 4$$
$$3y + 2 = -4$$
$$3y + 2 + (-2) = -4 + (-2)$$
$$3y = -6$$
$$\tfrac{1}{3}(3y) = \tfrac{1}{3}(-6)$$
$$y = -2$$

67. Using Method 2 to eliminate fractions:

$$\tfrac{3}{2}y + \tfrac{1}{3} = y - \tfrac{2}{3}$$
$$6\left(\tfrac{3}{2}y + \tfrac{1}{3}\right) = 6\left(y - \tfrac{2}{3}\right)$$
$$9y + 2 = 6y - 4$$
$$9y + (-6y) + 2 = 6y + (-6y) - 4$$
$$3y + 2 = -4$$
$$3y + 2 + (-2) = -4 + (-2)$$
$$3y = -6$$
$$\tfrac{1}{3}(3y) = \tfrac{1}{3}(-6)$$
$$y = -2$$

69. Solving the equation:
$$7.5x = 1500$$
$$\frac{7.5x}{7.5} = \frac{1500}{7.5}$$
$$x = 200$$
The break-even point is 200 tickets.

71. Solving the equation:
$$1 + 3(2) + 3x = 13$$
$$7 + 3x = 13$$
$$3x = 6$$
$$x = 2$$
Laura made 2 three-point shots.

73. Solving the proportion:
$$\frac{AX}{12} = \frac{15}{20}$$
$$20(AX) = 180$$
$$AX = 9$$

75. Solving the equation:
$$3x + 2 = 19$$
$$3x + 2 - 2 = 19 - 2$$
$$3x = 17$$
$$x = \frac{17}{3}$$

77. Solving the equation:
$$2(x + 10) = 40$$
$$2x + 20 = 40$$
$$2x + 20 - 20 = 40 - 20$$
$$2x = 20$$
$$x = 10$$

79. Using the distributive property and combining like terms: $5(2x - 8) - 3 = 10x - 40 - 3 = 10x - 43$

81. Using the distributive property and combining like terms:
$$-2(3x + 5) + 3(x - 1) = -6x - 10 + 3x - 3 = -6x + 3x - 10 - 3 = -3x - 13$$

83. Using the distributive property and combining like terms: $7 - 3(2y + 1) = 7 - 6y - 3 = -6y + 7 - 3 = -6y + 4$

2.4 Solving Linear Equations

1. Solving the equation:
$$2(x + 3) = 12$$
$$2x + 6 = 12$$
$$2x + 6 + (-6) = 12 + (-6)$$
$$2x = 6$$
$$\tfrac{1}{2}(2x) = \tfrac{1}{2}(6)$$
$$x = 3$$

3. Solving the equation:
$$6(x - 1) = -18$$
$$6x - 6 = -18$$
$$6x - 6 + 6 = -18 + 6$$
$$6x = -12$$
$$\tfrac{1}{6}(6x) = \tfrac{1}{6}(-12)$$
$$x = -2$$

5. Solving the equation:
$$2(4a + 1) = -6$$
$$8a + 2 = -6$$
$$8a + 2 + (-2) = -6 + (-2)$$
$$8a = -8$$
$$\tfrac{1}{8}(8a) = \tfrac{1}{8}(-8)$$
$$a = -1$$

7. Solving the equation:
$$14 = 2(5x - 3)$$
$$14 = 10x - 6$$
$$14 + 6 = 10x - 6 + 6$$
$$20 = 10x$$
$$\tfrac{1}{10}(20) = \tfrac{1}{10}(10x)$$
$$x = 2$$

9. Solving the equation:
$$-2(3y + 5) = 14$$
$$-6y - 10 = 14$$
$$-6y - 10 + 10 = 14 + 10$$
$$-6y = 24$$
$$-\tfrac{1}{6}(-6y) = -\tfrac{1}{6}(24)$$
$$y = -4$$

11. Solving the equation:
$$-5(2a + 4) = 0$$
$$-10a - 20 = 0$$
$$-10a - 20 + 20 = 0 + 20$$
$$-10a = 20$$
$$-\tfrac{1}{10}(-10a) = -\tfrac{1}{10}(20)$$
$$a = -2$$

13. Solving the equation:
$$1 = \tfrac{1}{2}(4x+2)$$
$$1 = 2x+1$$
$$1+(-1) = 2x+1+(-1)$$
$$0 = 2x$$
$$\tfrac{1}{2}(0) = \tfrac{1}{2}(2x)$$
$$x = 0$$

15. Solving the equation:
$$3(t-4)+5 = -4$$
$$3t-12+5 = -4$$
$$3t-7 = -4$$
$$3t-7+7 = -4+7$$
$$3t = 3$$
$$\tfrac{1}{3}(3t) = \tfrac{1}{3}(3)$$
$$t = 1$$

17. Solving the equation:
$$4(2y+1)-7 = 1$$
$$8y+4-7 = 1$$
$$8y-3 = 1$$
$$8y-3+3 = 1+3$$
$$8y = 4$$
$$\tfrac{1}{8}(8y) = \tfrac{1}{8}(4)$$
$$y = \tfrac{1}{2}$$

19. Solving the equation:
$$\tfrac{1}{2}(x-3) = \tfrac{1}{4}(x+1)$$
$$\tfrac{1}{2}x - \tfrac{3}{2} = \tfrac{1}{4}x + \tfrac{1}{4}$$
$$4\left(\tfrac{1}{2}x - \tfrac{3}{2}\right) = 4\left(\tfrac{1}{4}x + \tfrac{1}{4}\right)$$
$$2x-6 = x+1$$
$$2x+(-x)-6 = x+(-x)+1$$
$$x-6 = 1$$
$$x-6+6 = 1+6$$
$$x = 7$$

21. Solving the equation:
$$-0.7(2x-7) = 0.3(11-4x)$$
$$-1.4x+4.9 = 3.3-1.2x$$
$$-1.4x+1.2x+4.9 = 3.3-1.2x+1.2x$$
$$-0.2x+4.9 = 3.3$$
$$-0.2x+4.9+(-4.9) = 3.3+(-4.9)$$
$$-0.2x = -1.6$$
$$\frac{-0.2x}{-0.2} = \frac{-1.6}{-0.2}$$
$$x = 8$$

23. Solving the equation:
$$-2(3y+1) = 3(1-6y)-9$$
$$-6y-2 = 3-18y-9$$
$$-6y-2 = -18y-6$$
$$-6y+18y-2 = -18y+18y-6$$
$$12y-2 = -6$$
$$12y-2+2 = -6+2$$
$$12y = -4$$
$$\tfrac{1}{12}(12y) = \tfrac{1}{12}(-4)$$
$$y = -\tfrac{1}{3}$$

25. Solving the equation:
$$\tfrac{3}{4}(8x-4)+3 = \tfrac{2}{5}(5x+10)-1$$
$$6x-3+3 = 2x+4-1$$
$$6x = 2x+3$$
$$6x+(-2x) = 2x+(-2x)+3$$
$$4x = 3$$
$$\tfrac{1}{4}(4x) = \tfrac{1}{4}(3)$$
$$x = \tfrac{3}{4}$$

27. Solving the equation:
$$0.06x+0.08(100-x) = 6.5$$
$$0.06x+8-0.08x = 6.5$$
$$-0.02x+8 = 6.5$$
$$-0.02x+8+(-8) = 6.5+(-8)$$
$$-0.02x = -1.5$$
$$\frac{-0.02x}{-0.02} = \frac{-1.5}{-0.02}$$
$$x = 75$$

29. Solving the equation:
$$6-5(2a-3) = 1$$
$$6-10a+15 = 1$$
$$-10a+21 = 1$$
$$-10a+21+(-21) = 1+(-21)$$
$$-10a = -20$$
$$-\tfrac{1}{10}(-10a) = -\tfrac{1}{10}(-20)$$
$$a = 2$$

31. Solving the equation:
$$0.2x-0.5 = 0.5-0.2(2x-13)$$
$$0.2x-0.5 = 0.5-0.4x+2.6$$
$$0.2x-0.5 = -0.4x+3.1$$
$$0.2x+0.4x-0.5 = -0.4x+0.4x+3.1$$
$$0.6x-0.5 = 3.1$$
$$0.6x-0.5+0.5 = 3.1+0.5$$
$$0.6x = 3.6$$
$$\frac{0.6x}{0.6} = \frac{3.6}{0.6}$$
$$x = 6$$

33. Solving the equation:

$$2(t-3)+3(t-2)=28$$
$$2t-6+3t-6=28$$
$$5t-12=28$$
$$5t-12+12=28+12$$
$$5t=40$$
$$\tfrac{1}{5}(5t)=\tfrac{1}{5}(40)$$
$$t=8$$

35. Solving the equation:

$$5(x-2)-(3x+4)=3(6x-8)+10$$
$$5x-10-3x-4=18x-24+10$$
$$2x-14=18x-14$$
$$2x+(-18x)-14=18x+(-18x)-14$$
$$-16x-14=-14$$
$$-16x-14+14=-14+14$$
$$-16x=0$$
$$-\tfrac{1}{16}(-16x)=-\tfrac{1}{16}(0)$$
$$x=0$$

37. Solving the equation:

$$2(5x-3)-(2x-4)=5-(6x+1)$$
$$10x-6-2x+4=5-6x-1$$
$$8x-2=-6x+4$$
$$8x+6x-2=-6x+6x+4$$
$$14x-2=4$$
$$14x-2+2=4+2$$
$$14x=6$$
$$\tfrac{1}{14}(14x)=\tfrac{1}{14}(6)$$
$$x=\tfrac{3}{7}$$

39. Solving the equation:

$$-(3x+1)-(4x-7)=4-(3x+2)$$
$$-3x-1-4x+7=4-3x-2$$
$$-7x+6=-3x+2$$
$$-7x+3x+6=-3x+3x+2$$
$$-4x+6=2$$
$$-4x+6+(-6)=2+(-6)$$
$$-4x=-4$$
$$-\tfrac{1}{4}(-4x)=-\tfrac{1}{4}(-4)$$
$$x=1$$

41. Multiplying the fractions: $\tfrac{1}{2}(3)=\tfrac{1}{2}\cdot\tfrac{3}{1}=\tfrac{3}{2}$

43. Multiplying the fractions: $\tfrac{2}{3}(6)=\tfrac{2}{3}\cdot\tfrac{6}{1}=\tfrac{12}{3}=4$

45. Multiplying the fractions: $\tfrac{5}{9}\cdot\tfrac{9}{5}=\tfrac{45}{45}=1$

47. Applying the distributive property: $2(3x-5)=2\cdot 3x-2\cdot 5=6x-10$

49. Applying the distributive property: $\tfrac{1}{2}(3x+6)=\tfrac{1}{2}\cdot 3x+\tfrac{1}{2}\cdot 6=\tfrac{3}{2}x+3$

51. Applying the distributive property: $\tfrac{1}{3}(-3x+6)=\tfrac{1}{3}\cdot(-3x)+\tfrac{1}{3}\cdot 6=-x+2$

2.5 Formulas

1. Using the perimeter formula:

$$P=2l+2w$$
$$300=2l+2(50)$$
$$300=2l+100$$
$$200=2l$$
$$l=100$$

The length is 100 feet.

3. Substituting $x=3$:

$$2(3)+3y=6$$
$$6+3y=6$$
$$6+(-6)+3y=6+(-6)$$
$$3y=0$$
$$y=0$$

5. Substituting $x=0$:

$$2(0)+3y=6$$
$$0+3y=6$$
$$3y=6$$
$$y=2$$

7. Substituting $y=2$:

$$2x-5(2)=20$$
$$2x-10=20$$
$$2x-10+10=20+10$$
$$2x=30$$
$$x=15$$

9. Substituting $y=0$:

$$2x-5(0)=20$$
$$2x-0=20$$
$$2x=20$$
$$x=10$$

11. Substituting $y=7$:

$$7=2x-1$$
$$7+1=2x-1+1$$
$$2x=8$$
$$x=4$$

13. Substituting $y = 3$:
$$3 = 2x - 1$$
$$3 + 1 = 2x - 1 + 1$$
$$2x = 4$$
$$x = 2$$

15. Solving for l:
$$lw = A$$
$$\frac{lw}{w} = \frac{A}{w}$$
$$l = \frac{A}{w}$$

17. Solving for r:
$$rt = d$$
$$\frac{rt}{t} = \frac{d}{t}$$
$$r = \frac{d}{t}$$

19. Solving for h:
$$lwh = V$$
$$\frac{lwh}{lw} = \frac{V}{lw}$$
$$h = \frac{V}{lw}$$

21. Solving for P:
$$PV = nRT$$
$$\frac{PV}{V} = \frac{nRT}{V}$$
$$P = \frac{nRT}{V}$$

23. Solving for a:
$$a + b + c = P$$
$$a + b + c - b - c = P - b - c$$
$$a = P - b - c$$

25. Solving for x:
$$x - 3y = -1$$
$$x - 3y + 3y = -1 + 3y$$
$$x = 3y - 1$$

27. Solving for y:
$$-3x + y = 6$$
$$-3x + 3x + y = 6 + 3x$$
$$y = 3x + 6$$

29. Solving for y:
$$2x + 3y = 6$$
$$-2x + 2x + 3y = -2x + 6$$
$$3y = -2x + 6$$
$$\tfrac{1}{3}(3y) = \tfrac{1}{3}(-2x + 6)$$
$$y = -\tfrac{2}{3}x + 2$$

31. Solving for y:
$$6x + 3y = 12$$
$$-6x + 6x + 3y = -6x + 12$$
$$3y = -6x + 12$$
$$\tfrac{1}{3}(3y) = \tfrac{1}{3}(-6x + 12)$$
$$y = -2x + 4$$

33. Solving for y:
$$5x - 2y = 3$$
$$-5x + 5x - 2y = -5x + 3$$
$$-2y = -5x + 3$$
$$-\tfrac{1}{2}(-2y) = -\tfrac{1}{2}(-5x + 3)$$
$$y = \tfrac{5}{2}x - \tfrac{3}{2}$$

35. Solving for w:
$$2l + 2w = P$$
$$2l - 2l + 2w = P - 2l$$
$$2w = P - 2l$$
$$\frac{2w}{2} = \frac{P - 2l}{2}$$
$$w = \frac{P - 2l}{2}$$

37. Solving for v:
$$vt + 16t^2 = h$$
$$vt + 16t^2 - 16t^2 = h - 16t^2$$
$$vt = h - 16t^2$$
$$\frac{vt}{t} = \frac{h - 16t^2}{t}$$
$$v = \frac{h - 16t^2}{t}$$

39. Solving for h:
$$\pi r^2 + 2\pi rh = A$$
$$\pi r^2 - \pi r^2 + 2\pi rh = A - \pi r^2$$
$$2\pi rh = A - \pi r^2$$
$$\frac{2\pi rh}{2\pi r} = \frac{A - \pi r^2}{2\pi r}$$
$$h = \frac{A - \pi r^2}{2\pi r}$$

41. Solving for y:
$$\frac{x}{2} + \frac{y}{3} = 1$$
$$-\frac{x}{2} + \frac{x}{2} + \frac{y}{3} = -\frac{x}{2} + 1$$
$$\frac{y}{3} = -\frac{x}{2} + 1$$
$$3\left(\frac{y}{3}\right) = 3\left(-\frac{x}{2} + 1\right)$$
$$y = -\frac{3}{2}x + 3$$

43. Solving for y:
$$\frac{x}{7} - \frac{y}{3} = 1$$
$$-\frac{x}{7} + \frac{x}{7} - \frac{y}{3} = -\frac{x}{7} + 1$$
$$-\frac{y}{3} = -\frac{x}{7} + 1$$
$$-3\left(-\frac{y}{3}\right) = -3\left(-\frac{x}{7} + 1\right)$$
$$y = \frac{3}{7}x - 3$$

45. Solving for y:
$$-\frac{1}{4}x + \frac{1}{8}y = 1$$
$$-\frac{1}{4}x + \frac{1}{4}x + \frac{1}{8}y = 1 + \frac{1}{4}x$$
$$\frac{1}{8}y = \frac{1}{4}x + 1$$
$$8\left(\frac{1}{8}y\right) = 8\left(\frac{1}{4}x + 1\right)$$
$$y = 2x + 8$$

47. The complement of $30°$ is $90° - 30° = 60°$, and the supplement is $180° - 30° = 150°$.

49. The complement of $45°$ is $90° - 45° = 45°$, and the supplement is $180° - 45° = 135°$.

51. Translating into an equation and solving:
$$x = 0.25 \cdot 40$$
$$x = 10$$
The number 10 is 25% of 40.

53. Translating into an equation and solving:
$$x = 0.12 \cdot 2000$$
$$x = 240$$
The number 240 is 12% of 2000.

55. Translating into an equation and solving:
$$x \cdot 28 = 7$$
$$28x = 7$$
$$\frac{1}{28}(28x) = \frac{1}{28}(7)$$
$$x = 0.25 = 25\%$$
The number 7 is 25% of 28.

57. Translating into an equation and solving:
$$x \cdot 40 = 14$$
$$40x = 14$$
$$\frac{1}{40}(40x) = \frac{1}{40}(14)$$
$$x = 0.35 = 35\%$$
The number 14 is 35% of 40.

59. Translating into an equation and solving:
$$0.50 \cdot x = 32$$
$$\frac{0.50x}{0.50} = \frac{32}{0.50}$$
$$x = 64$$
The number 32 is 50% of 64.

61. Translating into an equation and solving:
$$0.12 \cdot x = 240$$
$$\frac{0.12x}{0.12} = \frac{240}{0.12}$$
$$x = 2000$$
The number 240 is 12% of 2000.

63. Substituting $F = 212$: $C = \frac{5}{9}(212 - 32) = \frac{5}{9}(180) = 100°C$. This value agrees with the information in Table 1.

65. Substituting $F = 68$: $C = \frac{5}{9}(68 - 32) = \frac{5}{9}(36) = 20°C$. This value agrees with the information in Table 1.

67. Solving for C:
$$\frac{9}{5}C + 32 = F$$
$$\frac{9}{5}C + 32 - 32 = F - 32$$
$$\frac{9}{5}C = F - 32$$
$$\frac{5}{9}\left(\frac{9}{5}C\right) = \frac{5}{9}(F - 32)$$
$$C = \frac{5}{9}(F - 32)$$

69. We need to find what percent of 150 is 90:
$$x \cdot 150 = 90$$
$$\frac{1}{150}(150x) = \frac{1}{150}(90)$$
$$x = 0.60 = 60\%$$
So 60% of the calories in one serving of vanilla ice cream are fat calories.

71. We need to find what percent of 98 is 26:

$$x \cdot 98 = 26$$

$$\tfrac{1}{98}(98x) = \tfrac{1}{98}(26)$$

$$x \approx 0.265 = 26.5\%$$

So 26.5% of one serving of frozen yogurt are carbohydrates.

73. Solving for r:

$$2\pi r = C$$

$$2 \cdot \tfrac{22}{7} r = 44$$

$$\tfrac{44}{7} r = 44$$

$$\tfrac{7}{44}\left(\tfrac{44}{7} r\right) = \tfrac{7}{44}(44)$$

$$r = 7 \text{ meters}$$

75. Solving for r:

$$2\pi r = C$$

$$2 \cdot 3.14 r = 9.42$$

$$6.28 r = 9.42$$

$$\frac{6.28 r}{6.28} = \frac{9.42}{6.28}$$

$$r = 1.5 \text{ inches}$$

77. Solving for h:

$$\pi r^2 h = V$$

$$\tfrac{22}{7}\left(\tfrac{7}{22}\right)^2 h = 42$$

$$\tfrac{7}{22} h = 42$$

$$\tfrac{22}{7}\left(\tfrac{7}{22} h\right) = \tfrac{22}{7}(42)$$

$$h = 132 \text{ feet}$$

79. Solving for h:

$$\pi r^2 h = V$$

$$3.14(3)^2 h = 6.28$$

$$28.26 h = 6.28$$

$$\frac{28.26 h}{28.26} = \frac{6.28}{28.26}$$

$$h = \tfrac{2}{9} \text{ centimeters}$$

81. Completing the table:

Pitcher, Team	W	L	Saves	Blown Saves	Rolaids Points
John Franco, New York	0	8	38	8	82
Billy Wagner, Houston	4	3	30	5	82
Gregg Olson, Arizona	3	4	30	4	80
Bob Wickman, Milwaukee	6	9	25	7	55

83. **a.** The percent of silver is: $\tfrac{12}{20} \cdot 100 = 60\%$ **b.** The percent of copper is: $\tfrac{8}{20} \cdot 100 = 40\%$

85. Completing the table:

Altitude (feet)	Temperature (°F)
0	56
9,000	24.5
15,000	3.5
21,000	−17.5
28,000	−42
40,000	−84

87. The sum of 4 and 1 is 5.

89. The difference of 6 and 2 is 4.

91. An equivalent expression is: $2(6+3) = 2(9) = 18$

93. An equivalent expression is: $2(5)+3 = 10+3 = 13$

2.6 Applications

1. Let x represent the number. The equation is:

$$x + 5 = 13$$
$$x = 8$$

The number is 8.

3. Let x represent the number. The equation is:
$$2x + 4 = 14$$
$$2x = 10$$
$$x = 5$$
The number is 5.

5. Let x represent the number. The equation is:
$$5(x + 7) = 30$$
$$5x + 35 = 30$$
$$5x = -5$$
$$x = -1$$
The number is -1.

7. Let x and $x + 2$ represent the two numbers. The equation is:
$$x + x + 2 = 8$$
$$2x + 2 = 8$$
$$2x = 6$$
$$x = 3$$
$$x + 2 = 5$$
The two numbers are 3 and 5.

9. Let x and $3x - 4$ represent the two numbers. The equation is:
$$(x + 3x - 4) + 5 = 25$$
$$4x + 1 = 25$$
$$4x = 24$$
$$x = 6$$
$$3x - 4 = 3(6) - 4 = 14$$
The two numbers are 6 and 14.

11. Completing the table:

	Five Years Ago	Now
Fred	$x + 4 - 5 = x - 1$	$x + 4$
Barney	$x - 5$	x

The equation is:
$$x - 1 + x - 5 = 48$$
$$2x - 6 = 48$$
$$2x = 54$$
$$x = 27$$
$$x + 4 = 31$$
Barney is 27 and Fred is 31.

13. Completing the table:

	Now	Three Years from Now
Jack	$2x$	$2x + 3$
Lacy	x	$x + 3$

The equation is:
$$2x + 3 + x + 3 = 54$$
$$3x + 6 = 54$$
$$3x = 48$$
$$x = 16$$
$$2x = 32$$
Lacy is 16 and Jack is 32.

15. Completing the table:

	Now	Two Years from Now
Pat	$x + 20$	$x + 20 + 2 = x + 22$
Patrick	x	$x + 2$

The equation is:
$$x + 22 = 2(x + 2)$$
$$x + 22 = 2x + 4$$
$$22 = x + 4$$
$$x = 18$$
$$x + 20 = 38$$
Patrick is 18 and Pat is 38.

17. Let w represent the width and $w + 5$ represent the length. The equation is:
$$2w + 2(w + 5) = 34$$
$$2w + 2w + 10 = 34$$
$$4w + 10 = 34$$
$$4w = 24$$
$$w = 6$$
$$w + 5 = 11$$
The length is 11 inches and the width is 6 inches.

19. Let s represent the side of the square. The equation is:
$$4s = 48$$
$$s = 12$$
The length of one side is 12 meters.

21. Let w represent the width and $2w - 3$ represent the length. The equation is:
$$2w + 2(2w - 3) = 54$$
$$2w + 4w - 6 = 54$$
$$6w - 6 = 54$$
$$6w = 60$$
$$w = 10$$
$$2w - 3 = 2(10) - 3 = 17$$
The length is 17 inches and the width is 10 inches.

23. Completing the table:

	Nickels	Dimes
Number	x	$x + 9$
Value (cents)	$5(x)$	$10(x + 9)$

The equation is:
$$5(x) + 10(x + 9) = 210$$
$$5x + 10x + 90 = 210$$
$$15x + 90 = 210$$
$$15x = 120$$
$$x = 8$$
$$x + 9 = 17$$
Sue has 8 nickels and 17 dimes.

25. Completing the table:

	Dimes	Quarters
Number	x	$2x$
Value (cents)	$10(x)$	$25(2x)$

The equation is:
$$10(x) + 25(2x) = 900$$
$$10x + 50x = 900$$
$$60x = 900$$
$$x = 15$$
$$2x = 30$$
You have 15 dimes and 30 quarters.

27. Completing the table:

	Nickels	Dimes	Quarters
Number	x	$x + 3$	$x + 5$
Value (cents)	$5(x)$	$10(x + 3)$	$25(x + 5)$

The equation is:
$$5(x) + 10(x + 3) + 25(x + 5) = 435$$
$$5x + 10x + 30 + 25x + 125 = 435$$
$$40x + 155 = 435$$
$$40x = 280$$
$$x = 7$$
$$x + 3 = 10$$
$$x + 5 = 12$$
Katie has 7 nickels, 10 dimes, and 12 quarters.

29. The statement is: 4 is less than 10

31. The statement is: 9 is greater than or equal to -5

33. The correct statement is: $12 < 20$

35. The correct statement is: $-8 < -6$

37. Simplifying: $|8 - 3| - |5 - 2| = |5| - |3| = 5 - 3 = 2$

39. Simplifying: $15 - |9 - 3(7 - 5)| = 15 - |9 - 3(2)| = 15 - |9 - 6| = 15 - |3| = 15 - 3 = 12$

2.7 More Applications

1.   Completing the table:

	Dollars Invested at 8%	Dollars Invested at 9%
Number of	x	$x + 2000$
Interest on	$0.08(x)$	$0.09(x + 2000)$

The equation is:
$$0.08(x) + 0.09(x + 2000) = 860$$
$$0.08x + 0.09x + 180 = 860$$
$$0.17x + 180 = 860$$
$$0.17x = 680$$
$$x = 4000$$
$$x + 2000 = 6000$$
You have $4,000 invested at 8% and $6,000 invested at 9%.

3.   Completing the table:

	Dollars Invested at 10%	Dollars Invested at 12%
Number of	x	$x + 500$
Interest on	$0.10(x)$	$0.12(x + 500)$

The equation is:
$$0.10(x) + 0.12(x + 500) = 214$$
$$0.10x + 0.12x + 60 = 214$$
$$0.22x + 60 = 214$$
$$0.22x = 154$$
$$x = 700$$
$$x + 500 = 1200$$
Tyler has $700 invested at 10% and $1,200 invested at 12%.

5.   Completing the table:

	Dollars Invested at 8%	Dollars Invested at 9%	Dollars Invested at 10%
Number of	x	$2x$	$3x$
Interest on	$0.08(x)$	$0.09(2x)$	$0.10(3x)$

The equation is:
$$0.08(x) + 0.09(2x) + 0.10(3x) = 280$$
$$0.08x + 0.18x + 0.30x = 280$$
$$0.56x = 280$$
$$x = 500$$
$$2x = 1000$$
$$3x = 1500$$
She has $500 invested at 8%, $1,000 invested at 9%, and $1,500 invested at 10%.

7.   Let x represent the measure of the two equal angles, so $x + x = 2x$ represents the measure of the third angle. Since the sum of the three angles is 180°, the equation is:
$$x + x + 2x = 180°$$
$$4x = 180°$$
$$x = 45°$$
$$2x = 90°$$
The measures of the three angles are 45°, 45°, and 90°.

9. Let x represent the measure of the largest angle. Then $\frac{1}{5}x$ represents the measure of the smallest angle, and $2\left(\frac{1}{5}x\right) = \frac{2}{5}x$ represents the measure of the other angle. Since the sum of the three angles is 180°, the equation is:

$$x + \frac{1}{5}x + \frac{2}{5}x = 180°$$
$$\frac{5}{5}x + \frac{1}{5}x + \frac{2}{5}x = 180°$$
$$\frac{8}{5}x = 180°$$
$$x = 112.5°$$
$$\frac{1}{5}x = 22.5°$$
$$\frac{2}{5}x = 45°$$

The measures of the three angles are 22.5°, 45°, and 112.5°.

11. Let x represent the measure of the other acute angle, and 90° is the measure of the right angle. Since the sum of the three angles is 180°, the equation is:

$$x + 37° + 90° = 180°$$
$$x + 127° = 180°$$
$$x = 53°$$

The other two angles are 53° and 90°.

13. Let x represent the total minutes for the call. Then $0.41 is charged for the first minute, and $0.32 is charged for the additional $x - 1$ minutes. The equation is:

$$0.41(1) + 0.32(x - 1) = 5.21$$
$$0.41 + 0.32x - 0.32 = 5.21$$
$$0.32x + 0.09 = 5.21$$
$$0.32x = 5.12$$
$$x = 16$$

The call was 16 minutes long.

15. Let x represent the hours Jo Ann worked that week. Then $12/hour is paid for the first 35 hours and $18/hour is paid for the additional $x - 35$ hours. The equation is:

$$12(35) + 18(x - 35) = 492$$
$$420 + 18x - 630 = 492$$
$$18x - 210 = 492$$
$$18x = 702$$
$$x = 39$$

Jo Ann worked 39 hours that week.

17. Let x represent the number of children's tickets Stacey sold, so $2x$ represents the number of adult tickets sold. The equation is:

$$6.00(2x) + 4.50(x) = 115.50$$
$$12x + 4.5x = 115.5$$
$$16.5x = 115.5$$
$$x = 7$$
$$2x = 14$$

Stacey sold 7 children's tickets and 14 adult tickets.

19. For Jeff, the total time traveled is $\dfrac{425 \text{ miles}}{55 \text{ miles / hour}} \approx 7.72 \text{ hours} \approx 463 \text{ minutes}$. Since he left at 11:00 AM, he will arrive at 6:43 PM. For Carla, the total time traveled is $\dfrac{425 \text{ miles}}{65 \text{ miles / hour}} \approx 6.54 \text{ hours} \approx 392 \text{ minutes}$. Since she left at 1:00 PM, she will arrive at 7:32 PM. Thus Jeff will arrive in Lake Tahoe first.

21. Since $\frac{1}{5}$ mile $= 0.2$ mile, the taxi charge is $1.25 for the first $\frac{1}{5}$ mile and $0.25 per fifth mile for the remaining 7.3 miles. Since 7.3 miles $= \dfrac{7.3}{0.2} = 36.5$ fifths, the total charge is: $\$1.25 + \$0.25(36.5) \approx \$10.38$

23. The first $\frac{1}{5}$ mile is \$1.25, and the remaining $12.4 - 0.2 = 12.2$ miles will be charged at \$0.25 per fifth mile. Since $12.2 \text{ miles} = \dfrac{12.2}{0.2} = 61 \text{ fifths}$, the total charge is: $\$1.25 + \$0.25(61) = \$16.50$. Yes, the meter is working correctly.

25. If all 36 people are Elk's Lodge members (which would be the least amount), the cost of the lessons would be $\$3(36) = \$108$. Since half of the money is paid to Ike and Nancy, the least amount they could make is $\frac{1}{2}(\$108) = \$54$.

27. Yes. The total receipts were \$160, which is possible if there were 10 Elk's members and 26 nonmembers. Computing the total receipts: $10(\$3) + 26(\$5) = \$30 + \$130 = \$160$

29. Let x represent the median starting salary for 1998 graduates. The equation is:
$$x + 0.07x = 85,200$$
$$1.07x = 85,200$$
$$x = \frac{85,200}{1.07} \approx \$79,626$$
The median starting salary for 1998 graduates was approximately \$79,626.

31. The pattern is to add -4, so the next number is: $-4 + (-4) = -8$

33. The pattern is to multiply by $-\frac{1}{2}$, so the next number is: $-\frac{3}{2}\left(-\frac{1}{2}\right) = \frac{3}{4}$

35. Each number is the square of a number, with alternating signs. For example, $1^2 = 1$, $-2^2 = -4$, $3^2 = 9$, and $-4^2 = -16$. Based on this pattern, the next number is: $5^2 = 25$

37. The pattern is to add $-\frac{1}{2}$, so the next number is: $\frac{1}{2} + \left(-\frac{1}{2}\right) = 0$

2.8 Linear Inequalities

1. Solving the inequality:
$$x - 5 < 7$$
$$x - 5 + 5 < 7 + 5$$
$$x < 12$$
Graphing the solution set:

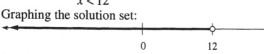

3. Solving the inequality:
$$a - 4 \le 8$$
$$a - 4 + 4 \le 8 + 4$$
$$a \le 12$$
Graphing the solution set:

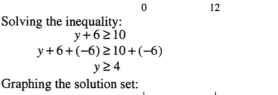

5. Solving the inequality:
$$x - 4.3 > 8.7$$
$$x - 4.3 + 4.3 > 8.7 + 4.3$$
$$x > 13$$
Graphing the solution set:

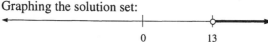

7. Solving the inequality:
$$y + 6 \ge 10$$
$$y + 6 + (-6) \ge 10 + (-6)$$
$$y \ge 4$$
Graphing the solution set:

9. Solving the inequality:
$$2 < x - 7$$
$$2 + 7 < x - 7 + 7$$
$$9 < x$$
$$x > 9$$
Graphing the solution set:

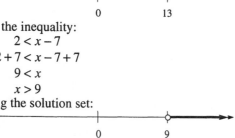

11. Solving the inequality:
$$3x < 6$$
$$\tfrac{1}{3}(3x) < \tfrac{1}{3}(6)$$
$$x < 2$$
Graphing the solution set:

13. Solving the inequality:

$$5a \le 25$$
$$\tfrac{1}{5}(5a) \le \tfrac{1}{5}(25)$$
$$a \le 5$$

Graphing the solution set:

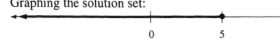

15. Solving the inequality:

$$\frac{x}{3} > 5$$
$$3\left(\frac{x}{3}\right) > 3(5)$$
$$x > 15$$

Graphing the solution set:

17. Solving the inequality:

$$-2x > 6$$
$$-\tfrac{1}{2}(-2x) < -\tfrac{1}{2}(6)$$
$$x < -3$$

Graphing the solution set:

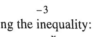

19. Solving the inequality:

$$-3x \ge -18$$
$$-\tfrac{1}{3}(-3x) \le -\tfrac{1}{3}(-18)$$
$$x \le 6$$

Graphing the solution set:

21. Solving the inequality:

$$-\frac{x}{5} \le 10$$
$$-5\left(-\frac{x}{5}\right) \ge -5(10)$$
$$x \ge -50$$

Graphing the solution set:

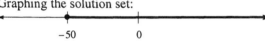

23. Solving the inequality:

$$-\tfrac{2}{3}y > 4$$
$$-\tfrac{3}{2}\left(-\tfrac{2}{3}y\right) < -\tfrac{3}{2}(4)$$
$$y < -6$$

Graphing the solution set:

25. Solving the inequality:

$$2x - 3 < 9$$
$$2x - 3 + 3 < 9 + 3$$
$$2x < 12$$
$$\tfrac{1}{2}(2x) < \tfrac{1}{2}(12)$$
$$x < 6$$

Graphing the solution set:

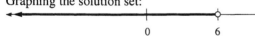

27. Solving the inequality:

$$-\tfrac{1}{5}y - \tfrac{1}{3} \le \tfrac{2}{3}$$
$$-\tfrac{1}{5}y - \tfrac{1}{3} + \tfrac{1}{3} \le \tfrac{2}{3} + \tfrac{1}{3}$$
$$-\tfrac{1}{5}y \le 1$$
$$-5\left(-\tfrac{1}{5}y\right) \ge -5(1)$$
$$y \ge -5$$

Graphing the solution set:

29. Solving the inequality:

$$-4x + 1 > -11$$
$$-4x + 1 + (-1) > -11 + (-1)$$
$$-4x > -12$$
$$-\tfrac{1}{4}(-4x) < -\tfrac{1}{4}(-12)$$
$$x < 3$$

Graphing the solution set:

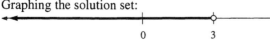

31. Solving the inequality:

$$\tfrac{2}{3}x - 5 \le 7$$
$$\tfrac{2}{3}x - 5 + 5 \le 7 + 5$$
$$\tfrac{2}{3}x \le 12$$
$$\tfrac{3}{2}\left(\tfrac{2}{3}x\right) \le \tfrac{3}{2}(12)$$
$$x \le 18$$

Graphing the solution set:

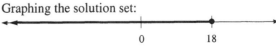

33. Solving the inequality:
$$-\tfrac{2}{5}a - 3 > 5$$
$$-\tfrac{2}{5}a - 3 + 3 > 5 + 3$$
$$-\tfrac{2}{5}a > 8$$
$$-\tfrac{5}{2}\left(-\tfrac{2}{5}a\right) < -\tfrac{5}{2}(8)$$
$$a < -20$$
Graphing the solution set:

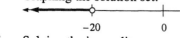

37. Solving the inequality:
$$0.3(a + 1) \le 1.2$$
$$0.3a + 0.3 \le 1.2$$
$$0.3a + 0.3 + (-0.3) \le 1.2 + (-0.3)$$
$$0.3a \le 0.9$$
$$\frac{0.3a}{0.3} \le \frac{0.9}{0.3}$$
$$a \le 3$$
Graphing the solution set:

41. Solving the inequality:

$$3x - 5 > 8x$$
$$-3x + 3x - 5 > -3x + 8x$$
$$-5 > 5x$$
$$\tfrac{1}{5}(-5) > \tfrac{1}{5}(5x)$$
$$-1 > x$$
$$x < -1$$

Graphing the solution set:

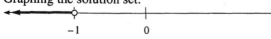

45. Solving the inequality:

$$-0.4x + 1.2 < -2x - 0.4$$
$$2x - 0.4x + 1.2 < 2x - 2x - 0.4$$
$$1.6x + 1.2 < -0.4$$
$$1.6x + 1.2 + (-1.2) < -0.4 + (-1.2)$$
$$1.6x < -1.6$$
$$\frac{1.6x}{1.6} < \frac{-1.6}{1.6}$$
$$x < -1$$

Graphing the solution set:

35. Solving the inequality:
$$5 - \tfrac{3}{5}y > -10$$
$$-5 + 5 - \tfrac{3}{5}y > -5 + (-10)$$
$$-\tfrac{3}{5}y > -15$$
$$-\tfrac{5}{3}\left(-\tfrac{3}{5}y\right) < -\tfrac{5}{3}(-15)$$
$$y < 25$$
Graphing the solution set:

39. Solving the inequality:
$$2(5 - 2x) \le -20$$
$$10 - 4x \le -20$$
$$-10 + 10 - 4x \le -10 + (-20)$$
$$-4x \le -30$$
$$-\tfrac{1}{4}(-4x) \ge -\tfrac{1}{4}(-30)$$
$$x \ge \tfrac{15}{2}$$
Graphing the solution set:

43. Solving the inequality:
$$\tfrac{1}{3}y - \tfrac{1}{2} \le \tfrac{5}{6}y + \tfrac{1}{2}$$
$$6\left(\tfrac{1}{3}y - \tfrac{1}{2}\right) \le 6\left(\tfrac{5}{6}y + \tfrac{1}{2}\right)$$
$$2y - 3 \le 5y + 3$$
$$-5y + 2y - 3 \le -5y + 5y + 3$$
$$-3y - 3 \le 3$$
$$-3y - 3 + 3 \le 3 + 3$$
$$-3y \le 6$$
$$-\tfrac{1}{3}(-3y) \ge -\tfrac{1}{3}(6)$$
$$y \ge -2$$
Graphing the solution set:

47. Solving the inequality:
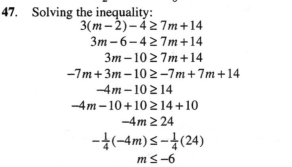
$$3(m - 2) - 4 \ge 7m + 14$$
$$3m - 6 - 4 \ge 7m + 14$$
$$3m - 10 \ge 7m + 14$$
$$-7m + 3m - 10 \ge -7m + 7m + 14$$
$$-4m - 10 \ge 14$$
$$-4m - 10 + 10 \ge 14 + 10$$
$$-4m \ge 24$$
$$-\tfrac{1}{4}(-4m) \le -\tfrac{1}{4}(24)$$
$$m \le -6$$
Graphing the solution set:

49. Solving the inequality:
$$3 - 4(x - 2) \le -5x + 6$$
$$3 - 4x + 8 \le -5x + 6$$
$$-4x + 11 \le -5x + 6$$
$$-4x + 5x + 11 \le -5x + 5x + 6$$
$$x + 11 \le 6$$
$$x + 11 + (-11) \le 6 + (-11)$$
$$x \le -5$$
Graphing the solution set:

51. Solving for y:
$$3x + 2y < 6$$
$$2y < -3x + 6$$
$$y < -\tfrac{3}{2}x + 3$$

53. Solving for y:
$$2x - 5y > 10$$
$$-5y > -2x + 10$$
$$y < \tfrac{2}{5}x - 2$$

55. Solving for y:
$$-3x + 7y \le 21$$
$$7y \le 3x + 21$$
$$y \le \tfrac{3}{7}x + 3$$

57. Solving for y:
$$2x - 4y \ge -4$$
$$-4y \ge -2x - 4$$
$$y \le \tfrac{1}{2}x + 1$$

59. The inequality is $x < 3$.

61. The inequality is $x \ge 3$.

63. Let x and $x + 1$ represent the integers. Solving the inequality:
$$x + x + 1 \ge 583$$
$$2x + 1 \ge 583$$
$$2x \ge 582$$
$$x \ge 291$$
The two numbers are at least 291.

65. Let x represent the number. Solving the inequality:
$$2x + 6 < 10$$
$$2x < 4$$
$$x < 2$$

67. Let x represent the number. Solving the inequality:
$$4x > x - 8$$
$$3x > -8$$
$$x > -\tfrac{8}{3}$$

69. Let w represent the width, so $3w$ represents the length. Using the formula for perimeter:
$$2(w) + 2(3w) \ge 48$$
$$2w + 6w \ge 48$$
$$8w \ge 48$$
$$w \ge 6$$
The width is at least 6 meters.

71. Let x, $x + 2$, and $x + 4$ represent the sides of the triangle. The inequality is:
$$x + (x + 2) + (x + 4) > 24$$
$$3x + 6 > 24$$
$$3x > 18$$
$$x > 6$$
The shortest side is an even number greater than 6 inches (greater than or equal to 8 inches).

73. The inequality is $t \ge 100$.

75. Let n represent the number of tickets they sell. The inequality is:
$$7.5n < 1500$$
$$n < 200$$
They will lose money if they sell less than 200 tickets. They will make a profit if they sell more than 200 tickets.

77. b (commutative property of addition)

79. a (distributive property)

81. b and c (commutative and associative properties of addition)

Chapter 2 Review

1. Simplifying the expression: $5x - 8x = (5-8)x = -3x$

3. Simplifying the expression: $-a + 2 + 5a - 9 = -a + 5a + 2 - 9 = 4a - 7$

5. Simplifying the expression: $6 - 2(3y+1) - 4 = 6 - 6y - 2 - 4 = -6y + 6 - 2 - 4 = -6y$

7. Evaluating when $x = 3$: $7x - 2 = 7(3) - 2 = 21 - 2 = 19$

9. Evaluating when $x = 3$: $-x - 2x - 3x = -6x = -6(3) = -18$

11. Evaluating when $x = -2$: $-3x + 2 = -3(-2) + 2 = 6 + 2 = 8$

13. Solving the equation:
$$x + 2 = -6$$
$$x + 2 + (-2) = -6 + (-2)$$
$$x = -8$$

15. Solving the equation:
$$10 - 3y + 4y = 12$$
$$10 + y = 12$$
$$-10 + 10 + y = -10 + 12$$
$$y = 2$$

17. Solving the equation:
$$2x = -10$$
$$\tfrac{1}{2}(2x) = \tfrac{1}{2}(-10)$$
$$x = -5$$

19. Solving the equation:
$$\frac{x}{3} = 4$$
$$3\left(\frac{x}{3}\right) = 3(4)$$
$$x = 12$$

21. Solving the equation:
$$3a - 2 = 5a$$
$$-3a + 3a - 2 = -3a + 5a$$
$$-2 = 2a$$
$$\tfrac{1}{2}(-2) = \tfrac{1}{2}(2a)$$
$$a = -1$$

23. Solving the equation:
$$3x + 2 = 5x - 8$$
$$3x + (-5x) + 2 = 5x + (-5x) - 8$$
$$-2x + 2 = -8$$
$$-2x + 2 + (-2) = -8 + (-2)$$
$$-2x = -10$$
$$-\tfrac{1}{2}(-2x) = -\tfrac{1}{2}(-10)$$
$$x = 5$$

25. Solving the equation:
$$0.7x - 0.1 = 0.5x - 0.1$$
$$0.7x + (-0.5x) - 0.1 = 0.5x + (-0.5x) - 0.1$$
$$0.2x - 0.1 = -0.1$$
$$0.2x - 0.1 + 0.1 = -0.1 + 0.1$$
$$0.2x = 0$$
$$\frac{0.2x}{0.2} = \frac{0}{0.2}$$
$$x = 0$$

27. Simplifying and then solving the equation:
$$2(x - 5) = 10$$
$$2x - 10 = 10$$
$$2x - 10 + 10 = 10 + 10$$
$$2x = 20$$
$$\tfrac{1}{2}(2x) = \tfrac{1}{2}(20)$$
$$x = 10$$

29. Simplifying and then solving the equation:
$$\tfrac{1}{2}(3t - 2) + \tfrac{1}{2} = \tfrac{5}{2}$$
$$\tfrac{3}{2}t - 1 + \tfrac{1}{2} = \tfrac{5}{2}$$
$$\tfrac{3}{2}t - \tfrac{1}{2} = \tfrac{5}{2}$$
$$\tfrac{3}{2}t - \tfrac{1}{2} + \tfrac{1}{2} = \tfrac{5}{2} + \tfrac{1}{2}$$
$$\tfrac{3}{2}t = 3$$
$$\tfrac{2}{3}\left(\tfrac{3}{2}t\right) = \tfrac{2}{3}(3)$$
$$t = 2$$

31. Simplifying and then solving the equation:
$$2(3x + 7) = 4(5x - 1) + 18$$
$$6x + 14 = 20x - 4 + 18$$
$$6x + 14 = 20x + 14$$
$$6x + (-20x) + 14 = 20x + (-20x) + 14$$
$$-14x + 14 = 14$$
$$-14x + 14 + (-14) = 14 + (-14)$$
$$-14x = 0$$
$$-\tfrac{1}{14}(-14x) = -\tfrac{1}{14}(0)$$
$$x = 0$$

33. Substituting $x = 5$:
$$4(5) - 5y = 20$$
$$20 - 5y = 20$$
$$-5y = 0$$
$$y = 0$$

35. Substituting $x = -5$:
$$4(-5) - 5y = 20$$
$$-20 - 5y = 20$$
$$-5y = 40$$
$$y = -8$$

37. Solving for y:

$$2x - 5y = 10$$
$$-5y = -2x + 10$$
$$y = \frac{2}{5}x - 2$$

39. Solving for h:

$$\pi r^2 h = V$$
$$\frac{\pi r^2 h}{\pi r^2} = \frac{V}{\pi r^2}$$
$$h = \frac{V}{\pi r^2}$$

41. Computing the amount:
$$0.86(240) = x$$
$$x = 206.4$$
So 86% of 240 is 206.4.

43. Let x represent the number. The equation is:
$$2x + 6 = 28$$
$$2x = 22$$
$$x = 11$$
The number is 11.

45. Completing the table:

	Dollars Invested at 9%	Dollars Invested at 10%
Number of	x	$x + 300$
Interest on	$0.09(x)$	$0.10(x + 300)$

The equation is:
$$0.09(x) + 0.10(x + 300) = 125$$
$$0.09x + 0.10x + 30 = 125$$
$$0.19x + 30 = 125$$
$$0.19x = 95$$
$$x = 500$$
$$x + 300 = 800$$
The man invested \$500 at 9% and \$800 at 10%.

47. Solving the inequality:

$$-2x < 4$$
$$-\frac{1}{2}(-2x) > -\frac{1}{2}(4)$$
$$x > -2$$

49. Solving the inequality:

$$-\frac{a}{2} \leq -3$$
$$-2\left(-\frac{a}{2}\right) \geq -2(-3)$$
$$a \geq 6$$

51. Solving the inequality:

$$-4x + 5 > 37$$
$$-4x > 32$$
$$x < -8$$

Graphing the solution set:

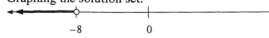

53. Solving the inequality:
$$2(3t + 1) + 6 \geq 5(2t + 4)$$
$$6t + 2 + 6 \geq 10t + 20$$
$$6t + 8 \geq 10t + 20$$
$$-4t + 8 \geq 20$$
$$-4t \geq 12$$
$$t \leq -3$$

Graphing the solution set:

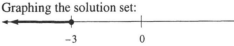

Chapters 1-2 Cumulative Review

1. Simplifying: $6+3(6+2)=6+3(8)=6+24=30$

3. Simplifying: $7-9-12=7+(-9)+(-12)=7+(-21)=-14$

5. Using the associative property: $\frac{1}{5}(10x)=\left(\frac{1}{5}\cdot10\right)x=2x$

7. Simplifying: $\left(-\frac{2}{3}\right)^3=\left(-\frac{2}{3}\right)\left(-\frac{2}{3}\right)\left(-\frac{2}{3}\right)=-\frac{8}{27}$

9. Simplifying: $-\frac{3}{4}\div\frac{15}{16}=-\frac{3}{4}\cdot\frac{16}{15}=-\frac{49}{60}=-\frac{4}{5}$

11. Simplifying: $\frac{-4(-6)}{-9}=\frac{24}{-9}=-\frac{8}{3}$

13. Simplifying: $\frac{(5-3)^2}{5^2-3^2}=\frac{2^2}{25-9}=\frac{4}{16}=\frac{1}{4}$

15. First factor the denominators to find the LCM:
$$21=3\cdot7$$
$$35=5\cdot7$$
$$LCM=3\cdot5\cdot7=105$$
Simplifying: $\frac{4}{21}-\frac{9}{35}=\frac{4\cdot5}{21\cdot5}-\frac{9\cdot3}{35\cdot3}=\frac{20}{105}-\frac{27}{105}=-\frac{7}{105}=-\frac{7}{3\cdot5\cdot7}=-\frac{1}{3\cdot5}=-\frac{1}{15}$

17. Solving the equation:
$$7x=6x+4$$
$$7x+(-6x)=6x+(-6x)+4$$
$$x=4$$

19. Solving the equation:
$$-\frac{3}{5}x=30$$
$$-\frac{5}{3}\left(-\frac{3}{5}x\right)=-\frac{5}{3}(30)$$
$$x=-50$$

21. Solving the equation:
$$5x-7=x-1$$
$$5x+(-x)-7=x+(-x)-1$$
$$4x-7=-1$$
$$4x-7+7=-1+7$$
$$4x=6$$
$$\frac{1}{4}(4x)=\frac{1}{4}(6)$$
$$x=\frac{3}{2}$$

23. Solving the equation:
$$15-3(2t+4)=1$$
$$15-6t-12=1$$
$$-6t+3=1$$
$$-6t+3+(-3)=1+(-3)$$
$$-6t=-2$$
$$-\frac{1}{6}(-6t)=-\frac{1}{6}(-2)$$
$$t=\frac{1}{3}$$

25. Solving the equation:
$$\frac{1}{3}(x-6)=\frac{1}{4}(x+8)$$
$$\frac{1}{3}x-2=\frac{1}{4}x+2$$
$$12\left(\frac{1}{3}x-2\right)=12\left(\frac{1}{4}x+2\right)$$
$$4x-24=3x+24$$
$$4x+(-3x)-24=3x+(-3x)+24$$
$$x-24=24$$
$$x-24+24=24+24$$
$$x=48$$

27. Solving for y:
$$3x+4y=12$$
$$3x+(-3x)+4y=12+(-3x)$$
$$4y=-3x+12$$
$$\frac{1}{4}(4y)=\frac{1}{4}(-3x+12)$$
$$y=-\frac{3}{4}x+3$$

29. Solving the inequality:
$$-\frac{1}{2}x<5$$
$$-2\left(-\frac{1}{2}x\right)>-2(5)$$
$$x>-10$$

Graphing the solution set:

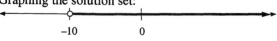

31. Solving the inequality:
$$-5x+9<-6$$
$$-5x+9+(-9)<-6+(-9)$$
$$-5x<-15$$
$$-\frac{1}{5}(-5x)>-\frac{1}{5}(-15)$$
$$x>3$$

Graphing the solution set:

33. The opposite of $-\frac{2}{3}$ is $\frac{2}{3}$, the reciprocal is $-\frac{3}{2}$, and the absolute value is $\left|-\frac{2}{3}\right|=\frac{2}{3}$.

35. The pattern is to add -3, so the next term is: $-5+(-3)=-8$

37. Using the distributive property: $\frac{1}{4}(8x-4) = \frac{1}{4} \cdot 8x - \frac{1}{4} \cdot 4 = 2x - 1$

39. Reducing the fraction: $\frac{234}{312} = \dfrac{2 \cdot 3 \cdot 3 \cdot 13}{2 \cdot 2 \cdot 2 \cdot 3 \cdot 13} = \dfrac{3}{2 \cdot 2} = \frac{3}{4}$

41. Evaluating when $a = 3$ and $b = -2$: $a^2 - 2ab + b^2 = (3)^2 - 2(3)(-2) + (-2)^2 = 9 + 12 + 4 = 25$

43. Let x represent the number. The equation is:
$$2x + 7 = 31$$
$$2x = 24$$
$$x = 12$$
The number is 12.

45. Let x represent the acute angle and $90°$ is the right angle. Since the sum of the three angles is $180°$, the equation is:
$$x + 42° + 90° = 180°$$
$$x + 132° = 180°$$
$$x = 48°$$
The other two angles are $48°$ and $90°$.

47. The other angle must be $90° - 25° = 65°$.

49. Completing the table:

	Dollars Invested at 5%	Dollars Invested at 6%
Number of	x	$x + 200$
Interest on	$0.05(x)$	$0.06(x+200)$

The equation is:
$$0.05(x) + 0.06(x + 200) = 56$$
$$0.05x + 0.06x + 12 = 56$$
$$0.11x + 12 = 56$$
$$0.11x = 44$$
$$x = 400$$
$$x + 200 = 600$$
You have $400 invested at 5% and $600 invested at 6%.

Chapter 2 Test

1. Simplifying: $3x + 2 - 7x + 3 = 3x - 7x + 2 + 3 = -4x + 5$

2. Simplifying: $4a - 5 - a + 1 = 4a - a - 5 + 1 = 3a - 4$

3. Simplifying: $7 - 3(y + 5) - 4 = 7 - 3y - 15 - 4 = -3y + 7 - 15 - 4 = -3y - 12$

4. Simplifying: $8(2x + 1) - 5(x - 4) = 16x + 8 - 5x + 20 = 16x - 5x + 8 + 20 = 11x + 28$

5. Evaluating when $x = -5$: $2x - 3 - 7x = -5x - 3 = -5(-5) - 3 = 25 - 3 = 22$

6. Evaluating when $x = 2$ and $y = 3$: $x^2 + 2xy + y^2 = (2)^2 + 2(2)(3) + (3)^2 = 4 + 12 + 9 = 25$

7. Solving the equation:
$$2x - 5 = 7$$
$$2x - 5 + 5 = 7 + 5$$
$$2x = 12$$
$$\tfrac{1}{2}(2x) = \tfrac{1}{2}(12)$$
$$x = 6$$

8. Solving the equation:
$$2y + 4 = 5y$$
$$-2y + 2y + 4 = -2y + 5y$$
$$4 = 3y$$
$$\tfrac{1}{3}(4) = \tfrac{1}{3}(3y)$$
$$y = \tfrac{4}{3}$$

9.   First clear the equation of fractions by multiplying by 10:
$$\tfrac{1}{2}x - \tfrac{1}{10} = \tfrac{1}{5}x + \tfrac{1}{2}$$
$$10\left(\tfrac{1}{2}x - \tfrac{1}{10}\right) = 10\left(\tfrac{1}{5}x + \tfrac{1}{2}\right)$$
$$5x - 1 = 2x + 5$$
$$5x + (-2x) - 1 = 2x + (-2x) + 5$$
$$3x - 1 = 5$$
$$3x - 1 + 1 = 5 + 1$$
$$3x = 6$$
$$\tfrac{1}{3}(3x) = \tfrac{1}{3}(6)$$
$$x = 2$$

10.   Solving the equation:
$$\tfrac{2}{5}(5x - 10) = -5$$
$$2x - 4 = -5$$
$$2x - 4 + 4 = -5 + 4$$
$$2x = -1$$
$$\tfrac{1}{2}(2x) = \tfrac{1}{2}(-1)$$
$$x = -\tfrac{1}{2}$$

11.   Solving the equation:
$$-5(2x + 1) - 6 = 19$$
$$-10x - 5 - 6 = 19$$
$$-10x - 11 = 19$$
$$-10x - 11 + 11 = 19 + 11$$
$$-10x = 30$$
$$-\tfrac{1}{10}(-10x) = -\tfrac{1}{10}(30)$$
$$x = -3$$

12.   Solving the equation:

$$0.04x + 0.06(100 - x) = 4.6$$
$$0.04x + 6 - 0.06x = 4.6$$
$$-0.02x + 6 = 4.6$$
$$-0.02x + 6 + (-6) = 4.6 + (-6)$$
$$-0.02x = -1.4$$
$$\frac{-0.02x}{-0.02} = \frac{-1.4}{-0.02}$$
$$x = 70$$

13.   Solving the equation:
$$2(t - 4) + 3(t + 5) = 2t - 2$$
$$2t - 8 + 3t + 15 = 2t - 2$$
$$5t + 7 = 2t - 2$$
$$5t + (-2t) + 7 = 2t + (-2t) - 2$$
$$3t + 7 = -2$$
$$3t + 7 + (-7) = -2 + (-7)$$
$$3t = -9$$
$$\tfrac{1}{3}(3t) = \tfrac{1}{3}(-9)$$
$$t = -3$$

14.   Solving the equation:
$$2x - 4(5x + 1) = 3x + 17$$
$$2x - 20x - 4 = 3x + 17$$
$$-18x - 4 = 3x + 17$$
$$-18x + (-3x) - 4 = 3x + (-3x) + 17$$
$$-21x - 4 = 17$$
$$-21x - 4 + 4 = 17 + 4$$
$$-21x = 21$$
$$-\tfrac{1}{21}(21x) = -\tfrac{1}{21}(21)$$
$$x = -1$$

15.   Finding the amount:

$$0.15(38) = x$$
$$5.7 = x$$

So 15% of 38 is 5.7.

16.   Finding the base:
$$0.12x = 240$$
$$\frac{0.12x}{0.12} = \frac{240}{0.12}$$
$$x = 2000$$
So 12% of 2,000 is 240.

17. Substituting $y = -2$:

$$2x - 3(-2) = 12$$
$$2x + 6 = 12$$
$$2x = 6$$
$$x = 3$$

18. Substituting $V = 88$, $\pi = \frac{22}{7}$, and $r = 3$:

$$\tfrac{1}{3} \cdot \tfrac{22}{7} \cdot (3)^2 h = 88$$
$$\tfrac{66}{7} h = 88$$
$$\tfrac{7}{66}\left(\tfrac{66}{7} h\right) = \tfrac{7}{66}(88)$$
$$h = \tfrac{28}{3} \text{ inches}$$

19. Solving for y:

$$2x + 5y = 20$$
$$5y = -2x + 20$$
$$y = -\tfrac{2}{5}x + 4$$

20. Solving for v:

$$x + vt + 16t^2 = h$$
$$vt = h - x - 16t^2$$
$$v = \frac{h - x - 16t^2}{t}$$

21. Completing the table:

	Ten Years Ago	Now
Dave	$2x - 10$	$2x$
Rick	$x - 10$	x

The equation is:
$$2x - 10 + x - 10 = 40$$
$$3x - 20 = 40$$
$$3x = 60$$
$$x = 20$$
$$2x = 40$$

Rick is 20 years old and Dave is 40 years old.

22. Let w represent the width and $2w$ represent the length. Using the perimeter formula:

$$2(w) + 2(2w) = 60$$
$$2w + 4w = 60$$
$$6w = 60$$
$$w = 10$$
$$2w = 20$$

The width is 10 inches and the length is 20 inches.

23. Completing the table:

	Dimes	Quarters
Number	$x + 7$	x
Value (cents)	$10(x + 7)$	$25(x)$

The equation is:
$$10(x + 7) + 25(x) = 350$$
$$10x + 70 + 25x = 350$$
$$35x + 70 = 350$$
$$35x = 280$$
$$x = 8$$
$$x + 7 = 15$$

He has 8 quarters and 15 dimes in his collection.

24. Completing the table:

	Dollars Invested at 7%	Dollars Invested at 9%
Number of	x	$x + 600$
Interest on	$0.07(x)$	$0.09(x + 600)$

The equation is:
$$0.07(x) + 0.09(x + 600) = 182$$
$$0.07x + 0.09x + 54 = 182$$
$$0.16x + 54 = 182$$
$$0.16x = 128$$
$$x = 800$$
$$x + 600 = 1400$$

She has \$800 invested at 7% and \$1,400 invested at 9%.

25.   Solving the inequality:

$$2x+3<5$$
$$2x+3+(-3)<5+(-3)$$
$$2x<2$$
$$\tfrac{1}{2}(2x)<\tfrac{1}{2}(2)$$
$$x<1$$

Graphing the solution set:

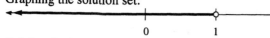

26.   Solving the inequality:

$$-5a>20$$
$$-\tfrac{1}{5}(-5a)<-\tfrac{1}{5}(20)$$
$$a<-4$$

Graphing the solution set:

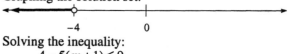

27.   Solving the inequality:

$$0.4-0.2x\geq1$$
$$-0.4+0.4-0.2x\geq-0.4+1$$
$$-0.2x\geq0.6$$
$$\frac{-0.2x}{-0.2}\leq\frac{0.6}{-0.2}$$
$$x\leq-3$$

Graphing the solution set:

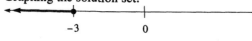

28.   Solving the inequality:

$$4-5(m+1)\leq9$$
$$4-5m-5\leq9$$
$$-5m-1\leq9$$
$$-5m-1+1\leq9+1$$
$$-5m\leq10$$
$$-\tfrac{1}{5}(-5m)\geq-\tfrac{1}{5}(10)$$
$$m\geq-2$$

Graphing the solution set:

Chapter 3
Graphing and Linear Systems

3.1 Paired Data and Graphing Ordered Pairs

1. Graphing the ordered pair:

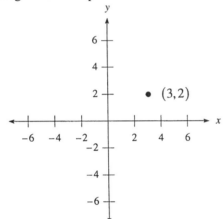

3. Graphing the ordered pair:

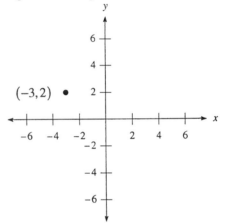

5. Graphing the ordered pair:

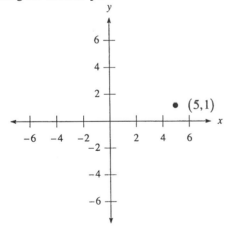

7. Graphing the ordered pair:

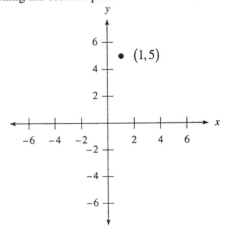

9.   Graphing the ordered pair:

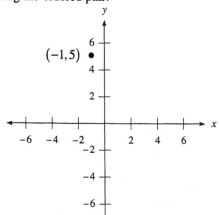

11.   Graphing the ordered pair:

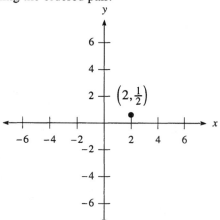

13.   Graphing the ordered pair:

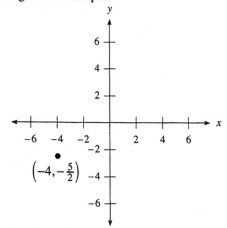

15.   Graphing the ordered pair:

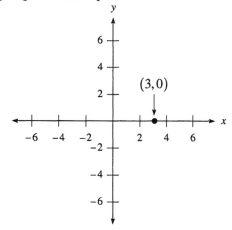

17.   Graphing the ordered pair:

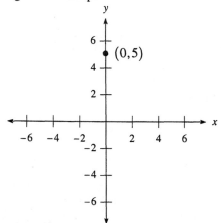

19.   The coordinates are $(-4, 4)$.
23.   The coordinates are $(-3, 0)$.
27.   The coordinates are $(-5, -5)$.

21.   The coordinates are $(-4, 2)$.
25.   The coordinates are $(2, -2)$.

29.   Graphing the line:

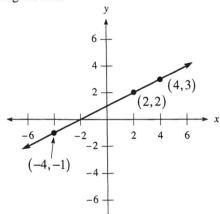

Yes, the point $(2,2)$ lies on the line.

31.   Graphing the line:

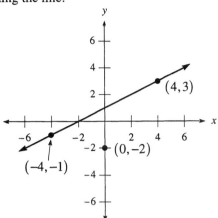

No, the point $(0,-2)$ does not lie on the line.

33.   Graphing the line:

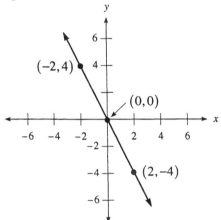

Yes, the point $(0,0)$ lies on the line.

34.   Graphing the line:

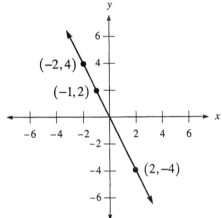

Yes, the point $(-1,2)$ lies on the line.

35.   Graphing the line:

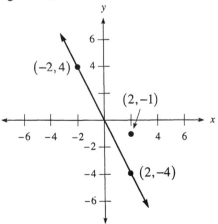

No, the point $(2,-1)$ does not lie on the line.

37.   Graphing the line:

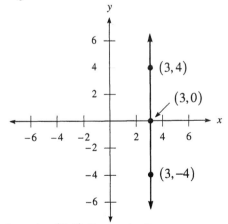

Yes, the point $(3,0)$ lies on the line.

39.   No, the x-coordinate of every point on this line is 3.

41.   Graphing the line:

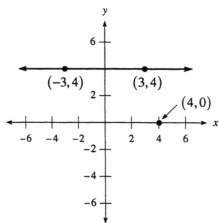

No, the point $(4,0)$ does not lie on the line.

43.   No, the y-coordinate of every point on this line is 4.

45.   **a.**   Three ordered pairs on the graph are (5,40), (10,80), and (20,160).
            b.   She will earn $320 for working 40 hours.
            c.   If her check is $240, she worked 30 hours that week.
            d.   No. She should be paid $280 for working 35 hours, not $260.

47.   Constructing a line graph:

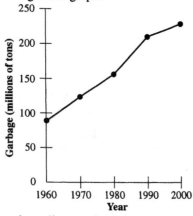

49.   Constructing a line graph:

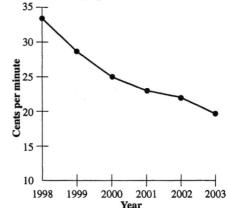

51.   Constructing a line graph:

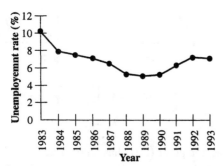

53.   Constructing a scatter diagram:

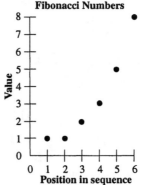

55.   The other three corners are the points $A(-2,-3)$, $B(-2,1)$, and $C(2,1)$.

57.   The pattern is to add 7, so the next number is: $24 + 7 = 31$

59. The pattern is to multiply by $\frac{1}{3}$, so the next number is: $\frac{1}{9} \cdot \frac{1}{3} = \frac{1}{27}$

61. The pattern is to add -3, so the next number is: $-2 + (-3) = -5$

63. The pattern is to multiply by 3, so the next number is: $189 \cdot 3 = 567$

65. The pattern is to add 1, add 2, add 3, so the next number is obtained by adding 4: $11 + 4 = 15$

3.2 Solutions to Linear Equations in Two Variables

1. Substituting $x = 0$, $y = 0$, and $y = -6$:

$$2(0) + y = 6 \qquad 2x + 0 = 6 \qquad 2x + (-6) = 6$$
$$0 + y = 6 \qquad 2x = 6 \qquad 2x = 12$$
$$y = 6 \qquad x = 3 \qquad x = 6$$

The ordered pairs are $(0,6)$, $(3,0)$, and $(6,-6)$.

3. Substituting $x = 0$, $y = 0$, and $x = -4$:

$$3(0) + 4y = 12 \qquad 3x + 4(0) = 12 \qquad 3(-4) + 4y = 12$$
$$0 + 4y = 12 \qquad 3x + 0 = 12 \qquad -12 + 4y = 12$$
$$4y = 12 \qquad 3x = 12 \qquad 4y = 24$$
$$y = 3 \qquad x = 4 \qquad y = 6$$

The ordered pairs are $(0,3)$, $(4,0)$, and $(-4,6)$.

5. Substituting $x = 1$, $y = 0$, and $x = 5$:

$$y = 4(1) - 3 \qquad 0 = 4x - 3 \qquad y = 4(5) - 3$$
$$y = 4 - 3 \qquad 3 = 4x \qquad y = 20 - 3$$
$$y = 1 \qquad x = \frac{3}{4} \qquad y = 17$$

The ordered pairs are $(1,1)$, $\left(\frac{3}{4},0\right)$, and $(5,17)$.

7. Substituting $x = 2$, $y = 6$, and $x = 0$:

$$y = 7(2) - 1 \qquad 6 = 7x - 1 \qquad y = 7(0) - 1$$
$$y = 14 - 1 \qquad 7 = 7x \qquad y = 0 - 1$$
$$y = 13 \qquad x = 1 \qquad y = -1$$

The ordered pairs are $(2,13)$, $(1,6)$, and $(0,-1)$.

9. Substituting $y = 4$, $y = -3$, and $y = 0$ results (in each case) in $x = -5$. The ordered pairs are $(-5,4)$, $(-5,-3)$, and $(-5,0)$.

11. Completing the table:

x	y
1	3
-3	-9
4	12
6	18

13. Completing the table:

x	y
0	0
$-1/2$	-2
-3	-12
3	12

15. Completing the table:

x	y
2	3
3	2
5	0
9	-4

17. Completing the table:

x	y
2	0
3	2
1	-2
-3	-10

19. Completing the table:

x	y
0	-1
-1	-7
-3	-19
3/2	8

21. Substituting each ordered pair into the equation:
$(2,3)$: $2(2) - 5(3) = 4 - 15 = -11 \neq 10$
$(0,-2)$: $2(0) - 5(-2) = 0 + 10 = 10$
$\left(\frac{5}{2},1\right)$: $2\left(\frac{5}{2}\right) - 5(1) = 5 - 5 = 0 \neq 10$
Only the ordered pair $(0,-2)$ is a solution.

23. Substituting each ordered pair into the equation:
$(1,5)$: $7(1) - 2 = 7 - 2 = 5$
$(0,-2)$: $7(0) - 2 = 0 - 2 = -2$
$(-2,-16)$: $7(-2) - 2 = -14 - 2 = -16$
All the ordered pairs $(1,5)$, $(0,-2)$ and $(-2,-16)$ are solutions.

25. Substituting each ordered pair into the equation:
$(1,6)$: $6(1) = 6$
$(-2,12)$: $6(-2) = -12 \neq 12$
$(0,0)$: $6(0) = 0$
The ordered pairs $(1,6)$ and $(0,0)$ are solutions.

27. Substituting each ordered pair into the equation:
$(1,1)$: $1 + 1 = 2 \neq 0$
$(2,-2)$: $2 + (-2) = 0$
$(3,3)$: $3 + 3 = 6 \neq 0$
Only the ordered pair $(2,-2)$ is a solution.

29. Since $x = 3$, the ordered pair $(5,3)$ cannot be a solution. The ordered pairs $(3,0)$ and $(3,-3)$ are solutions.

31. Substituting $w = 3$:
$2l + 2(3) = 30$
$2l + 6 = 30$
$2l = 24$
$l = 12$
The length is 12 inches.

33. Completing the table:

Minutes	0	10	20	30	40	50	60	70	80	90	100
Cost	$3	$4	$5	$6	$7	$8	$9	$10	$11	$12	$13

Creating a line graph:

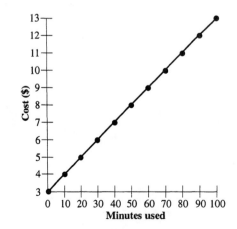

35. Completing the table:

Hours	0	1	2	3	4	5	6	7	8	9	10
Cost	$10	$13	$16	$19	$22	$25	$28	$31	$34	$37	$40

Creating a line graph:

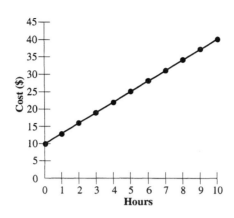

37. Substituting $x = 4$:
$$3(4) + 2y = 6$$
$$12 + 2y = 6$$
$$2y = -6$$
$$y = -3$$

39. Substituting $x = 0$: $y = -\frac{1}{3}(0) + 2 = 0 + 2 = 2$

41. Substituting $x = 2$: $y = \frac{3}{2}(2) - 3 = 3 - 3 = 0$

43. Solving for y:
$$5x + y = 4$$
$$y = -5x + 4$$

45. Solving for y:
$$3x - 2y = 6$$
$$-2y = -3x + 6$$
$$y = \frac{3}{2}x - 3$$

3.3 Graphing Linear Equations in Two Variables

1. The ordered pairs are $(0, 4)$, $(2, 2)$, and $(4, 0)$:

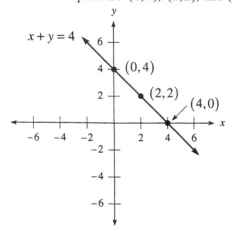

3. The ordered pairs are $(0, 3)$, $(2, 1)$, and $(4, -1)$:

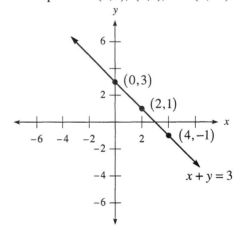

5.   The ordered pairs are $(0,0)$, $(-2,-4)$, and $(2,4)$:

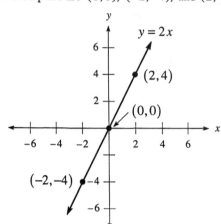

7.   The ordered pairs are $(-3,-1)$, $(0,0)$, and $(3,1)$:

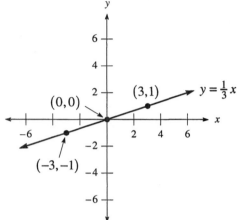

9.   The ordered pairs are $(0,1)$, $(-1,-1)$, and $(1,3)$:

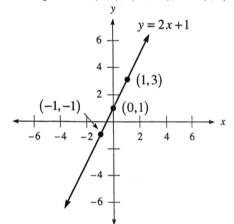

11.   The ordered pairs are $(0,4)$, $(-1,4)$, and $(2,4)$:

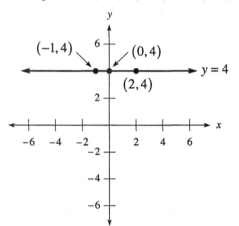

13.   The ordered pairs are $(-2,2)$, $(0,3)$, and $(2,4)$:

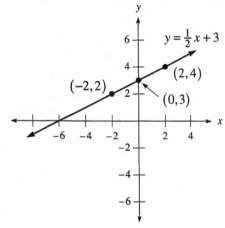

15.   The ordered pairs are $(-3,3)$, $(0,1)$, and $(3,-1)$:

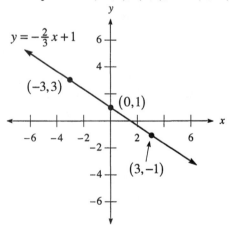

17.   Solving for y:

$$2x + y = 3$$
$$y = -2x + 3$$

The ordered pairs are $(-1, 5)$, $(0, 3)$, and $(1, 1)$:

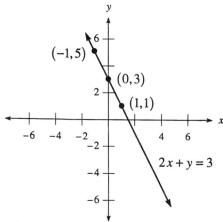

19.   Solving for y:

$$3x + 2y = 6$$
$$2y = -3x + 6$$
$$y = -\frac{3}{2}x + 3$$

The ordered pairs are $(0, 3)$, $(2, 0)$, and $(4, -3)$:

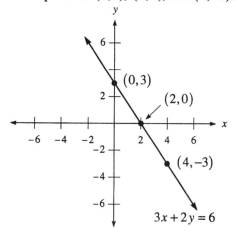

21.   Solving for y:

$$-x + 2y = 6$$
$$2y = x + 6$$
$$y = \frac{1}{2}x + 3$$

The ordered pairs are $(-2, 2)$, $(0, 3)$, and $(2, 4)$:

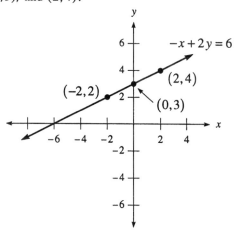

23.   Three solutions are $(-4,2)$, $(0,0)$, and $(4,-2)$:

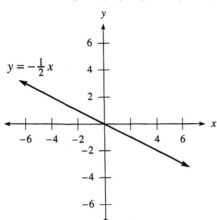

27.   Solving for y:

$$-2x + y = 1$$
$$y = 2x + 1$$

Three solutions are $(-2,-3)$, $(0,1)$, and $(2,5)$:

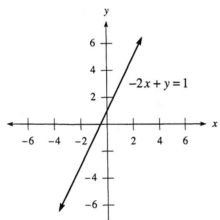

31.   Three solutions are $(-2,-4)$, $(-2,0)$, and $(-2,4)$:

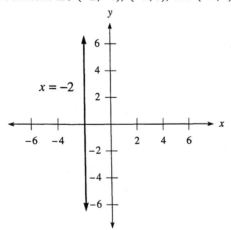

25.   Three solutions are $(-1,-4)$, $(0,-1)$, and $(1,2)$:

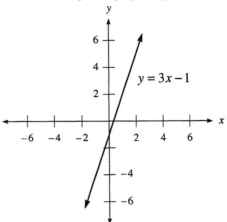

29.   Solving for y:

$$3x + 4y = 8$$
$$4y = -3x + 8$$
$$y = -\frac{3}{4}x + 2$$

Three solutions are $(-4,5)$, $(0,2)$, and $(4,-1)$:

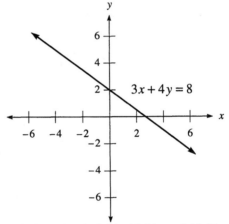

33.   Three solutions are $(-4,2)$, $(0,2)$, and $(4,2)$:

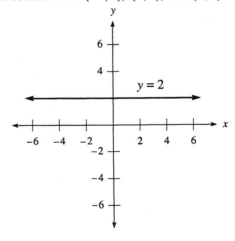

35. Completing the table:

x	y
−4	−3
−2	−2
0	−1
2	0
6	2

37. Graphing the three lines:

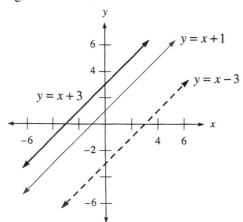

39. The ordered pairs are $(1,4)$, $(2,3)$, and $(3,2)$:

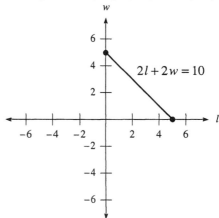

41. Solving the equation:

$$3(x-2)=9$$
$$3x-6=9$$
$$3x=15$$
$$x=5$$

43. Solving the equation:

$$2(3x-1)+4=-10$$
$$6x-2+4=-10$$
$$6x+2=-10$$
$$6x=-12$$
$$x=-2$$

45. Solving the equation:

$$6-2(4x-7)=-4$$
$$6-8x+14=-4$$
$$-8x+20=-4$$
$$-8x=-24$$
$$x=3$$

47. First multiply by 6 to clear the equation of fractions:

$$6\left(\tfrac{1}{2}x+4\right)=6\left(\tfrac{2}{3}x+5\right)$$
$$3x+24=4x+30$$
$$-x+24=30$$
$$-x=6$$
$$x=-6$$

3.4 More on Graphing: Intercepts

1.   To find the x-intercept, let $y = 0$:
$$2x + 0 = 4$$
$$2x = 4$$
$$x = 2$$
To find the y-intercept, let $x = 0$:
$$2(0) + y = 4$$
$$0 + y = 4$$
$$y = 4$$
Graphing the line:

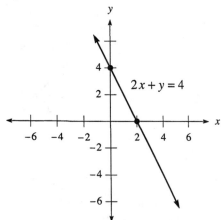

3.   To find the x-intercept, let $y = 0$:
$$-x + 0 = 3$$
$$-x = 3$$
$$x = -3$$
To find the y-intercept, let $x = 0$:
$$-0 + y = 3$$
$$y = 3$$
Graphing the line:

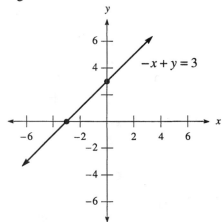

5.   To find the x-intercept, let $y = 0$:
$$-x + 2(0) = 2$$
$$-x = 2$$
$$x = -2$$
To find the y-intercept, let $x = 0$:
$$-0 + 2y = 2$$
$$2y = 2$$
$$y = 1$$
Graphing the line:

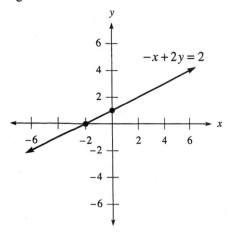

7.   To find the x-intercept, let $y = 0$:
$$5x + 2(0) = 10$$
$$5x = 10$$
$$x = 2$$
To find the y-intercept, let $x = 0$:
$$5(0) + 2y = 10$$
$$2y = 10$$
$$y = 5$$
Graphing the line:

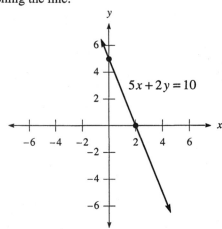

9.   To find the x-intercept, let $y = 0$:
$$4x - 2(0) = 8$$
$$4x = 8$$
$$x = 2$$
To find the y-intercept, let $x = 0$:
$$4(0) - 2y = 8$$
$$-2y = 8$$
$$y = -4$$
Graphing the line:

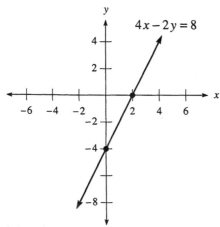

11.   To find the x-intercept, let $y = 0$:
$$-4x + 5(0) = 20$$
$$-4x = 20$$
$$x = -5$$
To find the y-intercept, let $x = 0$:
$$-4(0) + 5y = 20$$
$$5y = 20$$
$$y = 4$$
Graphing the line:

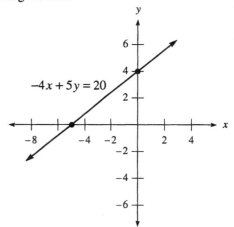

13.   To find the x-intercept, let $y = 0$:
$$2x - 6 = 0$$
$$2x = 6$$
$$x = 3$$
To find the y-intercept, let $x = 0$:
$$y = 2(0) - 6$$
$$y = -6$$
Graphing the line:

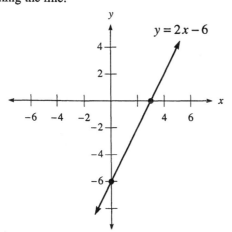

15.   To find the x-intercept, let $y = 0$:
$$2x + 2 = 0$$
$$2x = -2$$
$$x = -1$$
To find the y-intercept, let $x = 0$:
$$y = 2(0) + 2$$
$$y = 2$$
Graphing the line:

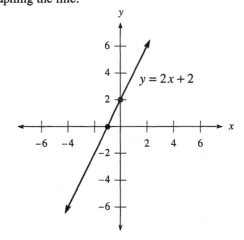

17. To find the x-intercept, let $y = 0$:

$$2x - 1 = 0$$
$$2x = 1$$
$$x = \tfrac{1}{2}$$

To find the y-intercept, let $x = 0$:

$$y = 2(0) - 1$$
$$y = -1$$

Graphing the line:

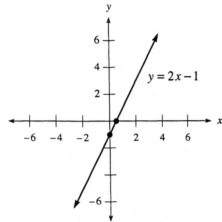

19. To find the x-intercept, let $y = 0$:

$$\tfrac{1}{2}x + 3 = 0$$
$$\tfrac{1}{2}x = -3$$
$$x = -6$$

To find the y-intercept, let $x = 0$:

$$y = \tfrac{1}{2}(0) + 3$$
$$y = 3$$

Graphing the line:

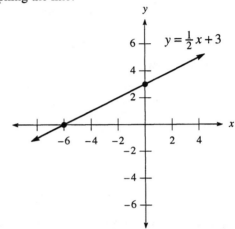

21. To find the x-intercept, let $y = 0$:

$$-\tfrac{1}{3}x - 2 = 0$$
$$-\tfrac{1}{3}x = 2$$
$$x = -6$$

Graphing the line:

To find the y-intercept, let $x = 0$:

$$y = -\tfrac{1}{3}(0) - 2$$
$$y = -2$$

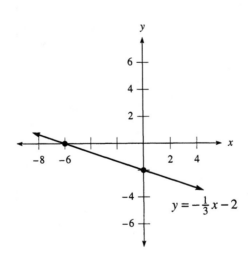

23.   Another point on the line is $(2, -4)$:

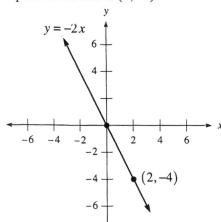

25.   Another point on the line is $(3, -1)$:

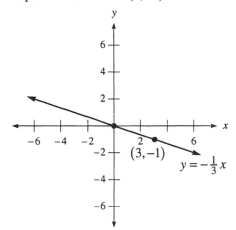

27.   Another point on the line is $(3, 2)$:

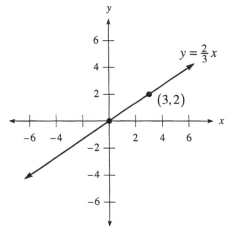

29.   The x-intercept is 3 and the y-intercept is 5.

31.   The x-intercept is -1 and the y-intercept is -3.

33.   Graphing the line:

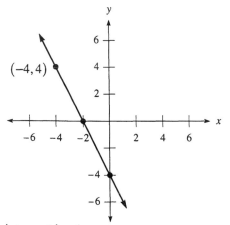

The y-intercept is -4.

35.   Graphing the line:

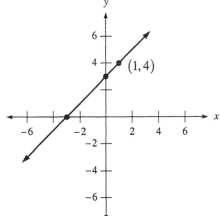

The x-intercept is -3.

37. Graphing the line:

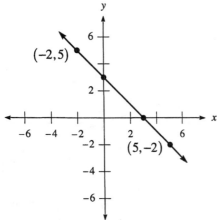

The x- and y-intercepts are both 3.

39. Completing the table:

x	y
−2	1
0	−1
−1	0
1	−2

41. Graphing the line:

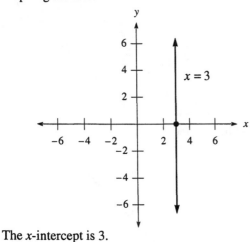

The x-intercept is 3.

43. Graphing the line:

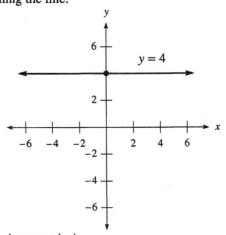

The y-intercept is 4.

45. Sketching the graph:

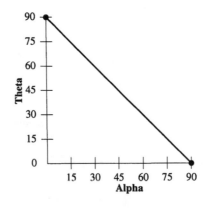

47.   Solving the inequality:
$$x - 3 < 2$$
$$x - 3 + 3 < 2 + 3$$
$$x < 5$$

49.   Solving the inequality:
$$-3x \geq 12$$
$$-\tfrac{1}{3}(-3x) \leq -\tfrac{1}{3}(12)$$
$$x \leq -4$$

51.   Solving the inequality:
$$-\frac{x}{3} \leq -1$$
$$-3\left(-\frac{x}{3}\right) \geq -3(-1)$$
$$x \geq 3$$

53.   Solving the inequality:
$$-4x + 1 < 17$$
$$-4x + 1 - 1 < 17 - 1$$
$$-4x < 16$$
$$-\tfrac{1}{4}(-4x) > -\tfrac{1}{4}(16)$$
$$x > -4$$

3.5 Solving Linear Systems by Graphing

1.   Graphing both lines:

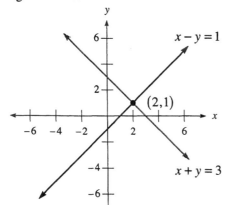

The intersection point is $(2,1)$.

3.   Graphing both lines:

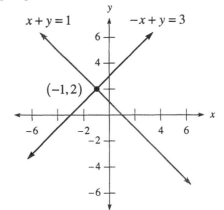

The intersection point is $(-1,2)$.

5.   Graphing both lines:

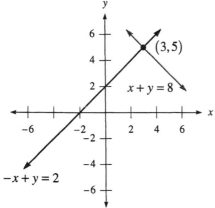

The intersection point is $(3,5)$.

7.   Graphing both lines:

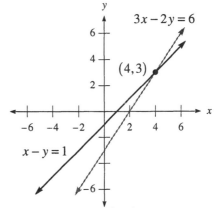

The intersection point is $(4,3)$.

9.   Graphing both lines:

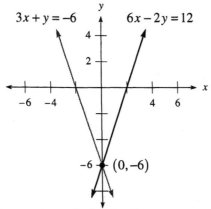

$3x + y = -6$ $6x - 2y = 12$

$(0, -6)$

The intersection point is $(0, -6)$.

11.   Graphing both lines:

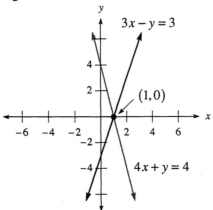

$3x - y = 3$

$(1, 0)$

$4x + y = 4$

The intersection point is $(1, 0)$.

13.   Graphing both lines:

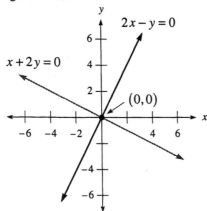

$2x - y = 0$

$x + 2y = 0$

$(0, 0)$

The intersection point is $(0, 0)$.

15.   Graphing both lines:

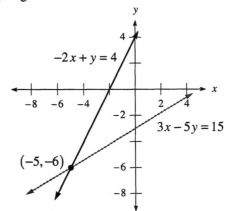

$-2x + y = 4$

$3x - 5y = 15$

$(-5, -6)$

The intersection point is $(-5, -6)$.

17.   Graphing both lines:

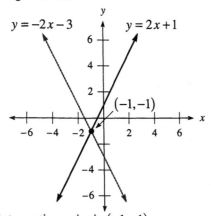

$y = -2x - 3$ $y = 2x + 1$

$(-1, -1)$

The intersection point is $(-1, -1)$.

19.   Graphing both lines:

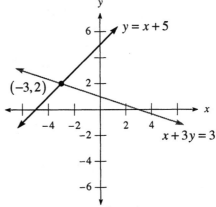

$y = x + 5$

$(-3, 2)$

$x + 3y = 3$

The intersection point is $(-3, 2)$.

21. Graphing both lines:

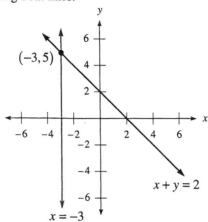

The intersection point is $(-3,5)$.

23. Graphing both lines:

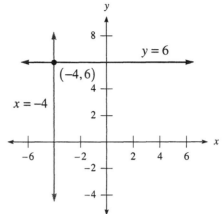

The intersection point is $(-4,6)$.

25. Graphing both lines:

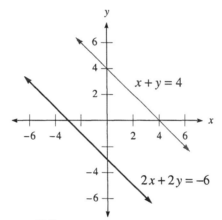

There is no intersection (the lines are parallel).

27. Graphing both lines:

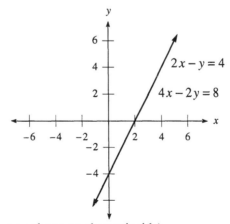

The system is dependent (both lines are the same, they coincide).

29.   Graphing both lines:

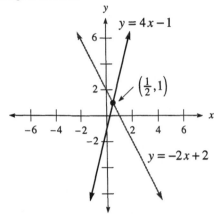

The intersection point is $\left(\frac{1}{2},1\right)$.

31.   Graphing both lines:

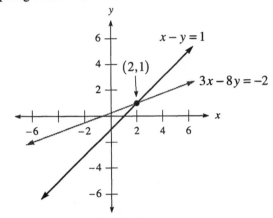

The intersection point is $(2,1)$.

33.   **a.**   If Jane worked 25 hours, she would earn the same amount at each position.
 b.   If Jane worked less than 20 hours, she should choose Gigi's since she earns more in that position.
 c.   If Jane worked more than 30 hours, she should choose Marcy's since she earns more in that position.

35.   **a.**   At 4 hours the cost of the two Internet providers will be the same.
 b.   Computer Service will be the cheaper plan if Patrice will be on the Internet for less than 3 hours.
 c.   ICM World will be the cheaper plan if Patrice will be on the Internet for more than 6 hours.

37.   Evaluating when $x = -3$: $2x - 9 = 2(-3) - 9 = -6 - 9 = -15$

39.   Evaluating when $x = -3$: $9 - 6x = 9 - 6(-3) = 9 + 18 = 27$

41.   Simplifying, then evaluating when $x = -3$: $4(3x + 2) + 1 = 12x + 8 + 1 = 12x + 9 = 12(-3) + 9 = -36 + 9 = -27$

43.   Evaluating when $x = -3$: $2x^2 + 3x + 4 = 2(-3)^2 + 3(-3) + 4 = 2(9) - 9 + 4 = 18 - 9 + 4 = 13$

3.6 The Addition Method

1.   Adding the two equations:
$$2x = 4$$
$$x = 2$$
Substituting into the first equation:
$$2 + y = 3$$
$$y = 1$$
The solution is $(2,1)$.

3.   Adding the two equations:
$$2y = 14$$
$$y = 7$$
Substituting into the first equation:
$$x + 7 = 10$$
$$x = 3$$
The solution is $(3,7)$.

5.   Adding the two equations:
$$-2y = 10$$
$$y = -5$$
Substituting into the first equation:
$$x - (-5) = 7$$
$$x + 5 = 7$$
$$x = 2$$
The solution is $(2,-5)$.

7.   Adding the two equations:
$$4x = -4$$
$$x = -1$$
Substituting into the first equation:
$$-1 + y = -1$$
$$y = 0$$
The solution is $(-1,0)$.

9.   Adding the two equations:
$$0 = 0$$
The lines coincide (the system is dependent).

11. Multiplying the first equation by 2:
$$6x - 2y = 8$$
$$2x + 2y = 24$$
Adding the two equations:
$$8x = 32$$
$$x = 4$$
Substituting into the first equation:
$$3(4) - y = 4$$
$$12 - y = 4$$
$$-y = -8$$
$$y = 8$$

The solution is $(4, 8)$.

15. Multiplying the second equation by 4:
$$11x - 4y = 11$$
$$20x + 4y = 20$$
Adding the two equations:
$$31x = 31$$
$$x = 1$$
Substituting into the second equation:
$$5(1) + y = 5$$
$$5 + y = 5$$
$$y = 0$$
The solution is $(1, 0)$.

19. Multiplying the first equation by –2:
$$2x + 16y = 2$$
$$-2x + 4y = 13$$
Adding the two equations:
$$20y = 15$$
$$y = \frac{3}{4}$$
Substituting into the first equation:
$$-x - 8\left(\frac{3}{4}\right) = -1$$
$$-x - 6 = -1$$
$$-x = 5$$
$$x = -5$$
The solution is $\left(-5, \frac{3}{4}\right)$.

23. Adding the two equations:
$$8x = -24$$
$$x = -3$$
Substituting into the second equation:
$$2(-3) + y = -16$$
$$-6 + y = -16$$
$$y = -10$$
The solution is $(-3, -10)$.

13. Multiplying the second equation by –3:
$$5x - 3y = -2$$
$$-30x + 3y = -3$$
Adding the two equations:
$$-25x = -5$$
$$x = \frac{1}{5}$$
Substituting into the first equation:
$$5\left(\frac{1}{5}\right) - 3y = -2$$
$$1 - 3y = -2$$
$$-3y = -3$$
$$y = 1$$
The solution is $\left(\frac{1}{5}, 1\right)$.

17. Multiplying the second equation by 3:
$$3x - 5y = 7$$
$$-3x + 3y = -3$$
Adding the two equations:
$$-2y = 4$$
$$y = -2$$
Substituting into the second equation:
$$-x - 2 = -1$$
$$-x = 1$$
$$x = -1$$
The solution is $(-1, -2)$.

21. Multiplying the first equation by 2:
$$-6x - 2y = 14$$
$$6x + 7y = 11$$
Adding the two equations:
$$5y = 25$$
$$y = 5$$
Substituting into the first equation:

$$-3x - 5 = 7$$
$$-3x = 12$$
$$x = -4$$

The solution is $(-4, 5)$.

25. Multiplying the second equation by 3:
$$x + 3y = 9$$
$$6x - 3y = 12$$
Adding the two equations:
$$7x = 21$$
$$x = 3$$
Substituting into the first equation:
$$3 + 3y = 9$$
$$3y = 6$$
$$y = 2$$

The solution is $(3, 2)$.

27. Multiplying the second equation by 2:
$$x - 6y = 3$$
$$8x + 6y = 42$$
Adding the two equations:
$$9x = 45$$
$$x = 5$$
Substituting into the second equation:
$$4(5) + 3y = 21$$
$$20 + 3y = 21$$
$$3y = 1$$
$$y = \frac{1}{3}$$
The solution is $\left(5, \frac{1}{3}\right)$.

29. Multiplying the second equation by –3:
$$2x + 9y = 2$$
$$-15x - 9y = 24$$
Adding the two equations:
$$-13x = 26$$
$$x = -2$$
Substituting into the first equation:
$$2(-2) + 9y = 2$$
$$-4 + 9y = 2$$
$$9y = 6$$
$$y = \frac{2}{3}$$
The solution is $\left(-2, \frac{2}{3}\right)$.

31. To clear each equation of fractions, multiply the first equation by 12 and the second equation by 6:
$$12\left(\tfrac{1}{3}x + \tfrac{1}{4}y\right) = 12\left(\tfrac{7}{6}\right) \qquad\qquad 6\left(\tfrac{3}{2}x - \tfrac{1}{3}y\right) = 6\left(\tfrac{7}{3}\right)$$
$$4x + 3y = 14 \qquad\qquad\qquad 9x - 2y = 14$$
The system of equations is:
$$4x + 3y = 14$$
$$9x - 2y = 14$$
Multiplying the first equation by 2 and the second equation by 3:
$$8x + 6y = 28$$
$$27x - 6y = 42$$
Adding the two equations:
$$35x = 70$$
$$x = 2$$
Substituting into $4x + 3y = 14$:
$$4(2) + 3y = 14$$
$$8 + 3y = 14$$
$$3y = 6$$
$$y = 2$$
The solution is $(2, 2)$.

33. Multiplying the first equation by –2:
$$-6x - 4y = 2$$
$$6x + 4y = 0$$
Adding the two equations:
$$0 = 2$$
Since this statement is false, the two lines are parallel, so the system has no solution.

35.   Multiplying the first equation by 2 and the second equation by 3:
$$22x + 12y = 34$$
$$15x - 12y = 3$$
Adding the two equations:
$$37x = 37$$
$$x = 1$$
Substituting into the second equation:
$$5(1) - 4y = 1$$
$$5 - 4y = 1$$
$$-4y = -4$$
$$y = 1$$
The solution is $(1,1)$.

37.   To clear each equation of fractions, multiply the first equation by 6 and the second equation by 6:
$$6\left(\tfrac{1}{2}x + \tfrac{1}{6}y\right) = 6\left(\tfrac{1}{3}\right) \qquad\qquad 6\left(-x - \tfrac{1}{3}y\right) = 6\left(-\tfrac{1}{6}\right)$$
$$3x + y = 2 \qquad\qquad\qquad\qquad -6x - 2y = -1$$
The system of equations is:
$$3x + y = 2$$
$$-6x - 2y = -1$$
Multiplying the first equation by 2:
$$6x + 2y = 4$$
$$-6x - 2y = -1$$
Adding the two equations:
$$0 = 3$$
Since this statement is false, the two lines are parallel, so the system has no solution.

39.   Adding $2x$ to each side of the first equation and $-3x$ to each side of the second equation:
$$6x - 5y = 17$$
$$-3x + 5y = 4$$
Adding the two equations:
$$3x = 21$$
$$x = 7$$
Substituting into $5y = 3x + 4$:
$$5y = 3(7) + 4$$
$$5y = 21 + 4$$
$$5y = 25$$
$$y = 5$$
The solution is $(7,5)$.

41.   Multiplying the second equation by 100 (to eliminate decimals):
$$x + y = 22$$
$$5x + 10y = 170$$
Multiplying the first equation by –5:
$$-5x - 5y = -110$$
$$5x + 10y = 170$$
Adding the two equations:
$$5y = 60$$
$$y = 12$$
Substituting into the first equation:
$$x + 12 = 22$$
$$x = 10$$
The solution is $(10,12)$.

43. Finding the amount:

$$(0.25)(300) = x$$
$$x = 75$$

So 75 is 25% of 300.

47. Finding the base:
$$0.15x = 60$$
$$x = \frac{60}{0.15}$$
$$x = 400$$
So 60 is 15% of 400.

45. Finding the percent:
$$p \cdot 125 = 25$$
$$p = \frac{25}{125}$$
$$p = 0.2 = 20\%$$

So 25 is 20% of 125.

3.7 The Substitution Method

1. Substituting into the first equation:
$$x + (2x - 1) = 11$$
$$3x - 1 = 11$$
$$3x = 12$$
$$x = 4$$
$$y = 2(4) - 1 = 7$$
The solution is $(4, 7)$.

2. Substituting into the first equation:
$$x + (5x + 2) = 20$$
$$6x + 2 = 20$$
$$6x = 18$$
$$x = 3$$
$$y = 5(3) + 2 = 17$$
The solution is $(3, 17)$.

5. Substituting into the first equation:
$$-2x + (-4x + 8) = -1$$
$$-6x + 8 = -1$$
$$-6x = -9$$
$$x = \frac{3}{2}$$
$$y = -4\left(\frac{3}{2}\right) + 8 = -6 + 8 = 2$$
The solution is $\left(\frac{3}{2}, 2\right)$.

7. Substituting into the first equation:
$$3(-y + 6) - 2y = -2$$
$$-3y + 18 - 2y = -2$$
$$-5y + 18 = -2$$
$$-5y = -20$$
$$y = 4$$
$$x = -4 + 6 = 2$$
The solution is $(2, 4)$.

9. Substituting into the first equation:
$$5x - 4(4) = -16$$
$$5x - 16 = -16$$
$$5x = 0$$
$$x = 0$$
The solution is $(0, 4)$.

11. Substituting into the first equation:
$$5x + 4(-3x) = 7$$
$$5x - 12x = 7$$
$$-7x = 7$$
$$x = -1$$
$$y = -3(-1) = 3$$
The solution is $(-1, 3)$.

13. Solving the second equation for x:
$$x - 2y = -1$$
$$x = 2y - 1$$
Substituting into the first equation:
$$(2y - 1) + 3y = 4$$
$$5y - 1 = 4$$
$$5y = 5$$
$$y = 1$$
$$x = 2(1) - 1 = 1$$
The solution is $(1, 1)$.

15. Solving the first equation for x:
$$x - 5y = 17$$
$$x = 5y + 17$$
Substituting into the first equation:
$$2(5y + 17) + y = 1$$
$$10y + 34 + y = 1$$
$$11y + 34 = 1$$
$$11y = -33$$
$$y = -3$$
$$x = 5(-3) + 17 = 2$$
The solution is $(2, -3)$.

17. Solving the second equation for x:
$$x - 5y = -5$$
$$x = 5y - 5$$
Substituting into the first equation:
$$3(5y - 5) + 5y = -3$$
$$15y - 15 + 5y = -3$$
$$20y - 15 = -3$$
$$20y = 12$$
$$y = \tfrac{3}{5}$$
$$x = 5\left(\tfrac{3}{5}\right) - 5 = 3 - 5 = -2$$
The solution is $\left(-2, \tfrac{3}{5}\right)$.

19. Solving the second equation for x:
$$x - 3y = -18$$
$$x = 3y - 18$$
Substituting into the first equation:
$$5(3y - 18) + 3y = 0$$
$$15y - 90 + 3y = 0$$
$$18y - 90 = 0$$
$$18y = 90$$
$$y = 5$$
$$x = 3(5) - 18 = -3$$
The solution is $(-3, 5)$.

21. Solving the second equation for x:
$$x + 3y = 12$$
$$x = -3y + 12$$
Substituting into the first equation:
$$-3(-3y + 12) - 9y = 7$$
$$9y - 36 - 9y = 7$$
$$-36 = 7$$
Since this statement is false, there is no solution to the system. The two lines are parallel.

23. Substituting into the first equation:
$$5x - 8(2x - 5) = 7$$
$$5x - 16x + 40 = 7$$
$$-11x + 40 = 7$$
$$-11x = -33$$
$$x = 3$$
$$y = 2(3) - 5 = 1$$
The solution is $(3, 1)$.

25. Substituting into the first equation:
$$7(2y - 1) - 6y = -1$$
$$14y - 7 - 6y = -1$$
$$8y - 7 = -1$$
$$8y = 6$$
$$y = \tfrac{3}{4}$$
$$x = 2\left(\tfrac{3}{4}\right) - 1 = \tfrac{3}{2} - 1 = \tfrac{1}{2}$$
The solution is $\left(\tfrac{1}{2}, \tfrac{3}{4}\right)$.

27. Substituting into the first equation:
$$-3x + 2(3x) = 6$$
$$-3x + 6x = 6$$
$$3x = 6$$
$$x = 2$$
$$y = 3(2) = 6$$
The solution is $(2, 6)$.

29. Substituting into the first equation:
$$5(y) - 6y = -4$$
$$-y = -4$$
$$y = 4$$
$$x = 4$$
The solution is $(4, 4)$.

31. Substituting into the first equation:
$$3x + 3(2x - 12) = 9$$
$$3x + 6x - 36 = 9$$
$$9x - 36 = 9$$
$$9x = 45$$
$$x = 5$$
$$y = 2(5) - 12 = -2$$
The solution is $(5, -2)$.

33. Substituting into the first equation:
$$7x - 11(10) = 16$$
$$7x - 110 = 16$$
$$7x = 126$$
$$x = 18$$
$$y = 10$$
The solution is $(18, 10)$.

35. Substituting into the first equation:
$$-4x + 4(x - 2) = -8$$
$$-4x + 4x - 8 = -8$$
$$-8 = -8$$
Since this statement is true, the system is dependent. The two lines coincide.

37. Substituting into the first equation:
$$0.05x + 0.10(22 - x) = 1.70$$
$$0.05x + 2.2 - 0.10x = 1.70$$
$$-0.05x + 2.2 = 1.7$$
$$-0.05x = -0.5$$
$$x = 10$$
$$y = 22 - 10 = 12$$
The solution is $(10, 12)$.

39.
   a. At 1,000 miles the car and truck cost the same to operate.
   b. If Daniel drives more than 1,200 miles, the car will be cheaper to operate.
   c. If Daniel drives less than 800 miles, the truck will be cheaper to operate.
   d. The graphs appear in the first quadrant only because all quantities are positive.

41. Let x represent the number of gallons consumed. For the charges for each company to be the same, the equation is:
$$7.00 + 1.10x = 5.00 + 1.15x$$
$$7.00 - 0.05x = 5.00$$
$$-0.05x = -2$$
$$x = 40$$
If 40 gallons are used in a month, the two companies charge the same amount.

43. Let w represent the width, and $3w$ represent the length. Using the perimeter formula:
$$2(w) + 2(3w) = 24$$
$$2w + 6w = 24$$
$$8w = 24$$
$$w = 3$$
$$3w = 9$$
The length is 9 meters and the width is 3 meters.

45. Completing the table:

	Nickels	Dimes
Number	x	$x + 3$
Value (cents)	$5(x)$	$10(x+3)$

The equation is:
$$5(x) + 10(x + 3) = 210$$
$$5x + 10x + 30 = 210$$
$$15x + 30 = 210$$
$$15x = 180$$
$$x = 12$$
$$x + 3 = 15$$
The collection consists of 12 nickels and 15 dimes.

47. Finding the amount:
$$0.08(6000) = x$$
$$x = 480$$
So 8% of 6,000 is 480.

49. Completing the table:

	Dollars Invested at 8%	Dollars Invested at 10%
Number of	x	$2x$
Interest on	$0.08(x)$	$0.10(2x)$

The equation is:
$$0.08(x) + 0.10(2x) = 224$$
$$0.08x + 0.20x = 224$$
$$0.28x = 224$$
$$x = 800$$
$$2x = 1600$$

The man invested $800 at 8% interest and $1,600 at 10% interest.

3.8 Applications

1. Let x and y represent the two numbers. The system of equations is:
$$x + y = 25$$
$$y = 5 + x$$
Substituting into the first equation:
$$x + (5 + x) = 25$$
$$2x + 5 = 25$$
$$2x = 20$$
$$x = 10$$
$$y = 5 + 10 = 15$$
The two numbers are 10 and 15.

3. Let x and y represent the two numbers. The system of equations is:
$$x + y = 15$$
$$y = 4x$$
Substituting into the first equation:
$$x + 4x = 15$$
$$5x = 15$$
$$x = 3$$
$$y = 4(3) = 12$$
The two numbers are 3 and 12.

5. Let x represent the larger number and y represent the smaller number. The system of equations is:
$$x - y = 5$$
$$x = 2y + 1$$
Substituting into the first equation:
$$2y + 1 - y = 5$$
$$y + 1 = 5$$
$$y = 4$$
$$x = 2(4) + 1 = 9$$
The two numbers are 4 and 9.

7. Let x and y represent the two numbers. The system of equations is:
$$y = 4x + 5$$
$$x + y = 35$$
Substituting into the second equation:
$$x + 4x + 5 = 35$$
$$5x + 5 = 35$$
$$5x = 30$$
$$x = 6$$
$$y = 4(6) + 5 = 29$$
The two numbers are 6 and 29.

9. Let x represent the amount invested at 6% and y represent the amount invested at 8%. The system of equations is:
$$x + y = 20000$$
$$0.06x + 0.08y = 1380$$
Multiplying the first equation by –0.06:
$$-0.06x - 0.06y = -1200$$
$$0.06x + 0.08y = 1380$$
Adding the two equations:
$$0.02y = 180$$
$$y = 9000$$
Substituting into the first equation:
$$x + 9000 = 20000$$
$$x = 11000$$
Mr. Wilson invested $9,000 at 8% and $11,000 at 6%.

11. Let x represent the amount invested at 5% and y represent the amount invested at 6%. The system of equations is:
$$x = 4y$$
$$0.05x + 0.06y = 520$$
Substituting into the second equation:
$$0.05(4y) + 0.06y = 520$$
$$0.20y + 0.06y = 520$$
$$0.26y = 520$$
$$y = 2000$$
$$x = 4(2000) = 8000$$
She invested $8,000 at 5% and $2,000 at 6%.

13. Let x represent the number of nickels and y represent the number of quarters. The system of equations is:
$$x + y = 14$$
$$0.05x + 0.25y = 2.30$$
Multiplying the first equation by –0.05:
$$-0.05x - 0.05y = -0.7$$
$$0.05x + 0.25y = 2.30$$
Adding the two equations:
$$0.20y = 1.6$$
$$y = 8$$
Substituting into the first equation:
$$x + 8 = 14$$
$$x = 6$$
Ron has 6 nickels and 8 quarters.

15.   Let x represent the number of dimes and y represent the number of quarters. The system of equations is:
$$x + y = 21$$
$$0.10x + 0.25y = 3.45$$
Multiplying the first equation by −0.10:
$$-0.10x - 0.10y = -2.10$$
$$0.10x + 0.25y = 3.45$$
Adding the two equations:
$$0.15y = 1.35$$
$$y = 9$$
Substituting into the first equation:
$$x + 9 = 21$$
$$x = 12$$
Tom has 12 dimes and 9 quarters.

17.   Let x represent the liters of 50% alcohol solution and y represent the liters of 20% alcohol solution. The system of equations is:
$$x + y = 18$$
$$0.50x + 0.20y = 0.30(18)$$
Multiplying the first equation by −0.20:
$$-0.20x - 0.20y = -3.6$$
$$0.50x + 0.20y = 5.4$$
Adding the two equations:
$$0.30x = 1.8$$
$$x = 6$$
Substituting into the first equation:
$$6 + y = 18$$
$$y = 12$$
The mixture contains 6 liters of 50% alcohol solution and 12 liters of 20% alcohol solution.

19.   Let x represent the gallons of 10% disinfectant and y represent the gallons of 7% disinfectant. The system of equations is:
$$x + y = 30$$
$$0.10x + 0.07y = 0.08(30)$$
Multiplying the first equation by −0.07:
$$-0.07x - 0.07y = -2.1$$
$$0.10x + 0.07y = 2.4$$
Adding the two equations:
$$0.03x = 0.3$$
$$x = 10$$
Substituting into the first equation:
$$10 + y = 30$$
$$y = 20$$
The mixture contains 10 gallons of 10% disinfectant and 20 gallons of 7% disinfectant.

21.   Let x represent the number of adult tickets and y represent the number of kids tickets. The system of equations is:
$$x + y = 70$$
$$5.50x + 4.00y = 310$$
Multiplying the first equation by −4:
$$-4.00x - 4.00y = -280$$
$$5.50x + 4.00y = 310$$
Adding the two equations:
$$1.5x = 30$$
$$x = 20$$
Substituting into the first equation:
$$20 + y = 70$$
$$y = 50$$
The matinee had 20 adult tickets sold and 50 kids tickets sold.

23.   Let x represent the width and y represent the length. The system of equations is:
$$2x + 2y = 96$$
$$y = 2x$$
Substituting into the first equation:
$$2x + 2(2x) = 96$$
$$2x + 4x = 96$$
$$6x = 96$$
$$x = 16$$
$$y = 2(16) = 32$$
The width is 16 feet and the length is 32 feet.

25.   Let x represent the number of \$5 chips and y represent the number of \$25 chips. The system of equations is:
$$x + y = 45$$
$$5x + 25y = 465$$
Multiplying the first equation by -5:
$$-5x - 5y = -225$$
$$5x + 25y = 465$$
Adding the two equations:
$$20y = 240$$
$$y = 12$$
Substituting into the first equation:
$$x + 12 = 45$$
$$x = 33$$
The gambler has 33 \$5 chips and 12 \$25 chips.

27.   Let x represent the number of shares of \$11 stock and y represent the number of shares of \$20 stock. The system of equations is:
$$x + y = 150$$
$$11x + 20y = 2550$$
Multiplying the first equation by -11:
$$-11x - 11y = -1650$$
$$11x + 20y = 2550$$
Adding the two equations:
$$9y = 900$$
$$y = 100$$
Substituting into the first equation:
$$x + 100 = 150$$
$$x = 50$$
She bought 50 shares at \$11 and 100 shares at \$20.

29.   Simplifying the expression: $7 - 3(2x - 4) - 8 = 7 - 6x + 12 - 8 = -6x + 11$

31.   Solving the equation:
$$-\tfrac{3}{2}x = 12$$
$$-\tfrac{2}{3}\left(-\tfrac{3}{2}x\right) = -\tfrac{2}{3}(12)$$
$$x = -8$$

33.   Solving the equation:
$$8 - 2(x + 7) = 2$$
$$8 - 2x - 14 = 2$$
$$-2x - 6 = 2$$
$$-2x = 8$$
$$x = -4$$

35.   Solving for w:
$$2l + 2w = P$$
$$2w = P - 2l$$
$$w = \frac{P - 2l}{2}$$

37. Solving the inequality:
$$3 - 2x > 5$$
$$3 - 3 - 2x > 5 - 3$$
$$-2x > 2$$
$$-\tfrac{1}{2}(-2x) < -\tfrac{1}{2}(2)$$
$$x < -1$$

Graphing the solution set:

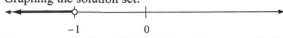

39. Solving for y:
$$3x - 2y \le 12$$
$$3x - 3x - 2y \le -3x + 12$$
$$-2y \le -3x + 12$$
$$-\tfrac{1}{2}(-2y) \ge -\tfrac{1}{2}(-3x + 12)$$
$$y \ge \tfrac{3}{2}x - 6$$

41. Let w represent the width and $3w + 5$ represent the length. Using the perimeter formula:
$$2(w) + 2(3w + 5) = 26$$
$$2w + 6w + 10 = 26$$
$$8w + 10 = 26$$
$$8w = 16$$
$$w = 2$$
$$3w + 5 = 3(2) + 5 = 11$$
The width is 2 inches and the length is 11 inches.

Chapter 3 Review

1. Substituting $x = 4$, $x = 0$, $y = 3$, and $y = 0$:

$3(4) + y = 6$	$3(0) + y = 6$	$3x + 3 = 6$	$3x + 0 = 6$
$12 + y = 6$	$0 + y = 6$	$3x = 3$	$3x = 6$
$y = -6$	$y = 6$	$x = 1$	$x = 2$

The ordered pairs are $(4, -6), (0, 6), (1, 3)$, and $(2, 0)$.

3. Substituting $x = 4$, $y = -2$, and $y = 3$:

$y = 2(4) - 6$	$-2 = 2x - 6$	$3 = 2x - 6$
$y = 8 - 6$	$4 = 2x$	$9 = 2x$
$y = 2$	$x = 2$	$x = \tfrac{9}{2}$

The ordered pairs are $(4, 2), (2, -2)$, and $\left(\tfrac{9}{2}, 3\right)$.

5. Substituting $x = 2$, $x = -1$, and $x = -3$ results (in each case) in $y = -3$. The ordered pairs are $(2, -3), (-1, -3)$, and $(-3, -3)$.

7. Substituting each ordered pair into the equation:
$$\left(-2, \tfrac{9}{2}\right): \quad 3(-2) - 4\left(\tfrac{9}{2}\right) = -6 - 18 = -24 \ne 12$$
$$(0, 3): \quad 3(0) - 4(3) = 0 - 12 = -12 \ne 12$$
$$\left(2, -\tfrac{3}{2}\right): \quad 3(2) - 4\left(-\tfrac{3}{2}\right) = 6 + 6 = 12$$

Only the ordered pair $\left(2, -\tfrac{3}{2}\right)$ is a solution.

9.   Graphing the ordered pair:

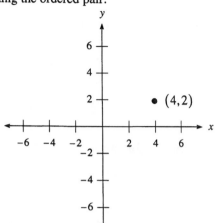

11.   Graphing the ordered pair:

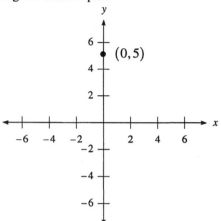

13.   Graphing the ordered pair:

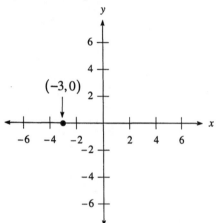

15.   The ordered pairs are $(-2,0), (0,-2)$, and $(1,-3)$:

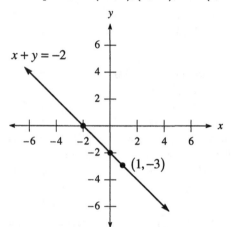

17.   The ordered pairs are $(1,1), (0,-1)$, and $(-1,-3)$:

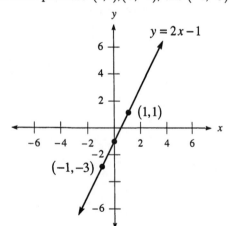

19.   Graphing the equation:

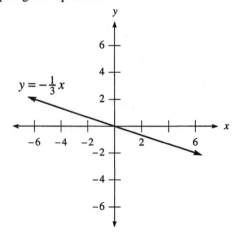

21.   Graphing the equation:

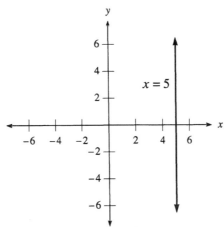

23.   To find the x-intercept, let $y = 0$:
$$3x - 0 = 6$$
$$3x = 6$$
$$x = 2$$

To find the y-intercept, let $x = 0$:
$$3(0) - y = 6$$
$$-y = 6$$
$$y = -6$$

25.   To find the x-intercept, let $y = 0$:
$$0 = x - 3$$
$$x = 3$$

To find the y-intercept, let $x = 0$:
$$y = 0 - 3$$
$$y = -3$$

27.   Graphing both lines:

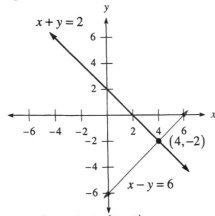

The intersection point is $(4, -2)$.

29.   Graphing both lines:

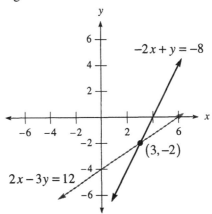

The intersection point is $(3, -2)$.

31.   Graphing both lines:

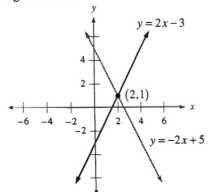

The intersection point is $(2, 1)$.

33.   Adding the two equations:
$$2x = 2$$
$$x = 1$$
Substituting into the second equation:
$$1 + y = -2$$
$$y = -3$$
The solution is $(1, -3)$.

35.   Multiplying the first equation by 2:
$$10x - 6y = 4$$
$$-10x + 6y = -4$$
Adding the two equations:
$$0 = 0$$
Since this statement is true, the system is dependent. The two lines coincide.

37.   Multiplying the second equation by -4:
$$-3x + 4y = 1$$
$$16x - 4y = 12$$
Adding the two equations:
$$13x = 13$$
$$x = 1$$
Substituting into the second equation:
$$-4(1) + y = -3$$
$$-4 + y = -3$$
$$y = 1$$
The solution is $(1, 1)$.

39.   Multiplying the first equation by 3 and the second equation by 5:
$$-6x + 15y = -33$$
$$35x - 15y = -25$$
Adding the two equations:
$$29x = -58$$
$$x = -2$$
Substituting into the first equation:
$$-2(-2) + 5y = -11$$
$$4 + 5y = -11$$
$$5y = -15$$
$$y = -3$$
The solution is $(-2, -3)$.

41.   Substituting into the first equation:
$$x + (-3x + 1) = 5$$
$$-2x + 1 = 5$$
$$-2x = 4$$
$$x = -2$$
Substituting into the second equation: $y = -3(-2) + 1 = 6 + 1 = 7$. The solution is $(-2, 7)$.

43.   Substituting into the first equation:
$$4x - 3(3x + 7) = -16$$
$$4x - 9x - 21 = -16$$
$$-5x - 21 = -16$$
$$-5x = 5$$
$$x = -1$$
Substituting into the second equation: $y = 3(-1) + 7 = -3 + 7 = 4$. The solution is $(-1, 4)$.

45. Solving the first equation for x:
$$x - 4y = 2$$
$$x = 4y + 2$$
Substituting into the second equation:
$$-3(4y + 2) + 12y = -8$$
$$-12y - 6 + 12y = -8$$
$$-6 = -8$$
Since this statement is false, there is no solution to the system. The two lines are parallel.

47. Solving the second equation for x:
$$x + 6y = -11$$
$$x = -6y - 11$$
Substituting into the first equation:
$$10(-6y - 11) - 5y = 20$$
$$-60y - 110 - 5y = 20$$
$$-65y - 110 = 20$$
$$-65y = 130$$
$$y = -2$$
Substituting into $x = -6y - 11$: $x = -6(-2) - 11 = 12 - 11 = 1$. The solution is $(1, -2)$.

49. Let x represent the smaller number and y represent the larger number. The system of equations is:
$$x + y = 18$$
$$2x = 6 + y$$
Solving the first equation for y:
$$x + y = 18$$
$$y = -x + 18$$
Substituting into the second equation:
$$2x = 6 + (-x + 18)$$
$$2x = -x + 24$$
$$3x = 24$$
$$x = 8$$
Substituting into the first equation:
$$8 + y = 18$$
$$y = 10$$
The two numbers are 8 and 10.

51. Let x represent the amount invested at 4% and y represent the amount invested at 5%. The system of equations is:
$$x + y = 12000$$
$$0.04x + 0.05y = 560$$
Multiplying the first equation by –0.04:
$$-0.04x - 0.04y = -480$$
$$0.04x + 0.05y = 560$$
Adding the two equations:
$$0.01y = 80$$
$$y = 8000$$
Substituting into the first equation:
$$x + 8000 = 12000$$
$$x = 4000$$
So $4,000 was invested at 4% and $8,000 was invested at 5%.

53. Let x represent the number of dimes and y represent the number of nickels. The system of equations is:
$$x + y = 17$$
$$0.10x + 0.05y = 1.35$$
Multiplying the first equation by –0.05:
$$-0.05x - 0.05y = -0.85$$
$$0.10x + 0.05y = 1.35$$
Adding the two equations:
$$0.05x = 0.50$$
$$x = 10$$
Substituting into the first equation:
$$10 + y = 17$$
$$y = 7$$
Barbara has 10 dimes and 7 nickels.

55. Let x represent the liters of 20% alcohol solution and y represent the liters of 10% alcohol solution. The system of equations is:
$$x + y = 50$$
$$0.20x + 0.10y = 0.12(50)$$
Multiplying the first equation by –0.10:
$$-0.10x - 0.10y = -5$$
$$0.20x + 0.10y = 6$$
Adding the two equations:
$$0.10x = 1$$
$$x = 10$$
Substituting into the first equation:
$$10 + y = 50$$
$$y = 40$$
The solution contains 40 liters of 10% alcohol solution and 10 liters of 20% alcohol solution.

Chapters 1-3 Cumulative Review

1. Simplifying using order of operations: $3 \cdot 4 + 5 = 12 + 5 = 17$
3. Simplifying using order of operations: $7[8 + (-5)] + 3(-7 + 12) = 7(3) + 3(5) = 21 + 15 = 36$
5. Simplifying using order of operations: $8 - 6(5 - 9) = 8 - 6(-4) = 8 + 24 = 32$
7. Simplifying: $\frac{2}{3} + \frac{3}{4} - \frac{1}{6} = \frac{2 \cdot 4}{3 \cdot 4} + \frac{3 \cdot 3}{4 \cdot 3} - \frac{1 \cdot 2}{6 \cdot 2} = \frac{8}{12} + \frac{9}{12} - \frac{2}{12} = \frac{15}{12} = \frac{5}{4}$
9. Solving the equation:

$$-5 - 6 = -y - 3 + 2y$$
$$-11 = y - 3$$
$$-11 + 3 = y - 3 + 3$$
$$y = -8$$

11. Solving the equation:
$$3(x - 4) = 9$$
$$3x - 12 = 9$$
$$3x - 12 + 12 = 9 + 12$$
$$3x = 21$$
$$\tfrac{1}{3}(3x) = \tfrac{1}{3}(21)$$
$$x = 7$$

13. Solving the inequality:
$$0.3x + 0.7 \le -2$$
$$0.3x + 0.7 - 0.7 \le -2 - 0.7$$
$$0.3x \le -2.7$$
$$\frac{0.3x}{0.3} \le \frac{-2.7}{0.3}$$
$$x \le -9$$
Graphing the solution set:

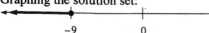

15. Graphing the line:

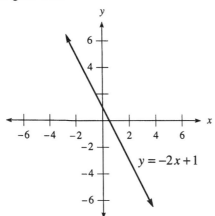

17. Graphing the line:

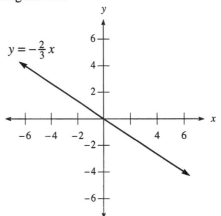

19. Graphing the two equations:

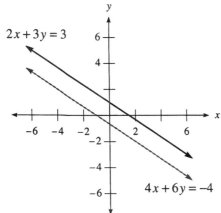

The two lines are parallel.

21. Multiplying the first equation by –2:
$$-2x - 2y = -14$$
$$2x + 2y = 14$$
Adding the two equations:
$$0 = 0$$
Since this statement is true, the system is dependent. The two lines coincide.

23. Multiplying the second equation by 3:
$$2x + 3y = 13$$
$$3x - 3y = -3$$
Adding the two equations:
$$5x = 10$$
$$x = 2$$
Substituting into the first equation:
$$2(2) + 3y = 13$$
$$4 + 3y = 13$$
$$3y = 9$$
$$y = 3$$
The solution is $(2, 3)$.

25. Multiplying the second equation by –2:
$$2x + 5y = 33$$
$$-2x + 6y = 0$$
Adding the two equations:
$$11y = 33$$
$$y = 3$$
Substituting into the second equation:
$$x - 3(3) = 0$$
$$x - 9 = 0$$
$$x = 9$$
The solution is $(9, 3)$.

27. Multiplying the second equation by 7:
$$3x - 7y = 12$$
$$14x + 7y = 56$$
Adding the two equations:
$$17x = 68$$
$$x = 4$$
Substituting into the second equation:
$$2(4) + y = 8$$
$$8 + y = 8$$
$$y = 0$$
The solution is $(4, 0)$.

29. Substituting into the first equation:
$$2x - 3(5x + 2) = 7$$
$$2x - 15x - 6 = 7$$
$$-13x - 6 = 7$$
$$-13x = 13$$
$$x = -1$$
Substituting into the second equation: $y = 5(-1) + 2 = -5 + 2 = -3$. The solution is $(-1, -3)$.

31. commutative property of addition

33. The quotient is: $\dfrac{-30}{6} = -5$

35. Finding the percent:
$$p \bullet 82 = 20.5$$
$$p = \frac{20.5}{82}$$
$$p = 0.25 = 25\%$$
So 25% of 82 is 20.5.

37. Simplifying, then evaluating when $x = 3$: $-3x + 7 + 5x = 2x + 7 = 2(3) + 7 = 6 + 7 = 13$

39. Evaluating when $x = -2$: $4x - 5 = 4(-2) - 5 = -8 - 5 = -13$

41. The expression is: $5 - (-8) = 5 + 8 = 13$

43. To find the x-intercept, let $y = 0$:
$$3x - 4(0) = 12$$
$$3x = 12$$
$$x = 4$$
To find the y-intercept, let $x = 0$:
$$3(0) - 4y = 12$$
$$-4y = 12$$
$$y = -3$$

45. Substituting $x = -1$:
$$2(-1) - 3y = 7$$
$$-2 - 3y = 7$$
$$-3y = 9$$
$$y = -3$$

47. Let w represent the width and $2w + 5$ represent the length. Using the perimeter formula:
$$2(w) + 2(2w + 5) = 44$$
$$2w + 4w + 10 = 44$$
$$6w + 10 = 44$$
$$6w = 34$$
$$w = \tfrac{17}{3}$$
$$2w + 5 = 2\left(\tfrac{17}{3}\right) + 5 = \tfrac{34}{3} + 5 = \tfrac{49}{3}$$
The width is $\tfrac{17}{3}$ cm and the length is $\tfrac{49}{3}$ cm.

49. Completing the table:

	Dollars Invested at 8%	Dollars Invested at 6%
Number of	$x + 900$	x
Interest on	$0.08(x + 900)$	$0.06(x)$

The equation is:

$$0.08(x + 900) + 0.06(x) = 240$$
$$0.08x + 72 + 0.06x = 240$$
$$0.14x + 72 = 240$$
$$0.14x = 168$$
$$x = 1200$$
$$x + 900 = 2100$$

Barbara invested $1,200 at 6% and $2,100 at 8%.

Chapter 3 Test

1. Substituting $x = 0$, $y = 0$, $x = 10$, and $y = -3$:

$$2(0) - 5y = 10$$
$$0 - 5y = 10$$
$$-5y = 10$$
$$y = -2$$

$$2x - 5(0) = 10$$
$$2x - 0 = 10$$
$$2x = 10$$
$$x = 5$$

$$2(10) - 5y = 10$$
$$20 - 5y = 10$$
$$-5y = -10$$
$$y = 2$$

$$2x - 5(-3) = 10$$
$$2x + 15 = 10$$
$$2x = -5$$
$$x = -\frac{5}{2}$$

The ordered pairs are $(0, -2), (5, 0), (10, 2),$ and $\left(-\frac{5}{2}, -3\right)$.

2. Substituting each ordered pair into the equation:

$(2, 5)$: $4(2) - 3 = 8 - 3 = 5$
$(0, -3)$: $4(0) - 3 = 0 - 3 = -3$
$(3, 0)$: $4(3) - 3 = 12 - 3 = 9 \neq 0$
$(-2, 11)$: $4(-2) - 3 = -8 - 3 = -11 \neq 11$

The ordered pairs $(2, 5)$ and $(0, -3)$ are solutions.

3. Graphing the line:

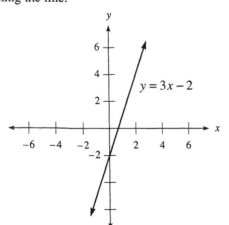

4. Graphing the line:

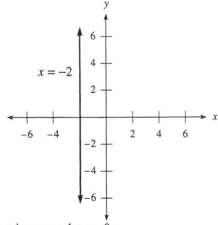

5. To find the x-intercept, let $y = 0$:

$$3x - 5(0) = 15$$
$$3x = 15$$
$$x = 5$$

To find the y-intercept, let $x = 0$:

$$3(0) - 5y = 15$$
$$-5y = 15$$
$$y = -3$$

6.   To find the x-intercept, let $y = 0$:   To find the y-intercept, let $x = 0$: $y = \frac{3}{2}(0) + 1 = 1$

$$0 = \frac{3}{2}x + 1$$
$$-1 = \frac{3}{2}x$$
$$x = -\frac{2}{3}$$

7.   Graphing the two equations:

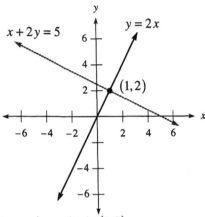

The intersection point is $(1, 2)$.

8.   Graphing the two equations:

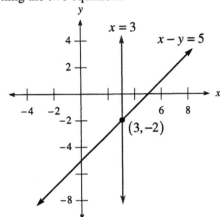

The intersection point is $(3, -2)$.

9.   Adding the two equations:

$$3x = -9$$
$$x = -3$$

Substituting into the first equation:

$$-3 - y = 1$$
$$-y = 4$$
$$y = -4$$

The solution is $(-3, -4)$.

10.   Multiplying the first equation by -1:

$$-2x - y = -7$$
$$3x + y = 12$$

Adding the two equations:

$$x = 5$$

Substituting into the first equation:

$$2(5) + y = 7$$
$$10 + y = 7$$
$$y = -3$$

The solution is $(5, -3)$.

11.   Multiplying the second equation by 4:

$$7x + 8y = -2$$
$$12x - 8y = 40$$

Adding the two equations:

$$19x = 38$$
$$x = 2$$

Substituting into the first equation:

$$7(2) + 8y = -2$$
$$14 + 8y = -2$$
$$8y = -16$$
$$y = -2$$

The solution is $(2, -2)$.

12.   Multiplying the first equation by -3 and the second equation by 2:

$$-18x + 30y = -18$$
$$18x - 30y = 18$$

Adding the two equations: $0 = 0$. Since this equation is true, the system is dependent. The two lines coincide.

13.   Substituting into the first equation:

$$3x + 2(2x + 3) = 20$$
$$3x + 4x + 6 = 20$$
$$7x + 6 = 20$$
$$7x = 14$$
$$x = 2$$

Substituting into the second equation: $y = 2(2) + 3 = 4 + 3 = 7$. The solution is $(2, 7)$.

14. Substituting into the first equation:
$$3(y+1)-6y=-6$$
$$3y+3-6y=-6$$
$$-3y+3=-6$$
$$-3y=-9$$
$$y=3$$
Substituting into the second equation: $x=3+1=4$. The solution is $(4,3)$.

15. Solving the second equation for y:
$$-3x+y=3$$
$$y=3x+3$$
Substituting into the first equation:
$$7x-2(3x+3)=-4$$
$$7x-6x-6=-4$$
$$x-6=-4$$
$$x=2$$
Substituting into $y=3x+3$: $y=3(2)+3=6+3=9$. The solution is $(2,9)$.

16. Solving the second equation for x:
$$x+3y=-8$$
$$x=-3y-8$$
Substituting into the first equation:
$$2(-3y-8)-3y=-7$$
$$-6y-16-3y=-7$$
$$-9y-16=-7$$
$$-9y=9$$
$$y=-1$$
Substituting into $x=-3y-8$: $x=-3(-1)-8=3-8=-5$. The solution is $(-5,-1)$.

17. Let x and y represent the two numbers. The system of equations is:
$$x+y=12$$
$$x-y=2$$
Adding the two equations:
$$2x=14$$
$$x=7$$
Substituting into the first equation:
$$7+y=12$$
$$y=5$$
The two numbers are 5 and 7.

18. Let x and y represent the two numbers. The system of equations is:
$$x+y=15$$
$$y=6+2x$$
Substituting into the first equation:
$$x+6+2x=15$$
$$3x+6=15$$
$$3x=9$$
$$x=3$$
Substituting into the second equation: $y=6+2(3)=6+6=12$. The two numbers are 3 and 12.

19. Let x represent the amount invested at 9% and y represent the amount invested at 11%. The system of equations is:
$$x + y = 10000$$
$$0.09x + 0.11y = 980$$
Multiplying the first equation by –0.09:
$$-0.09x - 0.09y = -900$$
$$0.09x + 0.11y = 980$$
Adding the two equations:
$$0.02y = 80$$
$$y = 4000$$
Substituting into the first equation:
$$x + 4000 = 10000$$
$$x = 6000$$
Dr. Stork should invest $6,000 at 9%.

20. Let x represent the number of nickels and y represent the number of quarters. The system of equations is:
$$x + y = 12$$
$$0.05x + 0.25y = 1.60$$
Multiplying the first equation by –0.05:
$$-0.05x - 0.05y = -0.60$$
$$0.05x + 0.25y = 1.60$$
Adding the two equations:
$$0.20y = 1.00$$
$$y = 5$$
Substituting into the first equation:
$$x + 5 = 12$$
$$x = 7$$
Diane has 7 nickels and 5 quarters.

Chapter 4
Exponents and Polynomials

4.1 Multiplication with Exponents

1. The base is 4 and the exponent is 2. Evaluating the expression: $4^2 = 4 \cdot 4 = 16$

3. The base is 0.3 and the exponent is 2. Evaluating the expression: $0.3^2 = 0.3 \cdot 0.3 = 0.09$

5. The base is 4 and the exponent is 3. Evaluating the expression: $4^3 = 4 \cdot 4 \cdot 4 = 64$

7. The base is −5 and the exponent is 2. Evaluating the expression: $(-5)^2 = (-5) \cdot (-5) = 25$

9. The base is 2 and the exponent is 3. Evaluating the expression: $-2^3 = -2 \cdot 2 \cdot 2 = -8$

11. The base is 3 and the exponent is 4. Evaluating the expression: $3^4 = 3 \cdot 3 \cdot 3 \cdot 3 = 81$

13. The base is $\frac{2}{3}$ and the exponent is 2. Evaluating the expression: $\left(\frac{2}{3}\right)^2 = \left(\frac{2}{3}\right) \cdot \left(\frac{2}{3}\right) = \frac{4}{9}$

15. The base is $\frac{1}{2}$ and the exponent is 4. Evaluating the expression: $\left(\frac{1}{2}\right)^4 = \left(\frac{1}{2}\right) \cdot \left(\frac{1}{2}\right) \cdot \left(\frac{1}{2}\right) \cdot \left(\frac{1}{2}\right) = \frac{1}{16}$

17. a. Completing the table:

Number (x)	1	2	3	4	5	6	7
Square (x^2)	1	4	9	16	25	36	49

   b. For numbers larger than 1, the square of the number is larger than the number.

19. Simplifying the expression: $x^4 \cdot x^5 = x^{4+5} = x^9$ 21. Simplifying the expression: $y^{10} \cdot y^{20} = y^{10+20} = y^{30}$

23. Simplifying the expression: $2^5 \cdot 2^4 \cdot 2^3 = 2^{5+4+3} = 2^{12}$

25. Simplifying the expression: $x^4 \cdot x^6 \cdot x^8 \cdot x^{10} = x^{4+6+8+10} = x^{28}$

27. Simplifying the expression: $\left(x^2\right)^5 = x^{2 \cdot 5} = x^{10}$ 29. Simplifying the expression: $\left(5^4\right)^3 = 5^{4 \cdot 3} = 5^{12}$

31. Simplifying the expression: $\left(y^3\right)^3 = y^{3 \cdot 3} = y^9$ 33. Simplifying the expression: $\left(2^5\right)^{10} = 2^{5 \cdot 10} = 2^{50}$

35. Simplifying the expression: $\left(a^3\right)^x = a^{3x}$ 37. Simplifying the expression: $\left(b^x\right)^y = b^{xy}$

39. Simplifying the expression: $(4x)^2 = 4^2 \cdot x^2 = 16x^2$ 41. Simplifying the expression: $(2y)^5 = 2^5 \cdot y^5 = 32y^5$

43. Simplifying the expression: $(-3x)^4 = (-3)^4 \cdot x^4 = 81x^4$

45. Simplifying the expression: $(0.5ab)^2 = (0.5)^2 \cdot a^2b^2 = 0.25a^2b^2$

47. Simplifying the expression: $(4xyz)^3 = 4^3 \cdot x^3y^3z^3 = 64x^3y^3z^3$

49. Simplifying using properties of exponents: $\left(2x^4\right)^3 = 2^3\left(x^4\right)^3 = 8x^{12}$

51. Simplifying using properties of exponents: $\left(4a^3\right)^2 = 4^2\left(a^3\right)^2 = 16a^6$

53. Simplifying using properties of exponents: $\left(x^2\right)^3\left(x^4\right)^2 = x^6 \cdot x^8 = x^{14}$

55. Simplifying using properties of exponents: $\left(a^3\right)^1\left(a^2\right)^4 = a^3 \cdot a^8 = a^{11}$

57. Simplifying using properties of exponents: $(2x)^3(2x)^4 = (2x)^7 = 2^7 x^7 = 128x^7$

59. Simplifying using properties of exponents: $\left(3x^2\right)^3(2x)^4 = 3^3 x^6 \cdot 2^4 x^4 = 27x^6 \cdot 16x^4 = 432x^{10}$

61. Simplifying using properties of exponents: $\left(4x^2 y^3\right)^2 = 4^2 x^4 y^6 = 16x^4 y^6$

63. Simplifying using properties of exponents: $\left(\frac{2}{3}a^4 b^5\right)^3 = \left(\frac{2}{3}\right)^3 a^{12} b^{15} = \frac{8}{27} a^{12} b^{15}$

65. **a.** Completing the table:

Number (x)	Square $\left(x^2\right)$
−3	9
−2	4
−1	1
0	0
1	1
2	4
3	9

b. Constructing a line graph:

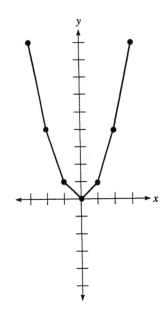

67. Completing the table:

Number (x)	Square (x^2)
−2.5	6.25
−1.5	2.25
−0.5	0.25
0	0
0.5	0.25
1.5	2.25
2.5	6.25

69. Writing in scientific notation: $43,200 = 4.32 \times 10^4$ **71.** Writing in scientific notation: $570 = 5.7 \times 10^2$

73. Writing in scientific notation: $238,000 = 2.38 \times 10^5$ **75.** Writing in expanded form: $2.49 \times 10^3 = 2,490$

77. Writing in expanded form: $3.52 \times 10^2 = 352$ **79.** Writing in expanded form: $2.8 \times 10^4 = 28,000$

81. The volume is given by: $V = (3 \text{ in.})^3 = 27 \text{ inches}^3$ **83.** The volume is given by: $V = (2.5 \text{ in.})^3 \approx 15.6 \text{ inches}^3$

85. The volume is given by: $V = (8 \text{ in.})(4.5 \text{ in.})(1 \text{ in.}) = 36 \text{ inches}^3$

87. Yes. If the box was 7 ft x 3 ft x 2 ft, you could fit inside of it.

89. Writing in scientific notation: $650,000,000 \text{ seconds} = 6.5 \times 10^8$ seconds

91. Writing in expanded form: 7.4×10^5 dollars $= \$740,000$

93. Writing in expanded form: 1.8×10^5 dollars $= \$180,000$

95. Substitute $c = 8$, $b = 3.35$, and $s = 3.11$: $d = \pi \cdot 3.11 \cdot 8 \cdot \left(\frac{1}{2} \cdot 3.35\right)^2 \approx 219 \text{ inches}^3$

97. Substitute $c = 6$, $b = 3.59$, and $s = 2.99$: $d = \pi \cdot 2.99 \cdot 6 \cdot \left(\frac{1}{2} \cdot 3.59\right)^2 \approx 182 \text{ inches}^3$

99. **a.** Each section is $\frac{1}{3}$ foot, so after stage 1 the length is $\frac{4}{3}$ feet.

   b. After stage 2 the length is $\left(\frac{4}{3}\right)^2 = \frac{16}{9}$ feet.   **c.** After stage 3 the length is $\left(\frac{4}{3}\right)^3 = \frac{64}{27}$ feet.

   d. After stage 10 the length is $\left(\frac{4}{3}\right)^{10}$ feet.

101. Subtracting: $4 - 7 = 4 + (-7) = -3$ **103.** Subtracting: $4 - (-7) = 4 + 7 = 11$

105. Subtracting: $15 - 20 = 15 + (-20) = -5$ **107.** Subtracting: $-15 - (-20) = -15 + 20 = 5$

4.2 Division with Exponents

1. Writing with positive exponents: $3^{-2} = \frac{1}{3^2} = \frac{1}{9}$ **3.** Writing with positive exponents: $6^{-2} = \frac{1}{6^2} = \frac{1}{36}$

5. Writing with positive exponents: $8^{-2} = \frac{1}{8^2} = \frac{1}{64}$ **7.** Writing with positive exponents: $5^{-3} = \frac{1}{5^3} = \frac{1}{125}$

9. Writing with positive exponents: $2x^{-3} = 2 \cdot \frac{1}{x^3} = \frac{2}{x^3}$

11. Writing with positive exponents: $(2x)^{-3} = \frac{1}{(2x)^3} = \frac{1}{8x^3}$

13. Writing with positive exponents: $(5y)^{-2} = \frac{1}{(5y)^2} = \frac{1}{25y^2}$

15. Writing with positive exponents: $10^{-2} = \frac{1}{10^2} = \frac{1}{100}$

17.　Completing the table:

Number (x)	Square (x^2)	Power of 2 (2^x)
-3	9	$\frac{1}{8}$
-2	4	$\frac{1}{4}$
-1	1	$\frac{1}{2}$
0	0	1
1	1	2
2	4	4
3	9	8

19.　Simplifying: $\dfrac{5^1}{5^3} = 5^{1-3} = 5^{-2} = \dfrac{1}{5^2} = \dfrac{1}{25}$

21.　Simplifying: $\dfrac{x^{10}}{x^4} = x^{10-4} = x^6$

23.　Simplifying: $\dfrac{4^3}{4^0} = 4^{3-0} = 4^3 = 64$

25.　Simplifying: $\dfrac{(2x)^7}{(2x)^4} = (2x)^{7-4} = (2x)^3 = 2^3 \, x^3 = 8x^3$

27.　Simplifying: $\dfrac{6^{11}}{6} = \dfrac{6^{11}}{6^1} = 6^{11-1} = 6^{10} \quad (= 60,466,176)$

29.　Simplifying: $\dfrac{6}{6^{11}} = \dfrac{6^1}{6^{11}} = 6^{1-11} = 6^{-10} = \dfrac{1}{6^{10}} \quad \left(= \dfrac{1}{60,466,176} \right)$

31.　Simplifying: $\dfrac{2^{-5}}{2^3} = 2^{-5-3} = 2^{-8} = \dfrac{1}{2^8} = \dfrac{1}{256}$

33.　Simplifying: $\dfrac{2^5}{2^{-3}} = 2^{5-(-3)} = 2^{5+3} = 2^8 = 256$

35.　Simplifying: $\dfrac{(3x)^{-5}}{(3x)^{-8}} = (3x)^{-5-(-8)} = (3x)^{-5+8} = (3x)^3 = 3^3 \, x^3 = 27x^3$

37.　Simplifying: $(3xy)^4 = 3^4 \, x^4 y^4 = 81x^4 y^4$

39.　Simplifying: $10^0 = 1$

41.　Simplifying: $\left(2a^2 b\right)^1 = 2a^2 b$

43.　Simplifying: $\left(7y^3\right)^{-2} = \dfrac{1}{\left(7y^3\right)^2} = \dfrac{1}{49y^6}$

45.　Simplifying: $x^{-3} \bullet x^{-5} = x^{-3-5} = x^{-8} = \dfrac{1}{x^8}$

47.　Simplifying: $y^7 \bullet y^{-10} = y^{7-10} = y^{-3} = \dfrac{1}{y^3}$

49.　Simplifying: $\dfrac{\left(x^2\right)^3}{x^4} = \dfrac{x^6}{x^4} = x^{6-4} = x^2$

51.　Simplifying: $\dfrac{\left(a^4\right)^3}{\left(a^3\right)^2} = \dfrac{a^{12}}{a^6} = a^{12-6} = a^6$

53.　Simplifying: $\dfrac{y^7}{\left(y^2\right)^8} = \dfrac{y^7}{y^{16}} = y^{7-16} = y^{-9} = \dfrac{1}{y^9}$

55.　Simplifying: $\left(\dfrac{y^7}{y^2}\right)^8 = \left(y^{7-2}\right)^8 = \left(y^5\right)^8 = y^{40}$

57.　Simplifying: $\dfrac{\left(x^{-2}\right)^3}{x^{-5}} = \dfrac{x^{-6}}{x^{-5}} = x^{-6-(-5)} = x^{-6+5} = x^{-1} = \dfrac{1}{x}$

59.　Simplifying: $\left(\dfrac{x^{-2}}{x^{-5}}\right)^3 = \left(x^{-2+5}\right)^3 = \left(x^3\right)^3 = x^9$

61.　Simplifying: $\dfrac{\left(a^3\right)^2 \left(a^4\right)^5}{\left(a^5\right)^2} = \dfrac{a^6 \bullet a^{20}}{a^{10}} = \dfrac{a^{26}}{a^{10}} = a^{26-10} = a^{16}$

63.　Simplifying: $\dfrac{\left(a^{-2}\right)^3 \left(a^4\right)^2}{\left(a^{-3}\right)^{-2}} = \dfrac{a^{-6} \bullet a^8}{a^6} = \dfrac{a^2}{a^6} = a^{2-6} = a^{-4} = \dfrac{1}{a^4}$

65. Completing the table:

Number (x)	Power of 2 $\left(2^x\right)$
-3	$\frac{1}{8}$
-2	$\frac{1}{4}$
-1	$\frac{1}{2}$
0	1
1	2
2	4
3	8

Constructing the line graph:

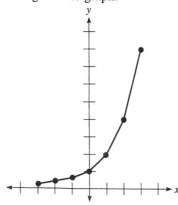

67. Writing in scientific notation: $0.0048 = 4.8 \times 10^{-3}$ **69.** Writing in scientific notation: $25 = 2.5 \times 10^1$

71. Writing in scientific notation: $0.000009 = 9 \times 10^{-6}$

73. Completing the table:

Expanded Form	Scientific Notation $\left(n \times 10^r\right)$
0.000357	3.57×10^{-4}
0.00357	3.57×10^{-3}
0.0357	3.57×10^{-2}
0.357	3.57×10^{-1}
3.57	3.57×10^0
35.7	3.57×10^1
357	3.57×10^2
$3,570$	3.57×10^3
$35,700$	3.57×10^4

75. Writing in expanded form: $4.23 \times 10^{-3} = 0.00423$ **77.** Writing in expanded form: $8 \times 10^{-5} = 0.00008$

79. Writing in expanded form: $4.2 \times 10^0 = 4.2$

81. Writing in expanded form: 2×10^{-3} seconds $= 0.002$ seconds

83. Writing in scientific notation: 0.006 inches $= 6 \times 10^{-3}$ inches

85. Writing in scientific notation: $25 \times 10^3 = 2.5 \times 10^4$ **87.** Writing in scientific notation: $23.5 \times 10^4 = 2.35 \times 10^5$

89. Writing in scientific notation: $0.82 \times 10^{-3} = 8.2 \times 10^{-4}$

91. The area of the smaller square is $(10 \text{ in.})^2 = 100 \text{ inches}^2$, while the area of the larger square is $(20 \text{ in.})^2 = 400 \text{ inches}^2$. It would take 4 smaller squares to cover the larger square.

93. The area of the smaller square is x^2, while the area of the larger square is $(2x)^2 = 4x^2$. It would take 4 smaller squares to cover the larger square.

95. The volume of the smaller box is $(6\text{ in.})^3 = 216$ inches^3, while the volume of the larger box is $(12\text{ in.})^3 = 1,728$ inches^3. Thus 8 smaller boxes will fit inside the larger box ($8 \cdot 216 = 1,728$).

97. The volume of the smaller box is x^3, while the volume of the larger box is $(2x)^3 = 8x^3$. Thus 8 smaller boxes will fit inside the larger box.

99. Simplifying by combining like terms: $4x + 3x = (4+3)x = 7x$

101. Simplifying by combining like terms: $5a - 3a = (5-3)a = 2a$

103. Simplifying by combining like terms: $4y + 5y + y = (4+5+1)y = 10y$

4.3 Operations with Monomials

1. Multiplying the monomials: $\left(3x^4\right)\left(4x^3\right) = 12x^{4+3} = 12x^7$

3. Multiplying the monomials: $\left(-2y^4\right)\left(8y^7\right) = -16y^{4+7} = -16y^{11}$

5. Multiplying the monomials: $(8x)(4x) = 32x^{1+1} = 32x^2$

7. Multiplying the monomials: $\left(10a^3\right)(10a)\left(2a^2\right) = 200a^{3+1+2} = 200a^6$

9. Multiplying the monomials: $\left(6ab^2\right)\left(-4a^2b\right) = -24a^{1+2}b^{2+1} = -24a^3b^3$

11. Multiplying the monomials: $\left(4x^2y\right)\left(3x^3y^3\right)\left(2xy^4\right) = 24x^{2+3+1}y^{1+3+4} = 24x^6y^8$

13. Dividing the monomials: $\dfrac{15x^3}{5x^2} = \dfrac{15}{5} \cdot \dfrac{x^3}{x^2} = 3x$

15. Dividing the monomials: $\dfrac{18y^9}{3y^{12}} = \dfrac{18}{3} \cdot \dfrac{y^9}{y^{12}} = 6 \cdot \dfrac{1}{y^3} = \dfrac{6}{y^3}$

17. Dividing the monomials: $\dfrac{32a^3}{64a^4} = \dfrac{32}{64} \cdot \dfrac{a^3}{a^4} = \dfrac{1}{2} \cdot \dfrac{1}{a} = \dfrac{1}{2a}$

19. Dividing the monomials: $\dfrac{21a^2b^3}{-7ab^5} = \dfrac{21}{-7} \cdot \dfrac{a^2}{a} \cdot \dfrac{b^3}{b^5} = -3 \cdot a \cdot \dfrac{1}{b^2} = -\dfrac{3a}{b^2}$

21. Dividing the monomials: $\dfrac{3x^3y^2z}{27xy^2z^3} = \dfrac{3}{27} \cdot \dfrac{x^3}{x} \cdot \dfrac{y^2}{y^2} \cdot \dfrac{z}{z^3} = \dfrac{1}{9} \cdot x^2 \cdot \dfrac{1}{z^2} = \dfrac{x^2}{9z^2}$

23. Completing the table:

a	b	ab	$\dfrac{a}{b}$	$\dfrac{b}{a}$
10	$5x$	$50x$	$\dfrac{2}{x}$	$\dfrac{x}{2}$
$20x^3$	$6x^2$	$120x^5$	$\dfrac{10x}{3}$	$\dfrac{3}{10x}$
$25x^5$	$5x^4$	$125x^9$	$5x$	$\dfrac{1}{5x}$
$3x^{-2}$	$3x^2$	9	$\dfrac{1}{x^4}$	x^4
$-2y^4$	$8y^7$	$-16y^{11}$	$-\dfrac{1}{4y^3}$	$-4y^3$

25. Finding the product: $\left(3 \times 10^3\right)\left(2 \times 10^5\right) = 6 \times 10^8$

27. Finding the product: $\left(3.5 \times 10^4\right)\left(5 \times 10^{-6}\right) = 17.5 \times 10^{-2} = 1.75 \times 10^{-1}$

29. Finding the product: $\left(5.5 \times 10^{-3}\right)\left(2.2 \times 10^{-4}\right) = 12.1 \times 10^{-7} = 1.21 \times 10^{-6}$

31. Finding the quotient: $\dfrac{8.4 \times 10^5}{2 \times 10^2} = 4.2 \times 10^3$

33. Finding the quotient: $\dfrac{6 \times 10^8}{2 \times 10^{-2}} = 3 \times 10^{10}$

35. Finding the quotient: $\dfrac{2.5 \times 10^{-6}}{5 \times 10^{-4}} = 0.5 \times 10^{-2} = 5.0 \times 10^{-3}$

37. Combining the monomials: $3x^2 + 5x^2 = (3+5)x^2 = 8x^2$

39. Combining the monomials: $8x^5 - 19x^5 = (8-19)x^5 = -11x^5$

41. Combining the monomials: $2a + a - 3a = (2+1-3)a = 0a = 0$

43. Combining the monomials: $10x^3 - 8x^3 + 2x^3 = (10-8+2)x^3 = 4x^3$

45. Combining the monomials: $20ab^2 - 19ab^2 + 30ab^2 = (20-19+30)ab^2 = 31ab^2$

47. Completing the table:

a	b	ab	$a+b$
$5x$	$3x$	$15x^2$	$8x$
$4x^2$	$2x^2$	$8x^4$	$6x^2$
$3x^3$	$6x^3$	$18x^6$	$9x^3$
$2x^4$	$-3x^4$	$-6x^8$	$-x^4$
x^5	$7x^5$	$7x^{10}$	$8x^5$

49. Simplifying the expression: $\dfrac{\left(3x^2\right)\left(8x^5\right)}{6x^4} = \dfrac{24x^7}{6x^4} = \dfrac{24}{6} \cdot \dfrac{x^7}{x^4} = 4x^3$

51. Simplifying the expression: $\dfrac{\left(9a^2b\right)\left(2a^3b^4\right)}{18a^5b^7} = \dfrac{18a^5b^5}{18a^5b^7} = \dfrac{18}{18} \cdot \dfrac{a^5}{a^5} \cdot \dfrac{b^5}{b^7} = 1 \cdot \dfrac{1}{b^2} = \dfrac{1}{b^2}$

53. Simplifying the expression: $\dfrac{\left(4x^3y^2\right)\left(9x^4y^{10}\right)}{\left(3x^5y\right)\left(2x^6y\right)} = \dfrac{36x^7y^{12}}{6x^{11}y^2} = \dfrac{36}{6} \cdot \dfrac{x^7}{x^{11}} \cdot \dfrac{y^{12}}{y^2} = 6 \cdot \dfrac{1}{x^4} \cdot y^{10} = \dfrac{6y^{10}}{x^4}$

55. Simplifying the expression: $\dfrac{\left(6 \times 10^8\right)\left(3 \times 10^5\right)}{9 \times 10^7} = \dfrac{18 \times 10^{13}}{9 \times 10^7} = 2 \times 10^6$

57. Simplifying the expression: $\dfrac{\left(5 \times 10^3\right)\left(4 \times 10^{-5}\right)}{2 \times 10^{-2}} = \dfrac{20 \times 10^{-2}}{2 \times 10^{-2}} = 10 = 1 \times 10^1$

59. Simplifying the expression: $\dfrac{\left(2.8 \times 10^{-7}\right)\left(3.6 \times 10^4\right)}{2.4 \times 10^3} = \dfrac{10.08 \times 10^{-3}}{2.4 \times 10^3} = 4.2 \times 10^{-6}$

61. Simplifying the expression: $\dfrac{18x^4}{3x} + \dfrac{21x^7}{7x^4} = 6x^3 + 3x^3 = 9x^3$

63. Simplifying the expression: $\dfrac{45a^6}{9a^4} - \dfrac{50a^8}{2a^6} = 5a^2 - 25a^2 = -20a^2$

65. Simplifying the expression: $\dfrac{6x^7y^4}{3x^2y^2} + \dfrac{8x^5y^8}{2y^6} = 2x^5y^2 + 4x^5y^2 = 6x^5y^2$

67. Solving the equation:
$$4^x \cdot 4^5 = 4^7$$
$$4^{x+5} = 4^7$$
$$x+5 = 7$$
$$x = 2$$

69. Solving the equation:
$$\left(7^3\right)^x = 7^{12}$$
$$7^{3x} = 7^{12}$$
$$3x = 12$$
$$x = 4$$

71. Simplifying each value:

$$(a+b)^2 = (4+5)^2 = 9^2 = 81$$
$$a^2 + b^2 = 4^2 + 5^2 = 16 + 25 = 41$$

Note that the values are not equal.

73. Simplifying each value:

$$(a+b)^2 = (3+4)^2 = 7^2 = 49 \qquad\qquad a^2 + 2ab + b^2 = 3^2 + 2(3)(4) + 4^2 = 9 + 24 + 16 = 49$$

Note that the values are equal.

75. Let x represent the width and $2x$ represent the length. The perimeter and area are given by:

$$P = 2(x) + 2(2x) = 2x + 4x = 6x \qquad\qquad A = x \cdot 2x = 2x^2$$

77. Let x represent the width and $2x$ represent the length. The volume is given by: $V = (x)(2x)(4) = 8x^2$ inches^3

79. **a.**   The volume is given by: $V = (8.5)(55)(10.1) \approx 4,700$ feet$^3 = 4.7 \times 10^3$ feet^3

      b.   Dividing: $\dfrac{4.7 \times 10^3}{0.15} \approx 31,000$ boxes or 3.1×10^3 boxes

81. Evaluating when $x = -2$: $4x = 4(-2) = -8$

83. Evaluating when $x = -2$: $-2x + 5 = -2(-2) + 5 = 4 + 5 = 9$

85. Evaluating when $x = -2$: $x^2 + 5x + 6 = (-2)^2 + 5(-2) + 6 = 4 - 10 + 6 = 0$

87. The ordered pairs are $(-2,-2)$, $(0,2)$, and $(2,6)$:   **89**.   The ordered pairs are $(-3,0)$, $(0,1)$, and $(3,2)$:

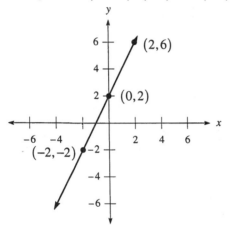
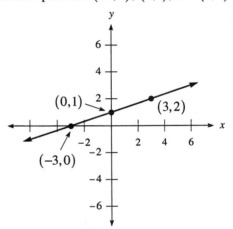

4.4 Addition and Subtraction of Polynomials

1.   This is a trinomial of degree 3.   **3.**   This is a trinomial of degree 3.

5.   This is a binomial of degree 1.   **7.**   This is a binomial of degree 2.

9.   This is a monomial of degree 2.   **11.**   This is a monomial of degree 0.

13. Combining the polynomials: $\left(2x^2 + 3x + 4\right) + \left(3x^2 + 2x + 5\right) = \left(2x^2 + 3x^2\right) + (3x + 2x) + (4 + 5) = 5x^2 + 5x + 9$

15. Combining the polynomials: $\left(3a^2 - 4a + 1\right) + \left(2a^2 - 5a + 6\right) = \left(3a^2 + 2a^2\right) + (-4a - 5a) + (1 + 6) = 5a^2 - 9a + 7$

17. Combining the polynomials: $x^2 + 4x + 2x + 8 = x^2 + (4x + 2x) + 8 = x^2 + 6x + 8$

19. Combining the polynomials: $6x^2 - 3x - 10x + 5 = 6x^2 + (-3x - 10x) + 5 = 6x^2 - 13x + 5$

21. Combining the polynomials: $x^2 - 3x + 3x - 9 = x^2 + (-3x + 3x) - 9 = x^2 - 9$

23. Combining the polynomials: $3y^2 - 5y - 6y + 10 = 3y^2 + (-5y - 6y) + 10 = 3y^2 - 11y + 10$

25. Combining the polynomials:

$$\left(6x^3 - 4x^2 + 2x\right) + \left(9x^2 - 6x + 3\right) = 6x^3 + \left(-4x^2 + 9x^2\right) + (2x - 6x) + 3 = 6x^3 + 5x^2 - 4x + 3$$

27. Combining the polynomials:

$$\left(\tfrac{2}{3}x^2 - \tfrac{1}{5}x - \tfrac{3}{4}\right) + \left(\tfrac{4}{3}x^2 - \tfrac{4}{5}x + \tfrac{7}{4}\right) = \left(\tfrac{2}{3}x^2 + \tfrac{4}{3}x^2\right) + \left(-\tfrac{1}{5}x - \tfrac{4}{5}x\right) + \left(-\tfrac{3}{4} + \tfrac{7}{4}\right) = 2x^2 - x + 1$$

29. Combining the polynomials: $\left(a^2 - a - 1\right) - \left(-a^2 + a + 1\right) = a^2 - a - 1 + a^2 - a - 1 = 2a^2 - 2a - 2$

31. Combining the polynomials:
$$\left(\tfrac{5}{9}x^3 + \tfrac{1}{3}x^2 - 2x + 1\right) - \left(\tfrac{2}{3}x^3 + x^2 + \tfrac{1}{2}x - \tfrac{3}{4}\right) = \tfrac{5}{9}x^3 + \tfrac{1}{3}x^2 - 2x + 1 - \tfrac{2}{3}x^3 - x^2 - \tfrac{1}{2}x + \tfrac{3}{4}$$
$$= -\tfrac{1}{9}x^3 - \tfrac{2}{3}x^2 - \tfrac{5}{2}x + \tfrac{7}{4}$$

33. Combining the polynomials:
$$\left(4y^2 - 3y + 2\right) + \left(5y^2 + 12y - 4\right) - \left(13y^2 - 6y + 20\right) = 4y^2 - 3y + 2 + 5y^2 + 12y - 4 - 13y^2 + 6y - 20$$
$$= \left(4y^2 + 5y^2 - 13y^2\right) + \left(-3y + 12y + 6y\right) + \left(2 - 4 - 20\right)$$
$$= -4y^2 + 15y - 22$$

35. Performing the subtraction:
$$\left(11x^2 - 10x + 13\right) - \left(10x^2 + 23x - 50\right) = 11x^2 - 10x + 13 - 10x^2 - 23x + 50$$
$$= \left(11x^2 - 10x^2\right) + \left(-10x - 23x\right) + \left(13 + 50\right)$$
$$= x^2 - 33x + 63$$

37. Performing the subtraction:
$$\left(11y^2 + 11y + 11\right) - \left(3y^2 + 7y - 15\right) = 11y^2 + 11y + 11 - 3y^2 - 7y + 15$$
$$= \left(11y^2 - 3y^2\right) + \left(11y - 7y\right) + \left(11 + 15\right)$$
$$= 8y^2 + 4y + 26$$

39. Performing the addition:
$$\left(25x^2 - 50x + 75\right) + \left(50x^2 - 100x - 150\right) = \left(25x^2 + 50x^2\right) + \left(-50x - 100x\right) + \left(75 - 150\right) = 75x^2 - 150x - 75$$

41. Performing the operations:
$$(3x - 2) + (11x + 5) - (2x + 1) = 3x - 2 + 11x + 5 - 2x - 1 = (3x + 11x - 2x) + (-2 + 5 - 1) = 12x + 2$$

43. Evaluating when $x = 3$: $x^2 - 2x + 1 = (3)^2 - 2(3) + 1 = 9 - 6 + 1 = 4$

45. Evaluating when $y = 10$: $(y - 5)^2 = (10 - 5)^2 = (5)^2 = 25$

47. Evaluating when $a = 2$: $a^2 + 4a + 4 = (2)^2 + 4(2) + 4 = 4 + 8 + 4 = 16$

49. Finding the volume of the cylinder and sphere:
$$V_{\text{cylinder}} = \pi\left(3^2\right)(6) = 54\pi \qquad\qquad V_{\text{sphere}} = \tfrac{4}{3}\pi\left(3^3\right) = 36\pi$$
Subtracting to find the amount of space to pack: $V = 54\pi - 36\pi = 18\pi \text{ inches}^3$

51. Multiplying the monomials: $3x(-5x) = -15x^2$

53. Multiplying the monomials: $2x\left(3x^2\right) = 6x^{1+2} = 6x^3$

55. Multiplying the monomials: $3x^2\left(2x^2\right) = 6x^{2+2} = 6x^4$

4.5 Multiplication with Polynomials

1. Using the distributive property: $2x(3x + 1) = 2x(3x) + 2x(1) = 6x^2 + 2x$

3. Using the distributive property: $2x^2\left(3x^2 - 2x + 1\right) = 2x^2\left(3x^2\right) - 2x^2(2x) + 2x^2(1) = 6x^4 - 4x^3 + 2x^2$

5. Using the distributive property: $2ab\left(a^2 - ab + 1\right) = 2ab\left(a^2\right) - 2ab(ab) + 2ab(1) = 2a^3b - 2a^2b^2 + 2ab$

7. Using the distributive property: $y^2\left(3y^2 + 9y + 12\right) = y^2\left(3y^2\right) + y^2(9y) + y^2(12) = 3y^4 + 9y^3 + 12y^2$

9. Using the distributive property:
$$4x^2y\left(2x^3y + 3x^2y^2 + 8y^3\right) = 4x^2y\left(2x^3y\right) + 4x^2y\left(3x^2y^2\right) + 4x^2y\left(8y^3\right) = 8x^5y^2 + 12x^4y^3 + 32x^2y^4$$

11. Multiplying using the FOIL method: $(x + 3)(x + 4) = x^2 + 3x + 4x + 12 = x^2 + 7x + 12$

13. Multiplying using the FOIL method: $(x + 6)(x + 1) = x^2 + 6x + 1x + 6 = x^2 + 7x + 6$

15. Multiplying using the FOIL method: $\left(x+\frac{1}{2}\right)\left(x+\frac{3}{2}\right)=x^2+\frac{1}{2}x+\frac{3}{2}x+\frac{3}{4}=x^2+2x+\frac{3}{4}$

17. Multiplying using the FOIL method: $(a+5)(a-3)=a^2+5a-3a-15=a^2+2a-15$

19. Multiplying using the FOIL method: $(x-a)(y+b)=xy-ay+bx-ab$

21. Multiplying using the FOIL method: $(x+6)(x-6)=x^2+6x-6x-36=x^2-36$

23. Multiplying using the FOIL method: $\left(y+\frac{5}{6}\right)\left(y-\frac{5}{6}\right)=y^2+\frac{5}{6}y-\frac{5}{6}y-\frac{25}{36}=y^2-\frac{25}{36}$

25. Multiplying using the FOIL method: $(2x-3)(x-4)=2x^2-3x-8x+12=2x^2-11x+12$

27. Multiplying using the FOIL method: $(a+2)(2a-1)=2a^2+4a-a-2=2a^2+3a-2$

29. Multiplying using the FOIL method: $(2x-5)(3x-2)=6x^2-15x-4x+10=6x^2-19x+10$

31. Multiplying using the FOIL method: $(2x+3)(a+4)=2ax+3a+8x+12$

33. Multiplying using the FOIL method: $(5x-4)(5x+4)=25x^2-20x+20x-16=25x^2-16$

35. Multiplying using the FOIL method: $\left(2x-\frac{1}{2}\right)\left(x+\frac{3}{2}\right)=2x^2-\frac{1}{2}x+3x-\frac{3}{4}=2x^2+\frac{5}{2}x-\frac{3}{4}$

37. Multiplying using the FOIL method: $(1-2a)(3-4a)=3-6a-4a+8a^2=3-10a+8a^2$

39. The product is $(x+2)(x+3)=x^2+5x+6$:

	x	3
x	x^2	$3x$
2	$2x$	6

41. The product is $(x+1)(2x+2)=2x^2+4x+2$:

	x	x	2
x	x^2	x^2	$2x$
1	x	x	2

43. Multiplying using the column method:

$$
\begin{array}{r}
a^2-3a+2 \\
a-3 \\
\hline
a^3-3a^2+2a \\
-3a^2+9a-6 \\
\hline
a^3-6a^2+11a-6
\end{array}
$$

45. Multiplying using the column method:

$$
\begin{array}{r}
x^2-2x+4 \\
x+2 \\
\hline
x^3-2x^2+4x \\
2x^2-4x+8 \\
\hline
x^3+8
\end{array}
$$

47. Multiplying using the column method:

$$
\begin{array}{r}
x^2+8x+9 \\
2x+1 \\
\hline
2x^3+16x^2+18x \\
x^2+8x+9 \\
\hline
2x^3+17x^2+26x+9
\end{array}
$$

49. Multiplying using the column method:

$$
\begin{array}{r}
5x^2+2x+1 \\
x^2-3x+5 \\
\hline
5x^4+2x^3+x^2 \\
-15x^3-6x^2-3x \\
25x^2+10x+5 \\
\hline
5x^4-13x^3+20x^2+7x+5
\end{array}
$$

51. Multiplying using the FOIL method: $\left(x^2+3\right)\left(2x^2-5\right)=2x^4-5x^2+6x^2-15=2x^4+x^2-15$

53. Multiplying using the FOIL method: $\left(3a^4+2\right)\left(2a^2+5\right)=6a^6+15a^4+4a^2+10$

55. First multiply two polynomials using the FOIL method: $(x+3)(x+4)=x^2+3x+4x+12=x^2+7x+12$
Now using the column method:

$$
\begin{array}{r}
x^2+7x+12 \\
x+5 \\
\hline
x^3+7x^2+12x \\
5x^2+35x+60 \\
\hline
x^3+12x^2+47x+60
\end{array}
$$

57. Let x represent the width and $2x+5$ represent the length. The area is given by: $A=x(2x+5)=2x^2+5x$

59. Let x and $x+1$ represent the width and length, respectively. The area is given by: $A=x(x+1)=x^2+x$

61. The revenue is: $R=xp=(1200-100p)p=1200p-100p^2$

63. The revenue is: $R = xp = (1700 - 100p)p = 1700p - 100p^2$

65. Graphing each line:

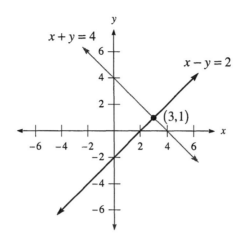

The intersection point is $(3,1)$.

67. Multiplying the first equation by 3 and the second equation by -2:

$$6x + 9y = -3$$
$$-6x - 10y = 4$$

Adding the two equations:

$$-y = 1$$
$$y = -1$$

Substituting into the first equation:

$$2x + 3(-1) = -1$$
$$2x - 3 = -1$$
$$2x = 2$$
$$x = 1$$

The solution is $(1, -1)$.

69. Substituting into the first equation:

$$2x - 6(3x + 1) = 2$$
$$2x - 18x - 6 = 2$$
$$-16x - 6 = 2$$
$$-16x = 8$$
$$x = -\tfrac{1}{2}$$

Substituting into the second equation: $y = 3\left(-\tfrac{1}{2}\right) + 1 = -\tfrac{3}{2} + 1 = -\tfrac{1}{2}$. The solution is $\left(-\tfrac{1}{2}, -\tfrac{1}{2}\right)$.

71. Let x represent the number of dimes and y represent the number of quarters. The system of equations is:

$$x + y = 11$$
$$0.10x + 0.25y = 1.85$$

Multiplying the first equation by -0.10:

$$-0.10x - 0.10y = -1.10$$
$$0.10x + 0.25y = 1.85$$

Adding the two equations:

$$0.15y = 0.75$$
$$y = 5$$

Substituting into the first equation:

$$x + 5 = 11$$
$$x = 6$$

Amy has 6 dimes and 5 quarters.

4.6 Binomial Squares and Other Special Products

1.   Multiplying using the FOIL method: $(x-2)^2 = (x-2)(x-2) = x^2 - 2x - 2x + 4 = x^2 - 4x + 4$

3.   Multiplying using the FOIL method: $(a+3)^2 = (a+3)(a+3) = a^2 + 3a + 3a + 9 = a^2 + 6a + 9$

5.   Multiplying using the FOIL method: $(x-5)^2 = (x-5)(x-5) = x^2 - 5x - 5x + 25 = x^2 - 10x + 25$

7.   Multiplying using the FOIL method: $\left(a-\frac{1}{2}\right)^2 = \left(a-\frac{1}{2}\right)\left(a-\frac{1}{2}\right) = a^2 - \frac{1}{2}a - \frac{1}{2}a + \frac{1}{4} = a^2 - a + \frac{1}{4}$

9.   Multiplying using the FOIL method: $(x+10)^2 = (x+10)(x+10) = x^2 + 10x + 10x + 100 = x^2 + 20x + 100$

11.   Multiplying using the square of binomial formula: $(a+0.8)^2 = a^2 + 2(a)(0.8) + (0.8)^2 = a^2 + 1.6a + 0.64$

13.   Multiplying using the square of binomial formula: $(2x-1)^2 = (2x)^2 - 2(2x)(1) + (1)^2 = 4x^2 - 4x + 1$

15.   Multiplying using the square of binomial formula: $(4a+5)^2 = (4a)^2 + 2(4a)(5) + (5)^2 = 16a^2 + 40a + 25$

17.   Multiplying using the square of binomial formula: $(3x-2)^2 = (3x)^2 - 2(3x)(2) + (2)^2 = 9x^2 - 12x + 4$

19.   Multiplying using the square of binomial formula: $(3a+5b)^2 = (3a)^2 + 2(3a)(5b) + (5b)^2 = 9a^2 + 30ab + 25b^2$

21.   Multiplying using the square of binomial formula: $(4x-5y)^2 = (4x)^2 - 2(4x)(5y) + (5y)^2 = 16x^2 - 40xy + 25y^2$

23.   Multiplying using the square of binomial formula: $(7m+2n)^2 = (7m)^2 + 2(7m)(2n) + (2n)^2 = 49m^2 + 28mn + 4n^2$

25.   Multiplying using the square of binomial formula:
$$(6x-10y)^2 = (6x)^2 - 2(6x)(10y) + (10y)^2 = 36x^2 - 120xy + 100y^2$$

27.   Multiplying using the square of binomial formula: $\left(x^2+5\right)^2 = \left(x^2\right)^2 + 2\left(x^2\right)(5) + (5)^2 = x^4 + 10x^2 + 25$

29.   Multiplying using the square of binomial formula: $\left(a^2+1\right)^2 = \left(a^2\right)^2 + 2\left(a^2\right)(1) + (1)^2 = a^4 + 2a^2 + 1$

31.   Completing the table:

x	$(x+3)^2$	x^2+9	x^2+6x+9
1	16	10	16
2	25	13	25
3	36	18	36
4	49	25	49

33.   Completing the table:

a	b	$(a+b)^2$	a^2+b^2	a^2+ab+b^2	$a^2+2ab+b^2$
1	1	4	2	3	4
3	5	64	34	49	64
3	4	49	25	37	49
4	5	81	41	61	81

35.   Multiplying using the FOIL method: $(a+5)(a-5) = a^2 + 5a - 5a - 25 = a^2 - 25$

37.   Multiplying using the FOIL method: $(y-1)(y+1) = y^2 - y + y - 1 = y^2 - 1$

39.   Multiplying using the difference of squares formula: $(9+x)(9-x) = (9)^2 - (x)^2 = 81 - x^2$

41.   Multiplying using the difference of squares formula: $(2x+5)(2x-5) = (2x)^2 - (5)^2 = 4x^2 - 25$

43.   Multiplying using the difference of squares formula: $\left(4x+\frac{1}{3}\right)\left(4x-\frac{1}{3}\right) = (4x)^2 - \left(\frac{1}{3}\right)^2 = 16x^2 - \frac{1}{9}$

45.   Multiplying using the difference of squares formula: $(2a+7)(2a-7) = (2a)^2 - (7)^2 = 4a^2 - 49$

47.   Multiplying using the difference of squares formula: $(6-7x)(6+7x) = (6)^2 - (7x)^2 = 36 - 49x^2$

49.   Multiplying using the difference of squares formula: $\left(x^2+3\right)\left(x^2-3\right) = \left(x^2\right)^2 - (3)^2 = x^4 - 9$

51. Multiplying using the difference of squares formula: $\left(a^2+4\right)\left(a^2-4\right)=\left(a^2\right)^2-(4)^2=a^4-16$

53. Multiplying using the difference of squares formula: $\left(5y^4-8\right)\left(5y^4+8\right)=\left(5y^4\right)^2-(8)^2=25y^8-64$

55. Multiplying and simplifying: $(x+3)(x-3)+(x+5)(x-5)=\left(x^2-9\right)+\left(x^2-25\right)=2x^2-34$

57. Multiplying and simplifying:
$$(2x+3)^2-(4x-1)^2=\left(4x^2+12x+9\right)-\left(16x^2-8x+1\right)=4x^2+12x+9-16x^2+8x-1=-12x^2+20x+8$$

59. Multiplying and simplifying:
$$(a+1)^2-(a+2)^2+(a+3)^2=\left(a^2+2a+1\right)-\left(a^2+4a+4\right)+\left(a^2+6a+9\right)$$
$$=a^2+2a+1-a^2-4a-4+a^2+6a+9$$
$$=a^2+4a+6$$

61. Multiplying and simplifying:
$$(2x+3)^3=(2x+3)(2x+3)^2$$
$$=(2x+3)\left(4x^2+12x+9\right)$$
$$=8x^3+24x^2+18x+12x^2+36x+27$$
$$=8x^3+36x^2+54x+27$$

63. Finding the product: $49(51)=(50-1)(50+1)=(50)^2-(1)^2=2,500-1=2,499$

65. Evaluating when $x=2$:
$$(x+3)^2=(2+3)^2=(5)^2=25$$
$$x^2+6x+9=(2)^2+6(2)+9=4+12+9=25$$

67. Let x and $x+1$ represent the two integers. The expression can be written as:
$$(x)^2+(x+1)^2=x^2+\left(x^2+2x+1\right)=2x^2+2x+1$$

69. Let x, $x+1$, and $x+2$ represent the three integers. The expression can be written as:
$$(x)^2+(x+1)^2+(x+2)^2=x^2+\left(x^2+2x+1\right)+\left(x^2+4x+4\right)=3x^2+6x+5$$

71. Verifying the areas: $(a+b)^2=a^2+ab+ab+b^2=a^2+2ab+b^2$

73. The product is $(2x+1)(2x+1)=4x^2+4x+1$:

	x	x	1
x	x^2	x^2	x
x	x^2	x^2	x
1	x	x	1

75. Simplifying: $\dfrac{15x^2y}{3xy}=\dfrac{15}{3}\cdot\dfrac{x^2}{x}\cdot\dfrac{y}{y}=5x$

77. Simplifying: $\dfrac{35a^6b^8}{70a^2b^{10}}=\dfrac{35}{70}\cdot\dfrac{a^6}{a^2}\cdot\dfrac{b^8}{b^{10}}=\dfrac{1}{2}\cdot a^4\cdot\dfrac{1}{b^2}=\dfrac{a^4}{2b^2}$

79.   Graphing both lines:

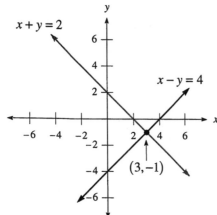

The intersection point is $(3, -1)$.

81.   Graphing both lines:

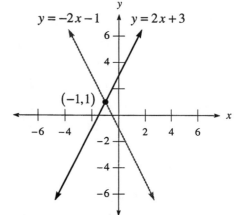

The intersection point is $(-1, 1)$.

4.7 Dividing a Polynomial by a Monomial

1.   Performing the division: $\dfrac{5x^2 - 10x}{5x} = \dfrac{5x^2}{5x} - \dfrac{10x}{5x} = x - 2$

3.   Performing the division: $\dfrac{15x - 10x^3}{5x} = \dfrac{15x}{5x} - \dfrac{10x^3}{5x} = 3 - 2x^2$

5.   Performing the division: $\dfrac{25x^2 y - 10xy}{5x} = \dfrac{25x^2 y}{5x} - \dfrac{10xy}{5x} = 5xy - 2y$

7.   Performing the division: $\dfrac{35x^5 - 30x^4 + 25x^3}{5x} = \dfrac{35x^5}{5x} - \dfrac{30x^4}{5x} + \dfrac{25x^3}{5x} = 7x^4 - 6x^3 + 5x^2$

9.   Performing the division: $\dfrac{50x^5 - 25x^3 + 5x}{5x} = \dfrac{50x^5}{5x} - \dfrac{25x^3}{5x} + \dfrac{5x}{5x} = 10x^4 - 5x^2 + 1$

11.   Performing the division: $\dfrac{8a^2 - 4a}{-2a} = \dfrac{8a^2}{-2a} + \dfrac{-4a}{-2a} = -4a + 2$

13.   Performing the division: $\dfrac{16a^5 + 24a^4}{-2a} = \dfrac{16a^5}{-2a} + \dfrac{24a^4}{-2a} = -8a^4 - 12a^3$

15.   Performing the division: $\dfrac{8ab + 10a^2}{-2a} = \dfrac{8ab}{-2a} + \dfrac{10a^2}{-2a} = -4b - 5a$

17.   Performing the division: $\dfrac{12a^3 b - 6a^2 b^2 + 14ab^3}{-2a} = \dfrac{12a^3 b}{-2a} + \dfrac{-6a^2 b^2}{-2a} + \dfrac{14ab^3}{-2a} = -6a^2 b + 3ab^2 - 7b^3$

19.   Performing the division: $\dfrac{a^2 + 2ab + b^2}{-2a} = \dfrac{a^2}{-2a} + \dfrac{2ab}{-2a} + \dfrac{b^2}{-2a} = -\dfrac{a}{2} - b - \dfrac{b^2}{2a}$

21.   Performing the division: $\dfrac{6x + 8y}{2} = \dfrac{6x}{2} + \dfrac{8y}{2} = 3x + 4y$

23.   Performing the division: $\dfrac{7y - 21}{-7} = \dfrac{7y}{-7} + \dfrac{-21}{-7} = -y + 3$

25.   Performing the division: $\dfrac{10xy - 8x}{2x} = \dfrac{10xy}{2x} - \dfrac{8x}{2x} = 5y - 4$

27.   Performing the division: $\dfrac{x^2 y - x^3 y^2}{x} = \dfrac{x^2 y}{x} - \dfrac{x^3 y^2}{x} = xy - x^2 y^2$

29.   Performing the division: $\dfrac{x^2 y - x^3 y^2}{-x^2 y} = \dfrac{x^2 y}{-x^2 y} + \dfrac{-x^3 y^2}{-x^2 y} = -1 + xy$

31. Performing the division: $\dfrac{a^2b^2 - ab^2}{-ab^2} = \dfrac{a^2b^2}{-ab^2} + \dfrac{-ab^2}{-ab^2} = -a + 1$

33. Performing the division: $\dfrac{x^3 - 3x^2y + xy^2}{x} = \dfrac{x^3}{x} - \dfrac{3x^2y}{x} + \dfrac{xy^2}{x} = x^2 - 3xy + y^2$

35. Performing the division: $\dfrac{10a^2 - 15a^2b + 25a^2b^2}{5a^2} = \dfrac{10a^2}{5a^2} - \dfrac{15a^2b}{5a^2} + \dfrac{25a^2b^2}{5a^2} = 2 - 3b + 5b^2$

37. Performing the division: $\dfrac{26x^2y^2 - 13xy}{-13xy} = \dfrac{26x^2y^2}{-13xy} + \dfrac{-13xy}{-13xy} = -2xy + 1$

39. Performing the division: $\dfrac{4x^2y^2 - 2xy}{4xy} = \dfrac{4x^2y^2}{4xy} - \dfrac{2xy}{4xy} = xy - \dfrac{1}{2}$

41. Performing the division: $\dfrac{5a^2x - 10ax^2 + 15a^2x^2}{20a^2x^2} = \dfrac{5a^2x}{20a^2x^2} - \dfrac{10ax^2}{20a^2x^2} + \dfrac{15a^2x^2}{20a^2x^2} = \dfrac{1}{4x} - \dfrac{1}{2a} + \dfrac{3}{4}$

43. Performing the division: $\dfrac{16x^5 + 8x^2 + 12x}{12x^3} = \dfrac{16x^5}{12x^3} + \dfrac{8x^2}{12x^3} + \dfrac{12x}{12x^3} = \dfrac{4x^2}{3} + \dfrac{2}{3x} + \dfrac{1}{x^2}$

45. Performing the division: $\dfrac{9a^{5m} - 27a^{3m}}{3a^{2m}} = \dfrac{9a^{5m}}{3a^{2m}} - \dfrac{27a^{3m}}{3a^{2m}} = 3a^{5m-2m} - 9a^{3m-2m} = 3a^{3m} - 9a^m$

47. Performing the division:
$$\dfrac{10x^{5m} - 25x^{3m} + 35x^m}{5x^m} = \dfrac{10x^{5m}}{5x^m} - \dfrac{25x^{3m}}{5x^m} + \dfrac{35x^m}{5x^m} = 2x^{5m-m} - 5x^{3m-m} + 7x^{m-m} = 2x^{4m} - 5x^{2m} + 7$$

49. Simplifying and then dividing:
$$\dfrac{2x^3(3x+2) - 3x^2(2x-4)}{2x^2} = \dfrac{6x^4 + 4x^3 - 6x^3 + 12x^2}{2x^2}$$
$$= \dfrac{6x^4 - 2x^3 + 12x^2}{2x^2}$$
$$= \dfrac{6x^4}{2x^2} - \dfrac{2x^3}{2x^2} + \dfrac{12x^2}{2x^2}$$
$$= 3x^2 - x + 6$$

51. Simplifying and then dividing:
$$\dfrac{(x+2)^2 - (x-2)^2}{2x} = \dfrac{\left(x^2 + 4x + 4\right) - \left(x^2 - 4x + 4\right)}{2x} = \dfrac{x^2 + 4x + 4 - x^2 + 4x - 4}{2x} = \dfrac{8x}{2x} = 4$$

53. Simplifying and then dividing:
$$\dfrac{(x+5)^2 + (x+5)(x-5)}{2x} = \dfrac{\left(x^2 + 10x + 25\right) + \left(x^2 - 25\right)}{2x} = \dfrac{2x^2 + 10x}{2x} = \dfrac{2x^2}{2x} + \dfrac{10x}{2x} = x + 5$$

55. Evaluating each expression when $x = 2$:
$$\dfrac{10x + 15}{5} = \dfrac{10(2) + 15}{5} = \dfrac{20 + 15}{5} = \dfrac{35}{5} = 7 \qquad\qquad 2x + 3 = 2(2) + 3 = 4 + 3 = 7$$

57. Evaluating each expression when $x = 10$:
$$\dfrac{3x + 8}{2} = \dfrac{3(10) + 8}{2} = \dfrac{30 + 8}{2} = \dfrac{38}{2} = 19 \qquad\qquad 3x + 4 = 3(10) + 4 = 30 + 4 = 34$$
Thus $\dfrac{3x + 8}{2} \neq 3x + 4$.

59. Adding the two equations:
$$2x = 14$$
$$x = 7$$
Substituting into the first equation:
$$7 + y = 6$$
$$y = -1$$
The solution is $(7, -1)$.

61. Multiplying the second equation by 3:
$$2x - 3y = -5$$
$$3x + 3y = 15$$
Adding the two equations:
$$5x = 10$$
$$x = 2$$
Substituting into the second equation:
$$2 + y = 5$$
$$y = 3$$
The solution is $(2, 3)$.

63. Substituting into the first equation:
$$x + 2x - 1 = 2$$
$$3x - 1 = 2$$
$$3x = 3$$
$$x = 1$$
Substituting into the second equation: $y = 2(1) - 1 = 2 - 1 = 1$. The solution is $(1, 1)$.

65. Substituting into the first equation:
$$4x + 2(-2x + 4) = 8$$
$$4x - 4x + 8 = 8$$
$$8 = 8$$
Since this statement is true, the system is dependent. The two lines coincide.

4.8 Dividing a Polynomial by a Polynomial

1. Using long division:

$$
\begin{array}{r}
x - 2 \\
x - 3 \overline{\smash{\big)}\, x^2 - 5x + 6} \\
\underline{x^2 - 3x} \\
-2x + 6 \\
\underline{-2x + 6} \\
0
\end{array}
$$

The quotient is $x - 2$.

3. Using long division:

$$
\begin{array}{r}
a + 4 \\
a + 5 \overline{\smash{\big)}\, a^2 + 9a + 20} \\
\underline{a^2 + 5a} \\
4a + 20 \\
\underline{4a + 20} \\
0
\end{array}
$$

The quotient is $a + 4$.

5. Using long division:

$$
\begin{array}{r}
x - 3 \\
x - 3 \overline{\smash{\big)}\, x^2 - 6x + 9} \\
\underline{x^2 - 3x} \\
-3x + 9 \\
\underline{-3x + 9} \\
0
\end{array}
$$

The quotient is $x - 3$.

7. Using long division:

$$
\begin{array}{r}
x + 3 \\
2x - 1 \overline{\smash{\big)}\, 2x^2 + 5x - 3} \\
\underline{2x^2 - x} \\
6x - 3 \\
\underline{6x - 3} \\
0
\end{array}
$$

The quotient is $x + 3$.

9. Using long division:

$$\begin{array}{r} a - 5 \\ 2a+1\overline{)2a^2 - 9a5} \\ \underline{2a^2 + a} \\ -10a - 5 \\ \underline{-10a - 5} \\ 0 \end{array}$$

The quotient is $a - 5$.

11. Using long division:

$$\begin{array}{r} x + 2 \\ x+3\overline{)x^2 + 5x + 8} \\ \underline{x^2 + 3x} \\ 2x + 8 \\ \underline{2x + 6} \\ 2 \end{array}$$

The quotient is $x + 2 + \dfrac{2}{x+3}$.

13. Using long division:

$$\begin{array}{r} a - 2 \\ a+5\overline{)a^2 + 3a + 2} \\ \underline{a^2 + 5a} \\ -2a + 2 \\ \underline{-2a - 10} \\ 12 \end{array}$$

The quotient is $a - 2 + \dfrac{12}{a+5}$.

15. Using long division:

$$\begin{array}{r} x + 4 \\ x-2\overline{)x^2 + 2x + 1} \\ \underline{x^2 - 2x} \\ 4x + 1 \\ \underline{4x - 8} \\ 9 \end{array}$$

The quotient is $x + 4 + \dfrac{9}{x-2}$.

17. Using long division:

$$\begin{array}{r} x + 4 \\ x+1\overline{)x^2 + 5x - 6} \\ \underline{x^2 + x} \\ 4x - 6 \\ \underline{4x + 4} \\ -10 \end{array}$$

The quotient is $x + 4 + \dfrac{-10}{x+1}$.

19. Using long division:

$$\begin{array}{r} a + 1 \\ a+2\overline{)a^2 + 3a + 1} \\ \underline{a^2 + 2a} \\ a + 1 \\ \underline{a + 2} \\ -1 \end{array}$$

The quotient is $a + 1 + \dfrac{-1}{a+2}$.

21. Using long division:

$$\begin{array}{r} x - 3 \\ 2x+4\overline{)2x^2 - 2x + 5} \\ \underline{2x^2 + 4x} \\ -6x + 5 \\ \underline{-6x - 12} \\ 17 \end{array}$$

The quotient is $x - 3 + \dfrac{17}{2x+4}$.

23. Using long division:

$$\begin{array}{r} 3a - 2 \\ 2a+3\overline{)6a^2 + 5a + 1} \\ \underline{6a^2 + 9a} \\ -4a + 1 \\ \underline{-4a - 6} \\ 7 \end{array}$$

The quotient is $3a - 2 + \dfrac{7}{2a+3}$.

25. Using long division:

$$\begin{array}{r} 2a^2 - a - 3 \\ 3a-5\overline{)6a^3 - 13a^2 - 4a + 15} \\ \underline{6a^3 - 10a^2} \\ -3a^2 - 4a \\ \underline{-3a^2 + 5a} \\ -9a + 15 \\ \underline{-9a + 15} \\ 0 \end{array}$$

The quotient is $2a^2 - a - 3$.

27. Using long division:

$$\begin{array}{r} x^2 - x + 5 \\ x+1\overline{)x^3 + 0x^2 + 4x + 5} \\ \underline{x^3 + x^2} \\ -x^2 + 4x \\ \underline{-x^2 - x} \\ 5x + 5 \\ \underline{5x + 5} \\ 0 \end{array}$$

The quotient is $x^2 - x + 5$.

29. Using long division:

$$\begin{array}{r}
x^2+x+1 \\
x-1\overline{\smash{\big)}\,x^3+0x^2+0x-1} \\
\underline{x^3-\ x^2} \\
x^2+0x \\
\underline{x^2-\ x} \\
x-1 \\
\underline{x-1} \\
0
\end{array}$$

The quotient is x^2+x+1.

31. Using long division:

$$\begin{array}{r}
x^2+2x+4 \\
x-2\overline{\smash{\big)}\,x^3+0x^2+0x-8} \\
\underline{x^3-2x^2} \\
2x^2+0x \\
\underline{2x^2-4x} \\
4x-8 \\
\underline{4x-8} \\
0
\end{array}$$

The quotient is x^2+2x+4.

33. Let x and y represent the two numbers. The system of equations is:
$$x+y=25$$
$$y=4x$$
Substituting into the first equation:
$$x+4x=25$$
$$5x=25$$
$$x=5$$
$$y=4(5)=20$$
The two numbers are 5 and 20.

35. Let x represent the amount invested at 8% and y represent the amount invested at 9%. The system of equations is:
$$x+y=1200$$
$$0.08x+0.09y=100$$
Multiplying the first equation by -0.08:
$$-0.08x-0.08y=-96$$
$$0.08x+0.09y=100$$
Adding the two equations:
$$0.01y=4$$
$$y=400$$
Substituting into the first equation:
$$x+400=1200$$
$$x=800$$
You have $800 invested at 8% and $400 invested at 9%.

37. Let x represent the number of $5 bills and $x+4$ represent the number of $10 bills. The equation is:
$$5(x)+10(x+4)=160$$
$$5x+10x+40=160$$
$$15x+40=160$$
$$15x=120$$
$$x=8$$
$$x+4=12$$
You have 8 $5 bills and 12 $10 bills.

39.   Let x represent the gallons of 20% antifreeze and y represent the gallons of 60% antifreeze. The system of equations is:
$$x + y = 16$$
$$0.20x + 0.60y = 0.35(16)$$
Multiplying the first equation by -0.20:
$$-0.20x - 0.20y = -3.2$$
$$0.20x + 0.60y = 5.6$$
Adding the two equations:
$$0.40y = 2.4$$
$$y = 6$$
Substituting into the first equation:
$$x + 6 = 16$$
$$x = 10$$
The mixture contains 10 gallons of 20% antifreeze and 6 gallons of 60% antifreeze.

Chapter 4 Review

1.   Simplifying the expression: $(-1)^3 = (-1)(-1)(-1) = -1$

3.   Simplifying the expression: $\left(\frac{3}{7}\right)^2 = \left(\frac{3}{7}\right) \cdot \left(\frac{3}{7}\right) = \frac{9}{49}$

5.   Simplifying the expression: $x^{15} \cdot x^7 \cdot x^5 \cdot x^3 = x^{15+7+5+3} = x^{30}$

7.   Simplifying the expression: $\left(2^6\right)^4 = 2^{6 \cdot 4} = 2^{24}$

9.   Simplifying the expression: $(-2xyz)^3 = (-2)^3 \cdot x^3 y^3 z^3 = -8x^3 y^3 z^3$

11.   Writing with positive exponents: $4x^{-5} = 4 \cdot \frac{1}{x^5} = \frac{4}{x^5}$

13.   Simplifying the expression: $\frac{a^9}{a^3} = a^{9-3} = a^6$

15.   Simplifying the expression: $\frac{x^9}{x^{-6}} = x^{9-(-6)} = x^{9+6} = x^{15}$

17.   Simplifying the expression: $(-3xy)^0 = 1$

19.   Simplifying the expression: $\left(3x^3 y^2\right)^2 = 3^2 x^6 y^4 = 9x^6 y^4$

21.   Simplifying the expression: $\left(-3xy^2\right)^{-3} = \frac{1}{\left(-3xy^2\right)^3} = \frac{1}{(-3)^3 x^3 y^6} = -\frac{1}{27x^3 y^6}$

23.   Simplifying the expression: $\frac{\left(x^{-3}\right)^3 \left(x^6\right)^{-1}}{\left(x^{-5}\right)^{-4}} = \frac{x^{-9} \cdot x^{-6}}{x^{20}} = \frac{x^{-15}}{x^{20}} = x^{-15-20} = x^{-35} = \frac{1}{x^{35}}$

25.   Simplifying the expression: $\frac{\left(10x^3 y^5\right)\left(21x^2 y^6\right)}{\left(7xy^3\right)\left(5x^9 y\right)} = \frac{210x^5 y^{11}}{35x^{10} y^4} = \frac{210}{35} \cdot \frac{x^5}{x^{10}} \cdot \frac{y^{11}}{y^4} = 6 \cdot \frac{1}{x^5} \cdot y^7 = \frac{6y^7}{x^5}$

27.   Simplifying the expression: $\frac{8x^8 y^3}{2x^3 y} - \frac{10x^6 y^9}{5xy^7} = 4x^5 y^2 - 2x^5 y^2 = 2x^5 y^2$

29.   Finding the quotient: $\frac{4.6 \times 10^5}{2 \times 10^{-3}} = 2.3 \times 10^8$

31.   Performing the operations: $\left(3a^2 - 5a + 5\right) + \left(5a^2 - 7a - 8\right) = \left(3a^2 + 5a^2\right) + (-5a - 7a) + (5 - 8) = 8a^2 - 12a - 3$

33. Performing the operations:
$$\left(4x^2 - 3x - 2\right) - \left(8x^2 + 3x - 2\right) = 4x^2 - 3x - 2 - 8x^2 - 3x + 2$$
$$= \left(4x^2 - 8x^2\right) + \left(-3x - 3x\right) + \left(-2 + 2\right)$$
$$= -4x^2 - 6x$$

35. Multiplying: $3x(4x - 7) = 3x(4x) - 3x(7) = 12x^2 - 21x$

37. Multiplying using the column method:

$$
\begin{array}{r}
a^2 + 5a - 4 \\
a + 1 \\
\hline
a^3 + 5a^2 - 4a \\
a^2 + 5a - 4 \\
\hline
a^3 + 6a^2 + a - 4
\end{array}
$$

39. Multiplying using the FOIL method: $(3x - 7)(2x - 5) = 6x^2 - 14x - 15x + 35 = 6x^2 - 29x + 35$

41. Multiplying using the difference of squares formula: $\left(a^2 - 3\right)\left(a^2 + 3\right) = \left(a^2\right)^2 - (3)^2 = a^4 - 9$

43. Multiplying using the square of binomial formula: $(3x + 4)^2 = (3x)^2 + 2(3x)(4) + (4)^2 = 9x^2 + 24x + 16$

45. Performing the division: $\dfrac{10ab + 20a^2}{-5a} = \dfrac{10ab}{-5a} + \dfrac{20a^2}{-5a} = -2b - 4a$

47. Using long division:

$$
\begin{array}{r}
x + 9 \\
x + 6 \overline{)\, x^2 + 15x + 54} \\
\underline{x^2 + 6x} \\
9x + 54 \\
\underline{9x + 54} \\
0
\end{array}
$$

The quotient is $x + 9$.

49. Using long division:

$$
\begin{array}{r}
x^2 - 4x + 16 \\
x + 4 \overline{)\, x^3 + 0x^2 + 0x + 64} \\
\underline{x^3 + 4x^2} \\
-4x^2 + 0x \\
\underline{-4x^2 - 16x} \\
16x + 64 \\
\underline{16x + 64} \\
0
\end{array}
$$

The quotient is $x^2 - 4x + 16$.

51. Using long division:

$$
\begin{array}{r}
x^2 - 4x + 5 \\
2x + 1 \overline{)\, 2x^3 - 7x^2 + 6x + 10} \\
\underline{2x^3 + x^2} \\
-8x^2 + 6x \\
\underline{-8x^2 - 4x} \\
10x + 10 \\
\underline{10x + 5} \\
5
\end{array}
$$

The quotient is $x^2 - 4x + 5 + \dfrac{5}{2x + 1}$.

Chapters 1-4 Cumulative Review

1. Simplifying the expression: $-\left(-\frac{3}{4}\right) = \frac{3}{4}$ **3.** Simplifying the expression: $6 \cdot 7 + 7 \cdot 9 = 42 + 63 = 105$

5. Simplifying the expression: $6(4a+2) - 3(5a-1) = 24a + 12 - 15a + 3 = 24a - 15a + 12 + 3 = 9a + 15$

7. Simplifying the expression: $-15 - (-3) = -15 + 3 = -12$

9. Simplifying the expression: $(-9)(-5) = 45$ **11.** Simplifying the expression: $(2y)^4 = 2^4 y^4 = 16y^4$

13. Simplifying the expression: $\dfrac{\left(12xy^5\right)^3 \left(16x^2 y^2\right)}{\left(8x^3 y^3\right)\left(3x^5 y\right)} = \dfrac{1728x^3 y^{15} \cdot 16x^2 y^2}{24x^8 y^4} = \dfrac{27648x^5 y^{17}}{24x^8 y^4} = \dfrac{1152y^{13}}{x^3}$

15. Multiplying using the square of binomial formula: $(5x-1)^2 = (5x)^2 - 2(5x)(1) + (1)^2 = 25x^2 - 10x + 1$

17. Multiplying using the column method:

$$
\begin{array}{r}
x^2 + x + 1 \\
x - 1 \\
\hline
x^3 + x^2 + x \\
-x^2 - x - 1 \\
\hline
x^3 - 1
\end{array}
$$

19. Solving the equation:

$$6a - 5 = 4a$$
$$-5 = -2a$$
$$a = \frac{5}{2}$$

21. Solving the equation:

$$2(3x+5) + 8 = 2x + 10$$
$$6x + 10 + 8 = 2x + 10$$
$$6x + 18 = 2x + 10$$
$$4x + 18 = 10$$
$$4x = -8$$
$$x = -2$$

23. Solving the inequality:

$$-4x > 28$$
$$-\tfrac{1}{4}(-4x) < -\tfrac{1}{4}(28)$$
$$x < -7$$

Graphing the solution set:

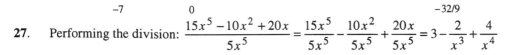

25. Solving the inequality:

$$3(2t-5) - 7 \le 5(3t+1) + 5$$
$$6t - 15 - 7 \le 15t + 5 + 5$$
$$6t - 22 \le 15t + 10$$
$$-9t - 22 \le 10$$
$$-9t \le 32$$
$$t \ge -\tfrac{32}{9}$$

Graphing the solution set:

27. Performing the division: $\dfrac{15x^5 - 10x^2 + 20x}{5x^5} = \dfrac{15x^5}{5x^5} - \dfrac{10x^2}{5x^5} + \dfrac{20x}{5x^5} = 3 - \dfrac{2}{x^3} + \dfrac{4}{x^4}$

28. Graphing the equation:

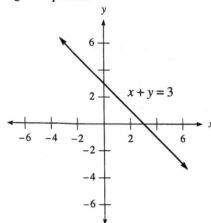

31. Graphing the two lines:

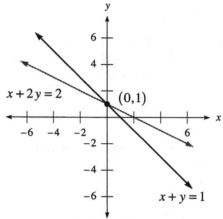

The intersection point is $(0,1)$.

33. Graphing the two lines:

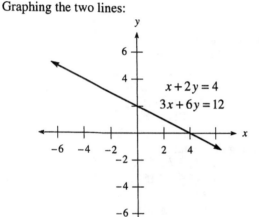

The system is dependent. The two lines coincide.

35. Multiplying the first equation by –3:
$$-3x - 6y = -15$$
$$3x + 6y = 14$$
Adding the two equations:
$$0 = -1$$
Since this statement is false, there is no solution to the system. The two lines are parallel.

37. To clear each equation of fractions, multiply the first equation by 12 and the second equation by 5:
$$12\left(\tfrac{1}{6}x + \tfrac{1}{4}y\right) = 12(1) \qquad\qquad 5\left(\tfrac{6}{5}x - y\right) = 5\left(\tfrac{8}{5}\right)$$
$$2x + 3y = 12 \qquad\qquad\qquad 6x - 5y = 8$$
The system of equations is:
$$2x + 3y = 12$$
$$6x - 5y = 8$$
Multiplying the first equation by –3:
$$-6x - 9y = -36$$
$$6x - 5y = 8$$
Adding the two equations:
$$-14y = -28$$
$$y = 2$$

Substituting into $2x + 3y = 12$:
$$2x + 3(2) = 12$$
$$2x + 6 = 12$$
$$2x = 6$$
$$x = 3$$
The solution is $(3, 2)$.

39. Solving the second equation for x:
$$2y = x - 3$$
$$x = 2y + 3$$
Substituting into the first equation:
$$4(2y + 3) + 5y = 25$$
$$8y + 12 + 5y = 25$$
$$13y + 12 = 25$$
$$13y = 13$$
$$y = 1$$
$$x = 2(1) + 3 = 5$$
The solution is $(5, 1)$.

41. Evaluating when $x = 3$: $8x - 3 = 8(3) - 3 = 24 - 3 = 21$ **43.** The irrational numbers are $-\sqrt{2}$ and π.

45. Substituting each ordered pair:
$$(0, 3): \quad 3(0) - 4(3) = 0 - 12 = -12 \neq 12$$
$$(4, 0): \quad 3(4) - 4(0) = 12 - 0 = 12$$
$$\left(\tfrac{16}{3}, 1\right): \quad 3\left(\tfrac{16}{3}\right) - 4(1) = 16 - 4 = 12$$

The ordered pairs $(4, 0)$ and $\left(\tfrac{16}{3}, 1\right)$ are solutions to the equation.

47. Writing in scientific notation: $186,000 = 1.86 \times 10^5$ **49.** Multiplying: $\left(2 \times 10^5\right)\left(3 \times 10^{-8}\right) = 6 \times 10^{-3} = 0.006$

Chapter 4 Test

1. Simplifying the expression: $(-3)^4 = (-3)(-3)(-3)(-3) = 81$

2. Simplifying the expression: $\left(\tfrac{3}{4}\right)^2 = \left(\tfrac{3}{4}\right)\left(\tfrac{3}{4}\right) = \tfrac{9}{16}$

3. Simplifying the expression: $\left(3x^3\right)^2 \left(2x^4\right)^3 = 9x^6 \bullet 8x^{12} = 72x^{18}$

4. Simplifying the expression: $3^{-2} = \dfrac{1}{3^2} = \tfrac{1}{9}$ **5.** Simplifying the expression: $\left(3a^4 b^2\right)^0 = 1$

6. Simplifying the expression: $\dfrac{a^{-3}}{a^{-5}} = a^{-3-(-5)} = a^{-3+5} = a^2$

7. Simplifying the expression: $\dfrac{\left(x^{-2}\right)^3 \left(x^{-3}\right)^{-5}}{\left(x^{-4}\right)^{-2}} = \dfrac{x^{-6} x^{15}}{x^8} = \dfrac{x^9}{x^8} = x^{9-8} = x$

8. Writing in scientific notation: $0.0278 = 2.78 \times 10^{-2}$ **9.** Writing in expanded form: $2.43 \times 10^5 = 243,000$

10. Simplifying the expression: $\dfrac{35x^2 y^4 z}{70x^6 y^2 z} = \dfrac{35}{70} \bullet \dfrac{x^2}{x^6} \bullet \dfrac{y^4}{y^2} \bullet \dfrac{z}{z} = \dfrac{1}{2} \bullet \dfrac{1}{x^4} \bullet y^2 = \dfrac{y^2}{2x^4}$

11. Simplifying the expression: $\dfrac{\left(6a^2 b\right)\left(9a^3 b^2\right)}{18a^4 b^3} = \dfrac{54a^5 b^3}{18a^4 b^3} = 3a$

12. Simplifying the expression: $\dfrac{24x^7}{3x^2} + \dfrac{14x^9}{7x^4} = 8x^5 + 2x^5 = 10x^5$

13. Simplifying the expression: $\dfrac{\left(2.4\times10^5\right)\left(4.5\times10^{-2}\right)}{1.2\times10^{-6}} = \dfrac{10.8\times10^3}{1.2\times10^{-6}} = 9.0\times10^9$

14. Performing the operations: $8x^2 - 4x + 6x + 2 = 8x^2 + 2x + 2$

15. Performing the operations: $\left(5x^2 - 3x + 4\right) - \left(2x^2 - 7x - 2\right) = 5x^2 - 3x + 4 - 2x^2 + 7x + 2 = 3x^2 + 4x + 6$

16. Performing the operations: $(6x - 8) - (3x - 4) = 6x - 8 - 3x + 4 = 3x - 4$

17. Evaluating when $y = -2$: $2y^2 - 3y - 4 = 2(-2)^2 - 3(-2) - 4 = 2(4) + 6 - 4 = 8 + 6 - 4 = 10$

18. Multiplying using the distributive property:
$$2a^2\left(3a^2 - 5a + 4\right) = 2a^2\left(3a^2\right) - 2a^2(5a) + 2a^2(4) = 6a^4 - 10a^3 + 8a^2$$

19. Multiplying using the FOIL method: $\left(x + \tfrac{1}{2}\right)\left(x + \tfrac{1}{3}\right) = x^2 + \tfrac{1}{2}x + \tfrac{1}{3}x + \tfrac{1}{6} = x^2 + \tfrac{5}{6}x + \tfrac{1}{6}$

20. Multiplying using the FOIL method: $(4x - 5)(2x + 3) = 8x^2 - 10x + 12x - 15 = 8x^2 + 2x - 15$

21. Multiplying using the column method:
$$\begin{array}{r} x^2 + 3x + 9 \\ x - 3 \\ \hline x^3 + 3x^2 + 9x \\ -3x^2 - 9x - 27 \\ \hline x^3 - 27 \end{array}$$

22. Multiplying using the square of binomial formula: $(x + 5)^2 = x^2 + 2(x)(5) + (5)^2 = x^2 + 10x + 25$

23. Multiplying using the square of binomial formula: $(3a - 2b)^2 = (3a)^2 - 2(3a)(2b) + (2b)^2 = 9a^2 - 12ab + 4b^2$

24. Multiplying using the difference of squares formula: $(3x - 4y)(3x + 4y) = (3x)^2 - (4y)^2 = 9x^2 - 16y^2$

25. Multiplying using the difference of squares formula: $\left(a^2 - 3\right)\left(a^2 + 3\right) = \left(a^2\right)^2 - (3)^2 = a^4 - 9$

26. Performing the division: $\dfrac{10x^3 + 15x^2 - 5x}{5x} = \dfrac{10x^3}{5x} + \dfrac{15x^2}{5x} - \dfrac{5x}{5x} = 2x^2 + 3x - 1$

27. Using long division:

$$\begin{array}{r} 4x + 3 \\ 2x - 3 \overline{\smash{\big)}\, 8x^2 - 6x - 5} \\ \underline{8x^2 - 12x} \\ 6x - 5 \\ \underline{6x - 9} \\ 4 \end{array}$$

The quotient is $4x + 3 + \dfrac{4}{2x - 3}$.

28. Using long division:

$$\begin{array}{r} 3x^2 + 9x + 25 \\ x - 3 \overline{\smash{\big)}\, 3x^3 + 0x^2 - 2x + 1} \\ \underline{3x^3 - 9x^2} \\ 9x^2 - 2x \\ \underline{9x^2 - 27x} \\ 25x + 1 \\ \underline{25x - 75} \\ 76 \end{array}$$

The quotient is $3x^2 + 9x + 25 + \dfrac{76}{x - 3}$.

29. Using the volume formula: $V = (2.5 \text{ cm})^3 = 15.625 \text{ cm}^3$

30. Let x represent the width, $5x$ represent the length, and $\tfrac{1}{5}x$ represent the height. The volume is given by:
$$V = (x)(5x)\left(\tfrac{1}{5}x\right) = x^3$$

Chapter 5
Factoring

5.1 The Greatest Common Factor and Factoring by Grouping

1. Factoring out the greatest common factor: $15x + 25 = 5(3x + 5)$

3. Factoring out the greatest common factor: $6a + 9 = 3(2a + 3)$

5. Factoring out the greatest common factor: $4x - 8y = 4(x - 2y)$

7. Factoring out the greatest common factor: $3x^2 - 6x - 9 = 3(x^2 - 2x - 3)$

9. Factoring out the greatest common factor: $3a^2 - 3a - 60 = 3(a^2 - a - 20)$

11. Factoring out the greatest common factor: $24y^2 - 52y + 24 = 4(6y^2 - 13y + 6)$

13. Factoring out the greatest common factor: $9x^2 - 8x^3 = x^2(9 - 8x)$

15. Factoring out the greatest common factor: $13a^2 - 26a^3 = 13a^2(1 - 2a)$

17. Factoring out the greatest common factor: $21x^2y - 28xy^2 = 7xy(3x - 4y)$

19. Factoring out the greatest common factor: $22a^2b^2 - 11ab^2 = 11ab^2(2a - 1)$

21. Factoring out the greatest common factor: $7x^3 + 21x^2 - 28x = 7x(x^2 + 3x - 4)$

23. Factoring out the greatest common factor: $121y^4 - 11x^4 = 11(11y^4 - x^4)$

25. Factoring out the greatest common factor: $100x^4 - 50x^3 + 25x^2 = 25x^2(4x^2 - 2x + 1)$

27. Factoring out the greatest common factor: $8a^2 + 16b^2 + 32c^2 = 8(a^2 + 2b^2 + 4c^2)$

29. Factoring out the greatest common factor: $4a^2b - 16ab^2 + 32a^2b^2 = 4ab(a - 4b + 8ab)$

31. Factoring out the greatest common factor: $121a^3b^2 - 22a^2b^3 + 33a^3b^3 = 11a^2b^2(11a - 2b + 3ab)$

33. Factoring out the greatest common factor: $12x^2y^3 - 72x^5y^3 - 36x^4y^4 = 12x^2y^3(1 - 6x^3 - 3x^2y)$

35. Factoring by grouping: $xy + 5x + 3y + 15 = x(y + 5) + 3(y + 5) = (y + 5)(x + 3)$

37. Factoring by grouping: $xy + 6x + 2y + 12 = x(y + 6) + 2(y + 6) = (y + 6)(x + 2)$

39. Factoring by grouping: $ab + 7a - 3b - 21 = a(b + 7) - 3(b + 7) = (b + 7)(a - 3)$

41. Factoring by grouping: $ax - bx + ay - by = x(a - b) + y(a - b) = (a - b)(x + y)$

43. Factoring by grouping: $2ax + 6x - 5a - 15 = 2x(a + 3) - 5(a + 3) = (a + 3)(2x - 5)$

45. Factoring by grouping: $3xb - 4b - 6x + 8 = b(3x - 4) - 2(3x - 4) = (3x - 4)(b - 2)$

47. Factoring by grouping: $x^2 + ax + 2x + 2a = x(x+a) + 2(x+a) = (x+a)(x+2)$

49. Factoring by grouping: $x^2 - ax - bx + ab = x(x-a) - b(x-a) = (x-a)(x-b)$

51. Factoring by grouping: $ax + ay + bx + by + cx + cy = a(x+y) + b(x+y) + c(x+y) = (x+y)(a+b+c)$

53. Factoring by grouping: $6x^2 + 9x + 4x + 6 = 3x(2x+3) + 2(2x+3) = (2x+3)(3x+2)$

55. Factoring by grouping: $20x^2 - 2x + 50x - 5 = 2x(10x-1) + 5(10x-1) = (10x-1)(2x+5)$

57. Factoring by grouping: $20x^2 + 4x + 25x + 5 = 4x(5x+1) + 5(5x+1) = (5x+1)(4x+5)$

59. Factoring by grouping: $x^3 + 2x^2 + 3x + 6 = x^2(x+2) + 3(x+2) = (x+2)(x^2+3)$

61. Factoring by grouping: $6x^3 - 4x^2 + 15x - 10 = 2x^2(3x-2) + 5(3x-2) = (3x-2)(2x^2+5)$

63. Its greatest common factor is $3 \cdot 2 = 6$.

65. The correct factoring is: $12x^2 + 6x + 3 = 3(4x^2 + 2x + 1)$

67. The factored form is: $A = 1000 + 1000r = 1000(1+r)$
Substituting $r = 0.12$: $A = 1000(1 + 0.12) = 1000(1.12) = \$1,120$

69. **a.**   Factoring: $A = 1,000,000 + 1,000,000r = 1,000,000(1+r)$
   b.   Substituting $r = 0.30$: $A = 1,000,000(1 + 0.3) = 1,300,000$ bacteria

71. Multiplying using the FOIL method: $(x-7)(x+2) = x^2 - 7x + 2x - 14 = x^2 - 5x - 14$

73. Multiplying using the FOIL method: $(x-3)(x+2) = x^2 - 3x + 2x - 6 = x^2 - x - 6$

75. Multiplying using the column method:

$$
\begin{array}{r}
x^2 - 3x + 9 \\
x + 3 \\
\hline
x^3 - 3x^2 + 9x \\
3x^2 - 9x + 27 \\
\hline
x^3 + 27
\end{array}
$$

77. Multiplying using the column method:

$$
\begin{array}{r}
x^2 + 4x - 3 \\
2x + 1 \\
\hline
2x^3 + 8x^2 - 6x \\
x^2 + 4x - 3 \\
\hline
2x^3 + 9x^2 - 2x - 3
\end{array}
$$

5.2 Factoring Trinomials

1. Factoring the trinomial: $x^2 + 7x + 12 = (x+3)(x+4)$ **3.** Factoring the trinomial: $x^2 + 3x + 2 = (x+2)(x+1)$

5. Factoring the trinomial: $a^2 + 10a + 21 = (a+7)(a+3)$ **7.** Factoring the trinomial: $x^2 - 7x + 10 = (x-5)(x-2)$

9. Factoring the trinomial: $y^2 - 10y + 21 = (y-7)(y-3)$ **11.** Factoring the trinomial: $x^2 - x - 12 = (x-4)(x+3)$

13. Factoring the trinomial: $y^2 + y - 12 = (y+4)(y-3)$ **15.** Factoring the trinomial: $x^2 + 5x - 14 = (x+7)(x-2)$

17. Factoring the trinomial: $r^2 - 8r - 9 = (r-9)(r+1)$ **19.** Factoring the trinomial: $x^2 - x - 30 = (x-6)(x+5)$

21. Factoring the trinomial: $a^2 + 15a + 56 = (a+7)(a+8)$ **23.** Factoring the trinomial: $y^2 - y - 42 = (y-7)(y+6)$

25. Factoring the trinomial: $x^2 + 13x + 42 = (x+7)(x+6)$

27. Factoring the trinomial: $2x^2 + 6x + 4 = 2(x^2 + 3x + 2) = 2(x+2)(x+1)$

29. Factoring the trinomial: $3a^2 - 3a - 60 = 3(a^2 - a - 20) = 3(a-5)(a+4)$

31. Factoring the trinomial: $100x^2 - 500x + 600 = 100(x^2 - 5x + 6) = 100(x-3)(x-2)$

33. Factoring the trinomial: $100p^2 - 1300p + 4000 = 100(p^2 - 13p + 40) = 100(p-8)(p-5)$

35. Factoring the trinomial: $x^4 - x^3 - 12x^2 = x^2(x^2 - x - 12) = x^2(x-4)(x+3)$

37. Factoring the trinomial: $2r^3 + 4r^2 - 30r = 2r(r^2 + 2r - 15) = 2r(r+5)(r-3)$

39. Factoring the trinomial: $2y^4 - 6y^3 - 8y^2 = 2y^2(y^2 - 3y - 4) = 2y^2(y-4)(y+1)$

41. Factoring the trinomial: $x^5 + 4x^4 + 4x^3 = x^3(x^2 + 4x + 4) = x^3(x+2)(x+2) = x^3(x+2)^2$

43.   Factoring the trinomial: $3y^4 - 12y^3 - 15y^2 = 3y^2\left(y^2 - 4y - 5\right) = 3y^2\left(y - 5\right)\left(y + 1\right)$

45.   Factoring the trinomial: $4x^4 - 52x^3 + 144x^2 = 4x^2\left(x^2 - 13x + 36\right) = 4x^2\left(x - 9\right)\left(x - 4\right)$

47.   Factoring the trinomial: $x^2 + 5xy + 6y^2 = \left(x + 2y\right)\left(x + 3y\right)$

49.   Factoring the trinomial: $x^2 - 9xy + 20y^2 = \left(x - 4y\right)\left(x - 5y\right)$

51.   Factoring the trinomial: $a^2 + 2ab - 8b^2 = \left(a + 4b\right)\left(a - 2b\right)$

53.   Factoring the trinomial: $a^2 - 10ab + 25b^2 = \left(a - 5b\right)\left(a - 5b\right) = \left(a - 5b\right)^2$

55.   Factoring the trinomial: $a^2 + 10ab + 25b^2 = \left(a + 5b\right)\left(a + 5b\right) = \left(a + 5b\right)^2$

57.   Factoring the trinomial: $x^2 + 2xa - 48a^2 = \left(x + 8a\right)\left(x - 6a\right)$

59.   Factoring the trinomial: $x^2 - 5xb - 36b^2 = \left(x - 9b\right)\left(x + 4b\right)$

61.   Factoring the trinomial: $x^4 - 5x^2 + 6 = \left(x^2 - 2\right)\left(x^2 - 3\right)$

63.   Factoring the trinomial: $x^2 - 80x - 2000 = \left(x - 100\right)\left(x + 20\right)$

65.   Factoring the trinomial: $x^2 - x + \frac{1}{4} = \left(x - \frac{1}{2}\right)\left(x - \frac{1}{2}\right) = \left(x - \frac{1}{2}\right)^2$

67.   Factoring the trinomial: $x^2 + 0.6x + 0.08 = \left(x + 0.4\right)\left(x + 0.2\right)$

69.   We can use long division to find the other factor:

$$\begin{array}{r} x + 250 \\ x + 10 \overline{\smash{\big)}\, x^2 + 260x + 2500} \\ \underline{x^2 + 10x} \\ 250x + 2500 \\ \underline{250x + 2500} \\ 0 \end{array}$$

The other factor is $x + 250$.

71.   Using FOIL to multiply out the factors: $\left(4x + 3\right)\left(x - 1\right) = 4x^2 + 3x - 4x - 3 = 4x^2 - x - 3$

73.   Multiplying using the FOIL method: $\left(6a + 1\right)\left(a + 2\right) = 6a^2 + a + 12a + 2 = 6a^2 + 13a + 2$

75.   Multiplying using the FOIL method: $\left(3a + 2\right)\left(2a + 1\right) = 6a^2 + 4a + 3a + 2 = 6a^2 + 7a + 2$

77.   Multiplying using the FOIL method: $\left(6a + 2\right)\left(a + 1\right) = 6a^2 + 2a + 6a + 2 = 6a^2 + 8a + 2$

79.   Subtracting the polynomials: $\left(5x^2 + 5x - 4\right) - \left(3x^2 - 2x + 7\right) = 5x^2 + 5x - 4 - 3x^2 + 2x - 7 = 2x^2 + 7x - 11$

81.   Subtracting the polynomials: $\left(7x + 3\right) - \left(4x - 5\right) = 7x + 3 - 4x + 5 = 3x + 8$

83.   Subtracting the polynomials: $\left(5x^2 - 5\right) - \left(2x^2 - 4x\right) = 5x^2 - 5 - 2x^2 + 4x = 3x^2 + 4x - 5$

5.3 More Trinomials to Factor

1.   Factoring the trinomial: $2x^2 + 7x + 3 = \left(2x + 1\right)\left(x + 3\right)$

3.   Factoring the trinomial: $2a^2 - a - 3 = \left(2a - 3\right)\left(a + 1\right)$

5.   Factoring the trinomial: $3x^2 + 2x - 5 = \left(3x + 5\right)\left(x - 1\right)$

7.   Factoring the trinomial: $3y^2 - 14y - 5 = \left(3y + 1\right)\left(y - 5\right)$

9.   Factoring the trinomial: $6x^2 + 13x + 6 = \left(3x + 2\right)\left(2x + 3\right)$

11.   Factoring the trinomial: $4x^2 - 12xy + 9y^2 = \left(2x - 3y\right)\left(2x - 3y\right) = \left(2x - 3y\right)^2$

13.   Factoring the trinomial: $4y^2 - 11y - 3 = \left(4y + 1\right)\left(y - 3\right)$

15.   Factoring the trinomial: $20x^2 - 41x + 20 = \left(4x - 5\right)\left(5x - 4\right)$

17.   Factoring the trinomial: $20a^2 + 48ab - 5b^2 = \left(10a - b\right)\left(2a + 5b\right)$

19. Factoring the trinomial: $20x^2 - 21x - 5 = (4x - 5)(5x + 1)$

21. Factoring the trinomial: $12m^2 + 16m - 3 = (6m - 1)(2m + 3)$

23. Factoring the trinomial: $20x^2 + 37x + 15 = (4x + 5)(5x + 3)$

25. Factoring the trinomial: $12a^2 - 25ab + 12b^2 = (3a - 4b)(4a - 3b)$

27. Factoring the trinomial: $3x^2 - xy - 14y^2 = (3x - 7y)(x + 2y)$

29. Factoring the trinomial: $14x^2 + 29x - 15 = (2x + 5)(7x - 3)$

31. Factoring the trinomial: $6x^2 - 43x + 55 = (3x - 5)(2x - 11)$

33. Factoring the trinomial: $15t^2 - 67t + 38 = (5t - 19)(3t - 2)$

35. Factoring the trinomial: $4x^2 + 2x - 6 = 2\left(2x^2 + x - 3\right) = 2(2x + 3)(x - 1)$

37. Factoring the trinomial: $24a^2 - 50a + 24 = 2\left(12a^2 - 25a + 12\right) = 2(4a - 3)(3a - 4)$

39. Factoring the trinomial: $10x^3 - 23x^2 + 12x = x\left(10x^2 - 23x + 12\right) = x(5x - 4)(2x - 3)$

41. Factoring the trinomial: $6x^4 - 11x^3 - 10x^2 = x^2\left(6x^2 - 11x - 10\right) = x^2(3x + 2)(2x - 5)$

43. Factoring the trinomial: $10a^3 - 6a^2 - 4a = 2a\left(5a^2 - 3a - 2\right) = 2a(5a + 2)(a - 1)$

45. Factoring the trinomial: $15x^3 - 102x^2 - 21x = 3x\left(5x^2 - 34x - 7\right) = 3x(5x + 1)(x - 7)$

47. Factoring the trinomial: $35y^3 - 60y^2 - 20y = 5y\left(7y^2 - 12y - 4\right) = 5y(7y + 2)(y - 2)$

49. Factoring the trinomial: $15a^4 - 2a^3 - a^2 = a^2\left(15a^2 - 2a - 1\right) = a^2(5a + 1)(3a - 1)$

51. Factoring the trinomial: $24x^2y - 6xy - 45y = 3y\left(8x^2 - 2x - 15\right) = 3y(4x + 5)(2x - 3)$

53. Factoring the trinomial: $12x^2y - 34xy^2 + 14y^3 = 2y\left(6x^2 - 17xy + 7y^2\right) = 2y(2x - y)(3x - 7y)$

55. Evaluating each expression when $x = 2$:
$$2x^2 + 7x + 3 = 2(2)^2 + 7(2) + 3 = 8 + 14 + 3 = 25$$
$$(2x + 1)(x + 3) = (2 \cdot 2 + 1)(2 + 3) = (5)(5) = 25$$

57. Multiplying using the difference of squares formula: $(2x + 3)(2x - 3) = (2x)^2 - (3)^2 = 4x^2 - 9$

59. Multiplying using the difference of squares formula:
$$(x + 3)(x - 3)\left(x^2 + 9\right) = \left(x^2 - 9\right)\left(x^2 + 9\right) = \left(x^2\right)^2 - (9)^2 = x^4 - 81$$

61. Factoring: $h = 8 + 62t - 16t^2 = 2\left(4 + 31t - 8t^2\right) = 2(4 - t)(1 + 8t)$. Now completing the table:

Time t (seconds)	Height h (feet)
0	8
1	54
2	68
3	50
4	0

63. **a.** Factoring: $V = 99x - 40x^2 + 4x^3 = x\left(99 - 40x + 4x^2\right) = x(9 - 2x)(11 - 2x)$

    b. The original dimensions were 9 inches by 11 inches.

65. Multiplying using the difference of squares formula: $(x + 3)(x - 3) = x^2 - (3)^2 = x^2 - 9$

67. Multiplying using the difference of squares formula: $(6a + 1)(6a - 1) = (6a)^2 - (1)^2 = 36a^2 - 1$

69. Multiplying using the square of binomial formula: $(x + 4)^2 = x^2 + 2(x)(4) + (4)^2 = x^2 + 8x + 16$

71. Multiplying using the square of binomial formula: $(2x + 3)^2 = (2x)^2 + 2(2x)(3) + (3)^2 = 4x^2 + 12x + 9$

5.4 The Difference of Two Squares

1. Factoring the binomial: $x^2 - 9 = (x + 3)(x - 3)$

5. Factoring the binomial: $x^2 - 49 = (x + 7)(x - 7)$

7. Factoring the binomial: $4a^2 - 16 = 4(a^2 - 4) = 4(a + 2)(a - 2)$

9. The expression $9x^2 + 25$ cannot be factored.

11. Factoring the binomial: $25x^2 - 169 = (5x + 13)(5x - 13)$

13. Factoring the binomial: $9a^2 - 16b^2 = (3a + 4b)(3a - 4b)$

15. Factoring the binomial: $9 - m^2 = (3 + m)(3 - m)$

17. Factoring the binomial: $25 - 4x^2 = (5 + 2x)(5 - 2x)$

19. Factoring the binomial: $2x^2 - 18 = 2(x^2 - 9) = 2(x + 3)(x - 3)$

21. Factoring the binomial: $32a^2 - 128 = 32(a^2 - 4) = 32(a + 2)(a - 2)$

23. Factoring the binomial: $8x^2y - 18y = 2y(4x^2 - 9) = 2y(2x + 3)(2x - 3)$

25. Factoring the binomial: $a^4 - b^4 = (a^2 + b^2)(a^2 - b^2) = (a^2 + b^2)(a + b)(a - b)$

27. Factoring the binomial: $16m^4 - 81 = (4m^2 + 9)(4m^2 - 9) = (4m^2 + 9)(2m + 3)(2m - 3)$

29. Factoring the binomial: $3x^3y - 75xy^3 = 3xy(x^2 - 25y^2) = 3xy(x + 5y)(x - 5y)$

31. Factoring the trinomial: $x^2 - 2x + 1 = (x - 1)(x - 1) = (x - 1)^2$

33. Factoring the trinomial: $x^2 + 2x + 1 = (x + 1)(x + 1) = (x + 1)^2$

35. Factoring the trinomial: $a^2 - 10a + 25 = (a - 5)(a - 5) = (a - 5)^2$

37. Factoring the trinomial: $y^2 + 4y + 4 = (y + 2)(y + 2) = (y + 2)^2$

39. Factoring the trinomial: $x^2 - 4x + 4 = (x - 2)(x - 2) = (x - 2)^2$

41. Factoring the trinomial: $m^2 - 12m + 36 = (m - 6)(m - 6) = (m - 6)^2$

43. Factoring the trinomial: $4a^2 + 12a + 9 = (2a + 3)(2a + 3) = (2a + 3)^2$

45. Factoring the trinomial: $49x^2 - 14x + 1 = (7x - 1)(7x - 1) = (7x - 1)^2$

47. Factoring the trinomial: $9y^2 - 30y + 25 = (3y - 5)(3y - 5) = (3y - 5)^2$

49. Factoring the trinomial: $x^2 + 10xy + 25y^2 = (x + 5y)(x + 5y) = (x + 5y)^2$

51. Factoring the trinomial: $9a^2 + 6ab + b^2 = (3a + b)(3a + b) = (3a + b)^2$

53. Factoring the trinomial: $3a^2 + 18a + 27 = 3(a^2 + 6a + 9) = 3(a + 3)(a + 3) = 3(a + 3)^2$

55. Factoring the trinomial: $2x^2 + 20xy + 50y^2 = 2(x^2 + 10xy + 25y^2) = 2(x + 5y)(x + 5y) = 2(x + 5y)^2$

57. Factoring the trinomial: $5x^3 + 30x^2y + 45xy^2 = 5x(x^2 + 6xy + 9y^2) = 5x(x + 3y)(x + 3y) = 5x(x + 3y)^2$

59. Factoring by grouping: $x^2 + 6x + 9 - y^2 = (x + 3)^2 - y^2 = (x + 3 + y)(x + 3 - y)$

61. Factoring by grouping: $x^2 + 2xy + y^2 - 9 = (x + y)^2 - 9 = (x + y + 3)(x + y - 3)$

63. Since $(x + 7)^2 = x^2 + 14x + 49$, the value is $b = 14$.

65. Since $(x + 5)^2 = x^2 + 10x + 25$, the value is $c = 25$.

67.   a.   Subtracting square areas, the area is $x^2 - 4^2 = x^2 - 16$.

      b.   Factoring: $x^2 - 16 = (x + 4)(x - 4)$

      c.   The square can be rearranged to be a rectangle with dimensions $x - 4$ by $x + 4$. Cut off the right "flap", then place it along the bottom of the left rectangle to produce the desired rectangle.

69.   The area is $a^2 - b^2 = (a+b)(a-b)$.

71.   Using long division:

$$
\begin{array}{r}
x-2 \\
x-3{\overline{\smash{\big)}\,x^2-5x+8}} \\
\underline{x^2-3x} \\
-2x+8 \\
\underline{-2x+6} \\
2
\end{array}
$$

The quotient is $x-2+\dfrac{2}{x-3}$.

73.   Using long division:

$$
\begin{array}{r}
3x-2 \\
2x+3{\overline{\smash{\big)}\,6x^2+5x+3}} \\
\underline{6x^2+9x} \\
-4x+3 \\
\underline{-4x-6} \\
9
\end{array}
$$

The quotient is $3x-2+\dfrac{9}{2x+3}$.

5.5 Factoring: A General Review

1.   Factoring the polynomial: $x^2 - 81 = (x+9)(x-9)$

3.   Factoring the polynomial: $x^2 + 2x - 15 = (x+5)(x-3)$

5.   Factoring the polynomial: $x^2 + 6x + 9 = (x+3)(x+3) = (x+3)^2$

7.   Factoring the polynomial: $y^2 - 10y + 25 = (y-5)(y-5) = (y-5)^2$

9.   Factoring the polynomial: $2a^3b + 6a^2b + 2ab = 2ab(a^2 + 3a + 1)$

11.   The polynomial $x^2 + x + 1$ cannot be factored.

13.   Factoring the polynomial: $12a^2 - 75 = 3(4a^2 - 25) = 3(2a+5)(2a-5)$

15.   Factoring the polynomial: $9x^2 - 12xy + 4y^2 = (3x-2y)(3x-2y) = (3x-2y)^2$

17.   Factoring the polynomial: $4x^3 + 16xy^2 = 4x(x^2 + 4y^2)$

19.   Factoring the polynomial: $2y^3 + 20y^2 + 50y = 2y(y^2 + 10y + 25) = 2y(y+5)(y+5) = 2y(y+5)^2$

21.   Factoring the polynomial: $a^6 + 4a^4b^2 = a^4(a^2 + 4b^2)$

23.   Factoring the polynomial: $xy + 3x + 4y + 12 = x(y+3) + 4(y+3) = (y+3)(x+4)$

25.   Factoring the polynomial: $x^4 - 16 = (x^2 + 4)(x^2 - 4) = (x^2 + 4)(x+2)(x-2)$

27.   Factoring the polynomial: $xy - 5x + 2y - 10 = x(y-5) + 2(y-5) = (y-5)(x+2)$

29.   Factoring the polynomial: $5a^2 + 10ab + 5b^2 = 5(a^2 + 2ab + b^2) = 5(a+b)(a+b) = 5(a+b)^2$

31.   The polynomial $x^2 + 49$ cannot be factored.

33.   Factoring the polynomial: $3x^2 + 15xy + 18y^2 = 3(x^2 + 5xy + 6y^2) = 3(x+2y)(x+3y)$

35.   Factoring the polynomial: $2x^2 + 15x - 38 = (2x+19)(x-2)$

37.   Factoring the polynomial: $100x^2 - 300x + 200 = 100(x^2 - 3x + 2) = 100(x-2)(x-1)$

39.   Factoring the polynomial: $x^2 - 64 = (x+8)(x-8)$

41.   Factoring the polynomial: $x^2 + 3x + ax + 3a = x(x+3) + a(x+3) = (x+3)(x+a)$

43.   Factoring the polynomial: $49a^7 - 9a^5 = a^5(49a^2 - 9) = a^5(7a+3)(7a-3)$

45.   The polynomial $49x^2 + 9y^2$ cannot be factored.

47.   Factoring the polynomial: $25a^3 + 20a^2 + 3a = a(25a^2 + 20a + 3) = a(5a+3)(5a+1)$

49.   Factoring the polynomial: $xa - xb + ay - by = x(a-b) + y(a-b) = (a-b)(x+y)$

51.   Factoring the polynomial: $48a^4b - 3a^2b = 3a^2b(16a^2 - 1) = 3a^2b(4a+1)(4a-1)$

53. Factoring the polynomial: $20x^4 - 45x^2 = 5x^2\left(4x^2 - 9\right) = 5x^2(2x + 3)(2x - 3)$

55. Factoring the polynomial: $3x^2 + 35xy - 82y^2 - (3x + 41y)(x - 2y)$

57. Factoring the polynomial: $16x^5 - 44x^4 + 30x^3 = 2x^3\left(8x^2 - 22x + 15\right) = 2x^3(2x - 3)(4x - 5)$

59. Factoring the polynomial: $2x^2 + 2ax + 3x + 3a = 2x(x + a) + 3(x + a) = (x + a)(2x + 3)$

61. Factoring the polynomial: $y^4 - 1 = \left(y^2 + 1\right)\left(y^2 - 1\right) = \left(y^2 + 1\right)(y + 1)(y - 1)$

63. Factoring the polynomial:
$$12x^4y^2 + 36x^3y^3 + 27x^2y^4 = 3x^2y^2\left(4x^2 + 12xy + 9y^2\right) = 3x^2y^2(2x + 3y)(2x + 3y) = 3x^2y^2(2x + 3y)^2$$

65. Solving the equation:
$$3x - 6 = 9$$
$$3x = 15$$
$$x = 5$$

67. Solving the equation:
$$2x + 3 = 0$$
$$2x = -3$$
$$x = -\frac{3}{2}$$

69. Solving the equation:
$$4x + 3 = 0$$
$$4x = -3$$
$$x = -\frac{3}{4}$$

71. Simplifying the expression: $x^8 \cdot x^7 = x^{8+7} = x^{15}$

73. Simplifying the expression: $\left(3x^3\right)^2\left(2x^4\right)^3 = 9x^6 \cdot 8x^{12} = 72x^{18}$

75. Writing in scientific notation: $57,600 = 5.76 \times 10^4$

5.6 Solving Equations by Factoring

1. Setting each factor equal to 0:
$$x + 2 = 0 \qquad\qquad x - 1 = 0$$
$$x = -2 \qquad\qquad x = 1$$
The solutions are –2 and 1.

3. Setting each factor equal to 0:
$$a - 4 = 0 \qquad\qquad a - 5 = 0$$
$$a = 4 \qquad\qquad a = 5$$
The solutions are 4 and 5.

5. Setting each factor equal to 0:
$$x = 0 \qquad x + 1 = 0 \qquad\qquad x - 3 = 0$$
$$x = -1 \qquad\qquad x = 3$$
The solutions are 0, –1 and 3.

7. Setting each factor equal to 0:
$$3x + 2 = 0 \qquad\qquad 2x + 3 = 0$$
$$3x = -2 \qquad\qquad 2x = -3$$
$$x = -\frac{2}{3} \qquad\qquad x = -\frac{3}{2}$$
The solutions are $-\frac{2}{3}$ and $-\frac{3}{2}$.

9. Setting each factor equal to 0:
$$m = 0 \qquad 3m + 4 = 0 \qquad\qquad 3m - 4 = 0$$
$$3m = -4 \qquad\qquad 3m = 4$$
$$m = -\frac{4}{3} \qquad\qquad m = \frac{4}{3}$$
The solutions are 0, $-\frac{4}{3}$ and $\frac{4}{3}$.

11.   Setting each factor equal to 0:

$$2y = 0 \qquad 3y + 1 = 0 \qquad\qquad 5y + 3 = 0$$
$$y = 0 \qquad\quad 3y = -1 \qquad\qquad\quad 5y = -3$$
$$y = -\tfrac{1}{3} \qquad\qquad\qquad y = -\tfrac{3}{5}$$

The solutions are 0, $-\tfrac{1}{3}$ and $-\tfrac{3}{5}$.

13.   Solving by factoring:

$$x^2 + 3x + 2 = 0$$
$$(x + 2)(x + 1) = 0$$
$$x = -2, -1$$

15.   Solving by factoring:

$$x^2 - 9x + 20 = 0$$
$$(x - 4)(x - 5) = 0$$
$$x = 4, 5$$

17.   Solving by factoring:

$$a^2 - 2a - 24 = 0$$
$$(a - 6)(a + 4) = 0$$
$$a = 6, -4$$

19.   Solving by factoring:

$$100x^2 - 500x + 600 = 0$$
$$100\left(x^2 - 5x + 6\right) = 0$$
$$100(x - 2)(x - 3) = 0$$
$$x = 2, 3$$

21.   Solving by factoring:

$$x^2 = -6x - 9$$
$$x^2 + 6x + 9 = 0$$
$$(x + 3)^2 = 0$$
$$x + 3 = 0$$
$$x = -3$$

23.   Solving by factoring:

$$a^2 - 16 = 0$$
$$(a + 4)(a - 4) = 0$$
$$a = -4, 4$$

25.   Solving by factoring:

$$2x^2 + 5x - 12 = 0$$
$$(2x - 3)(x + 4) = 0$$
$$x = \tfrac{3}{2}, -4$$

27.   Solving by factoring:

$$9x^2 + 12x + 4 = 0$$
$$(3x + 2)^2 = 0$$
$$3x + 2 = 0$$
$$x = -\tfrac{2}{3}$$

29.   Solving by factoring:

$$a^2 + 25 = 10a$$
$$a^2 - 10a + 25 = 0$$
$$(a - 5)^2 = 0$$
$$a - 5 = 0$$
$$a = 5$$

31.   Solving by factoring:

$$2x^2 = 3x + 20$$
$$2x^2 - 3x - 20 = 0$$
$$(2x + 5)(x - 4) = 0$$
$$x = -\tfrac{5}{2}, 4$$

33.   Solving by factoring:

$$3m^2 = 20 - 7m$$
$$3m^2 + 7m - 20 = 0$$
$$(3m - 5)(m + 4) = 0$$
$$m = \tfrac{5}{3}, -4$$

35.   Solving by factoring:

$$4x^2 - 49 = 0$$
$$(2x + 7)(2x - 7) = 0$$
$$x = -\tfrac{7}{2}, \tfrac{7}{2}$$

37.   Solving by factoring:

$$x^2 + 6x = 0$$
$$x(x + 6) = 0$$
$$x = 0, -6$$

39.   Solving by factoring:

$$x^2 - 3x = 0$$
$$x(x - 3) = 0$$
$$x = 0, 3$$

41.   Solving by factoring:

$$2x^2 = 8x$$
$$2x^2 - 8x = 0$$
$$2x(x - 4) = 0$$
$$x = 0, 4$$

43.   Solving by factoring:

$$3x^2 = 15x$$
$$3x^2 - 15x = 0$$
$$3x(x - 5) = 0$$
$$x = 0, 5$$

45.   Solving by factoring:
$$1400 = 400 + 700x - 100x^2$$
$$100x^2 - 700x + 1000 = 0$$
$$100\left(x^2 - 7x + 10\right) = 0$$
$$100(x - 5)(x - 2) = 0$$
$$x = 2, 5$$

47.   Solving by factoring:
$$6x^2 = -5x + 4$$
$$6x^2 + 5x - 4 = 0$$
$$(3x + 4)(2x - 1) = 0$$
$$x = -\frac{4}{3}, \frac{1}{2}$$

49.   Solving by factoring:
$$x(2x - 3) = 20$$
$$2x^2 - 3x = 20$$
$$2x^2 - 3x - 20 = 0$$
$$(2x + 5)(x - 4) = 0$$
$$x = -\frac{5}{2}, 4$$

51.   Solving by factoring:
$$t(t + 2) = 80$$
$$t^2 + 2t = 80$$
$$t^2 + 2t - 80 = 0$$
$$(t + 10)(t - 8) = 0$$
$$t = -10, 8$$

53.   Solving by factoring:
$$4000 = (1300 - 100p)p$$
$$4000 = 1300p - 100p^2$$
$$100p^2 - 1300p + 4000 = 0$$
$$100\left(p^2 - 13p + 40\right) = 0$$
$$100(p - 8)(p - 5) = 0$$
$$p = 5, 8$$

55.   Solving by factoring:
$$x(14 - x) = 48$$
$$14x - x^2 = 48$$
$$-x^2 + 14x - 48 = 0$$
$$x^2 - 14x + 48 = 0$$
$$(x - 6)(x - 8) = 0$$
$$x = 6, 8$$

57.   Solving by factoring:
$$(x + 5)^2 = 2x + 9$$
$$x^2 + 10x + 25 = 2x + 9$$
$$x^2 + 8x + 16 = 0$$
$$(x + 4)^2 = 0$$
$$x + 4 = 0$$
$$x = -4$$

59.   Solving by factoring:
$$(y - 6)^2 = y - 4$$
$$y^2 - 12y + 36 = y - 4$$
$$y^2 - 13y + 40 = 0$$
$$(y - 5)(y - 8) = 0$$
$$y = 5, 8$$

61.   Solving by factoring:
$$10^2 = (x + 2)^2 + x^2$$
$$100 = x^2 + 4x + 4 + x^2$$
$$100 = 2x^2 + 4x + 4$$
$$0 = 2x^2 + 4x - 96$$
$$0 = 2\left(x^2 + 2x - 48\right)$$
$$0 = 2(x + 8)(x - 6)$$
$$x = -8, 6$$

63.   Solving by factoring:
$$2x^3 + 11x^2 + 12x = 0$$
$$x\left(2x^2 + 11x + 12\right) = 0$$
$$x(2x + 3)(x + 4) = 0$$
$$x = 0, -\frac{3}{2}, -4$$

65.   Solving by factoring:
$$4y^3 - 2y^2 - 30y = 0$$
$$2y\left(2y^2 - y - 15\right) = 0$$
$$2y(2y + 5)(y - 3) = 0$$
$$y = 0, -\frac{5}{2}, 3$$

67.   Solving by factoring:
$$8x^3 + 16x^2 = 10x$$
$$8x^3 + 16x^2 - 10x = 0$$
$$2x\left(4x^2 + 8x - 5\right) = 0$$
$$2x(2x - 1)(2x + 5) = 0$$
$$x = 0, \frac{1}{2}, -\frac{5}{2}$$

69.   Solving by factoring:
$$20a^3 = -18a^2 + 18a$$
$$20a^3 + 18a^2 - 18a = 0$$
$$2a\left(10a^2 + 9a - 9\right) = 0$$
$$2a(5a - 3)(2a + 3) = 0$$
$$a = 0, \tfrac{3}{5}, -\tfrac{3}{2}$$

71.   Solving by factoring:
$$x^3 + 3x^2 - 4x - 12 = 0$$
$$x^2(x + 3) - 4(x + 3) = 0$$
$$(x + 3)\left(x^2 - 4\right) = 0$$
$$(x + 3)(x + 2)(x - 2) = 0$$
$$x = -3, -2, 2$$

73.   Solving by factoring:
$$x^3 + x^2 - 16x - 16 = 0$$
$$x^2(x + 1) - 16(x + 1) = 0$$
$$(x + 1)\left(x^2 - 16\right) = 0$$
$$(x + 1)(x + 4)(x - 4) = 0$$
$$x = -1, -4, 4$$

75.   Let x represent the cost of the suit and $5x$ represent the cost of the bicycle. The equation is:
$$x + 5x = 90$$
$$6x = 90$$
$$x = 15$$
$$5x = 75$$
The suit costs \$15 and the bicycle costs \$75.

77.   Let x represent the cost of the lot and $4x$ represent the cost of the house. The equation is:
$$x + 4x = 3000$$
$$5x = 3000$$
$$x = 600$$
$$4x = 2400$$
The lot cost \$600 and the house cost \$2,400.

79.   Simplifying using the properties of exponents: $2^{-3} = \dfrac{1}{2^3} = \dfrac{1}{8}$

81.   Simplifying using the properties of exponents: $\dfrac{x^5}{x^{-3}} = x^{5-(-3)} = x^{5+3} = x^8$

83.   Simplifying using the properties of exponents: $\dfrac{\left(x^2\right)^3}{\left(x^{-3}\right)^4} = \dfrac{x^6}{x^{-12}} = x^{6-(-12)} = x^{6+12} = x^{18}$

85.   Writing in scientific notation: $0.0056 = 5.6 \times 10^{-3}$

5.7 Applications

1. Let x and $x + 2$ represent the two integers. The equation is:
$$x(x+2) = 80$$
$$x^2 + 2x = 80$$
$$x^2 + 2x - 80 = 0$$
$$(x+10)(x-8) = 0$$
$$x = -10, 8$$
$$x + 2 = -8, 10$$
The two numbers are either −10 and −8, or 8 and 10.

3. Let x and $x + 2$ represent the two integers. The equation is:
$$x(x+2) = 99$$
$$x^2 + 2x = 99$$
$$x^2 + 2x - 99 = 0$$
$$(x+11)(x-9) = 0$$
$$x = -11, 9$$
$$x + 2 = -9, 11$$
The two numbers are either −11 and −9, or 9 and 11.

5. Let x and $x + 2$ represent the two integers. The equation is:
$$x(x+2) = 5(x + x + 2) - 10$$
$$x^2 + 2x = 5(2x + 2) - 10$$
$$x^2 + 2x = 10x + 10 - 10$$
$$x^2 + 2x = 10x$$
$$x^2 - 8x = 0$$
$$x(x-8) = 0$$
$$x = 0, 8$$
$$x + 2 = 2, 10$$
The two numbers are either 0 and 2, or 8 and 10.

7. Let x and $14 - x$ represent the two numbers. The equation is:
$$x(14 - x) = 48$$
$$14x - x^2 = 48$$
$$0 = x^2 - 14x + 48$$
$$0 = (x-8)(x-6)$$
$$x = 8, 6$$
$$14 - x = 6, 8$$
The two numbers are 6 and 8.

9. Let x and $5x + 2$ represent the two numbers. The equation is:
$$x(5x + 2) = 24$$
$$5x^2 + 2x = 24$$
$$5x^2 + 2x - 24 = 0$$
$$(5x+12)(x-2) = 0$$
$$x = -\frac{12}{5}, 2$$
$$5x + 2 = -10, 12$$
The two numbers are either $-\frac{12}{5}$ and −10, or 2 and 12.

11.   Let x and $4x$ represent the two numbers. The equation is:
$$x(4x) = 4(x + 4x)$$
$$4x^2 = 4(5x)$$
$$4x^2 = 20x$$
$$4x^2 - 20x = 0$$
$$4x(x - 5) = 0$$
$$x = 0, 5$$
$$4x = 0, 20$$
The two numbers are either 0 and 0, or 5 and 20.

13.   Let w represent the width and $w + 1$ represent the length. The equation is:
$$w(w + 1) = 12$$
$$w^2 + w = 12$$
$$w^2 + w - 12 = 0$$
$$(w + 4)(w - 3) = 0$$
$$w = 3 \quad (w = -4 \text{ is impossible})$$
$$w + 1 = 4$$
The width is 3 inches and the length is 4 inches.

15.   Let b represent the base and $2b$ represent the height. The equation is:
$$\tfrac{1}{2}b(2b) = 9$$
$$b^2 = 9$$
$$b^2 - 9 = 0$$
$$(b + 3)(b - 3) = 0$$
$$b = 3 \quad (b = -3 \text{ is impossible})$$
The base is 3 inches.

17.   Let x and $x + 2$ represent the two legs. The equation is:
$$x^2 + (x + 2)^2 = 10^2$$
$$x^2 + x^2 + 4x + 4 = 100$$
$$2x^2 + 4x + 4 = 100$$
$$2x^2 + 4x - 96 = 0$$
$$2(x^2 + 2x - 48) = 0$$
$$2(x + 8)(x - 6) = 0$$
$$x = 6 \quad (x = -8 \text{ is impossible})$$
$$x + 2 = 8$$
The legs are 6 inches and 8 inches.

19.   Let x represent the longer leg and $x + 1$ represent the hypotenuse. The equation is:
$$5^2 + x^2 = (x + 1)^2$$
$$25 + x^2 = x^2 + 2x + 1$$
$$25 = 2x + 1$$
$$24 = 2x$$
$$x = 12$$
The longer leg is 12 meters.

21. Setting $C = \$1,400$:

$$1400 = 400 + 700x - 100x^2$$
$$100x^2 - 700x + 1000 = 0$$
$$100\left(x^2 - 7x + 10\right) = 0$$
$$100(x - 5)(x - 2) = 0$$
$$x = 2, 5$$

The company can manufacture either 200 items or 500 items.

23. Setting $C = \$2,200$:

$$2200 = 600 + 1000x - 100x^2$$
$$100x^2 - 1000x + 1600 = 0$$
$$100\left(x^2 - 10x + 16\right) = 0$$
$$100(x - 2)(x - 8) = 0$$
$$x = 2, 8$$

The company can manufacture either 200 videotapes or 800 videotapes.

25. The revenue is given by: $R = xp = (1200 - 100p)p$. Setting $R = \$3,200$:

$$3200 = (1200 - 100p)p$$
$$3200 = 1200p - 100p^2$$
$$100p^2 - 1200p + 3200 = 0$$
$$100\left(p^2 - 12p + 32\right) = 0$$
$$100(p - 4)(p - 8) = 0$$
$$p = 4, 8$$

The company should sell the ribbons for either \$4 or \$8.

27. The revenue is given by: $R = xp = (1700 - 100p)p$. Setting $R = \$7,000$:

$$7000 = (1700 - 100p)p$$
$$7000 = 1700p - 100p^2$$
$$100p^2 - 1700p + 7000 = 0$$
$$100\left(p^2 - 17p + 70\right) = 0$$
$$100(p - 7)(p - 10) = 0$$
$$p = 7, 10$$

The calculators should be sold for either \$7 or \$10.

29. **a.** Let x represent the distance from the base to the wall, and $2x + 2$ represent the height on the wall. Using the Pythagorean theorem:

$$x^2 + (2x + 2)^2 = 13^2$$
$$x^2 + 4x^2 + 8x + 4 = 169$$
$$5x^2 + 8x - 165 = 0$$
$$(5x + 33)(x - 5) = 0$$
$$x = 5 \quad \left(x = -\tfrac{33}{5} \text{ is impossible}\right)$$

The base of the ladder is 5 feet from the wall.

b. Since $2x + 2 = 2 \cdot 5 + 2 = 12$, the ladder reaches a height of 12 feet.

31.   **a.**   Finding when $h = 0$:

$$0 = -16t^2 + 396t + 100$$
$$16t^2 - 396t - 100 = 0$$
$$4t^2 - 99t - 25 = 0$$
$$(4t + 1)(t - 25) = 0$$
$$t = 25 \quad \left(t = -\tfrac{1}{4} \text{ is impossible}\right)$$

The bullet will land on the ground after 25 seconds.

b.   Completing the table:

t (seconds)	h (feet)
0	100
5	1,680
10	2,460
15	2,440
20	1,620
25	0

33.   Simplifying the expression: $\left(5x^3\right)^2 \left(2x^6\right)^3 = 25x^6 \cdot 8x^{18} = 200x^{24}$

35.   Simplifying the expression: $\dfrac{x^4}{x^{-3}} = x^{4-(-3)} = x^{4+3} = x^7$

37.   Simplifying the expression: $\left(2 \times 10^{-4}\right)\left(4 \times 10^5\right) = 8 \times 10^1 = 80$

39.   Simplifying the expression: $20ab^2 - 16ab^2 + 6ab^2 = 10ab^2$

41.   Multiplying using the distributive property:

$$2x^2\left(3x^2 + 3x - 1\right) = 2x^2\left(3x^2\right) + 2x^2(3x) - 2x^2(1) = 6x^4 + 6x^3 - 2x^2$$

43.   Multiplying using the square of binomial formula: $(3y - 5)^2 = (3y)^2 - 2(3y)(5) + (5)^2 = 9y^2 - 30y + 25$

45.   Multiplying using the difference of squares formula: $\left(2a^2 + 7\right)\left(2a^2 - 7\right) = \left(2a^2\right)^2 - (7)^2 = 4a^4 - 49$

Chapter 5 Review

1.   Factoring the polynomial: $10x - 20 = 10(x - 2)$

3.   Factoring the polynomial: $5x - 5y = 5(x - y)$

5.   Factoring the polynomial: $8x + 4 = 4(2x + 1)$

7.   Factoring the polynomial: $24y^2 - 40y + 48 = 8\left(3y^2 - 5y + 6\right)$

9.   Factoring the polynomial: $49a^3 - 14b^3 = 7\left(7a^3 - 2b^3\right)$

11.   Factoring the polynomial: $xy + bx + ay + ab = x(y + b) + a(y + b) = (y + b)(x + a)$

13.   Factoring the polynomial: $2xy + 10x - 3y - 15 = 2x(y + 5) - 3(y + 5) = (y + 5)(2x - 3)$

15.   Factoring the polynomial: $y^2 + 9y + 14 = (y + 7)(y + 2)$

17.   Factoring the polynomial: $a^2 - 14a + 48 = (a - 8)(a - 6)$

19.   Factoring the polynomial: $y^2 + 20y + 99 = (y + 9)(y + 11)$

21.   Factoring the polynomial: $2x^2 + 13x + 15 = (2x + 3)(x + 5)$

23.   Factoring the polynomial: $5y^2 + 11y + 6 = (5y + 6)(y + 1)$

25.   Factoring the polynomial: $6r^2 + 5rt - 6t^2 = (3r - 2t)(2r + 3t)$

27.   Factoring the polynomial: $n^2 - 81 = (n + 9)(n - 9)$

29. The expression $x^2 + 49$ cannot be factored.

31. Factoring the polynomial: $64a^2 - 121b^2 = (8a + 11b)(8a - 11b)$

33. Factoring the polynomial: $y^2 + 20y + 100 = (y + 10)(y + 10) = (y + 10)^2$

35. Factoring the polynomial: $64t^2 + 16t + 1 = (8t + 1)(8t + 1) = (8t + 1)^2$

37. Factoring the polynomial: $4r^2 - 12rt + 9t^2 = (2r - 3t)(2r - 3t) = (2r - 3t)^2$

39. Factoring the polynomial: $2x^2 + 20x + 48 = 2(x^2 + 10x + 24) = 2(x + 4)(x + 6)$

41. Factoring the polynomial: $3m^3 - 18m^2 - 21m = 3m(m^2 - 6m - 7) = 3m(m - 7)(m + 1)$

43. Factoring the polynomial: $8x^2 + 16x + 6 = 2(4x^2 + 8x + 3) = 2(2x + 1)(2x + 3)$

45. Factoring the polynomial: $20m^3 - 34m^2 + 6m = 2m(10m^2 - 17m + 3) = 2m(5m - 1)(2m - 3)$

47. Factoring the polynomial: $4x^2 + 40x + 100 = 4(x^2 + 10x + 25) = 4(x + 5)(x + 5) = 4(x + 5)^2$

49. Factoring the polynomial: $5x^2 - 45 = 5(x^2 - 9) = 5(x + 3)(x - 3)$

51. Factoring the polynomial: $6a^3b + 33a^2b^2 + 15ab^3 = 3ab(2a^2 + 11ab + 5b^2) = 3ab(2a + b)(a + 5b)$

53. Factoring the polynomial: $4y^6 + 9y^4 = y^4(4y^2 + 9)$

55. Factoring the polynomial: $30a^4b + 35a^3b^2 - 15a^2b^3 = 5a^2b(6a^2 + 7ab - 3b^2) = 5a^2b(3a - b)(2a + 3b)$

57. Setting each factor equal to 0:
$$x - 5 = 0 \qquad\qquad x + 2 = 0$$
$$x = 5 \qquad\qquad\quad x = -2$$
The solutions are –2, 5.

59. Solving the equation by factoring:
$$m^2 + 3m = 10$$
$$m^2 + 3m - 10 = 0$$
$$(m + 5)(m - 2) = 0$$
$$m = -5, 2$$

61. Solving the equation by factoring:
$$m^2 - 9m = 0$$
$$m(m - 9) = 0$$
$$m = 0, 9$$

63. Solving the equation by factoring:
$$9x^4 + 9x^3 = 10x^2$$
$$9x^4 + 9x^3 - 10x^2 = 0$$
$$x^2(9x^2 + 9x - 10) = 0$$
$$x^2(3x - 2)(3x + 5) = 0$$
$$x = 0, \tfrac{2}{3}, -\tfrac{5}{3}$$

65. Let x and $x + 1$ represent the two integers. The equation is:
$$x(x + 1) = 110$$
$$x^2 + x = 110$$
$$x^2 + x - 110 = 0$$
$$(x + 11)(x - 10) = 0$$
$$x = -11, 10$$
$$x + 1 = -10, 11$$
The two integers are either –11 and –10, or 10 and 11.

67. Let x and $20 - x$ represent the two numbers. The equation is:

$$x(20 - x) = 75$$
$$20x - x^2 = 75$$
$$0 = x^2 - 20x + 75$$
$$0 = (x - 15)(x - 5)$$
$$x = 15, 5$$
$$20 - x = 5, 15$$

The two numbers are 5 and 15.

69. Let b represent the base, and $8b$ represent the height. The equation is:

$$\frac{1}{2}(b)(8b) = 16$$
$$4b^2 = 16$$
$$4b^2 - 16 = 0$$
$$4\left(b^2 - 4\right) = 0$$
$$4(b + 2)(b - 2) = 0$$
$$b = 2 \quad \left(b = -2 \text{ is impossible}\right)$$

The base is 2 inches.

Chapters 1-5 Cumulative Review

1. Simplifying the expression: $-|-9| = -9$

3. Simplifying the expression: $20 - (-9) = 20 + 9 = 29$

5. Simplifying the expression: $\dfrac{9(-2)}{-2} = \dfrac{-18}{-2} = 9$

7. Simplifying the expression: $\dfrac{-3(4 - 7) - 5(7 - 2)}{-5 - 2 - 1} = \dfrac{-3(-3) - 5(5)}{-8} = \dfrac{9 - 25}{-8} = \dfrac{-16}{-8} = 2$

9. Simplifying the expression: $6 - 2(4a + 2) - 5 = 6 - 8a - 4 - 5 = -8a - 3$

11. Simplifying using the rules of exponents: $(9xy)^0 = 1$

13. Simplifying using the rules of exponents: $\dfrac{50x^8 y^8}{25x^4 y^2} + \dfrac{28x^7 y^7}{14x^3 y} = 2x^4 y^6 + 2x^4 y^6 = 4x^4 y^6$

15. Solving the equation:

$$3x = -18$$
$$\tfrac{1}{3}(3x) = \tfrac{1}{3}(-18)$$
$$x = -6$$

17. Solving the equation:

$$-\frac{x}{3} = 7$$
$$-3\left(-\frac{x}{3}\right) = -3(7)$$
$$x = -21$$

19. Setting each factor equal to 0:

$$4m = 0 \qquad m - 7 = 0 \qquad 2m - 7 = 0$$
$$m = 0 \qquad\quad m = 7 \qquad\quad 2m = 7$$
$$m = \tfrac{7}{2}$$

The solutions are $0, 7,$ and $\tfrac{7}{2}$.

21. Solving the inequality:

$$x - 5 < -3$$
$$x - 5 + 5 < -3 + 5$$
$$x < 2$$

23. Solving the inequality:

$$7x + 1 \geq 2x - 5$$
$$5x + 1 \geq -5$$
$$5x \geq -6$$
$$x \geq -\tfrac{6}{5}$$

25.   Graphing the line:

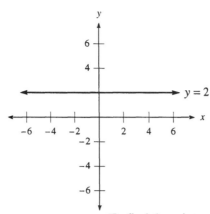

27.   To find the x-intercept, let $y = 0$:   To find the y-intercept, let $x = 0$:
$$10x - 3(0) = 30 \qquad\qquad 10(0) - 3y = 30$$
$$10x = 30 \qquad\qquad\qquad -3y = 30$$
$$x = 3 \qquad\qquad\qquad\qquad y = -10$$

29.   Substituting the points:
$$(0, 2): \ 2(0) + 5(2) = 0 + 10 = 10$$
$$(-5, 0): \ 2(-5) + 5(0) = -10 + 0 = -10 \neq 10$$
$$\left(4, \tfrac{2}{5}\right): \ 2(4) + 5\left(\tfrac{2}{5}\right) = 8 + 2 = 10$$

The points $(0, 2)$ and $\left(4, \tfrac{2}{5}\right)$ are solutions to the equation.

31.   Graphing both lines:

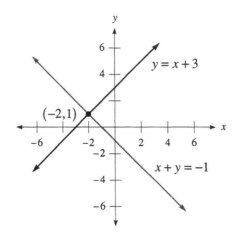

The intersection point is $(-2, 1)$.

33.   Multiplying the first equation by -3 and the second equation by 7:
$$-15x - 21y = 54$$
$$56x + 21y = 28$$
Adding the two equations:
$$41x = 82$$
$$x = 2$$
Substituting into the second equation:
$$8(2) + 3y = 4$$
$$16 + 3y = 4$$
$$3y = -12$$
$$y = -4$$
The solution is $(2, -4)$.

35.   Multiplying the first equation by –0.04:
$$-0.04x - 0.04y = -200$$
$$0.04x + 0.06y = 270$$
Adding the two equations:
$$0.02y = 70$$
$$y = 3500$$
Substituting into the first equation:
$$x + 3500 = 5000$$
$$x = 1500$$
The solution is $(1500, 3500)$.

37.   Factoring the polynomial: $n^2 - 5n - 36 = (n-9)(n+4)$

39.   Factoring the polynomial: $16 - a^2 = (4+a)(4-a)$

41.   Factoring the polynomial: $45x^2y - 30xy^2 + 5y^3 = 5y(9x^2 - 6xy + y^2) = 5y(3x-y)(3x-y) = 5y(3x-y)^2$

43.   Factoring the polynomial: $3xy + 15x - 2y - 10 = 3x(y+5) - 2(y+5) = (y+5)(3x-2)$

45.   Commutative property of addition

47.   Dividing by the monomial: $\dfrac{28x^4y^4 - 14x^2y^3 + 21xy^2}{-7xy^2} = \dfrac{28x^4y^4}{-7xy^2} + \dfrac{-14x^2y^3}{-7xy^2} + \dfrac{21xy^2}{-7xy^2} = -4x^3y^2 + 2xy - 3$

49.   Let x and $x + 4$ represent the length of each piece. The equation is:
$$x + x + 4 = 72$$
$$2x + 4 = 72$$
$$2x = 68$$
$$x = 34$$
$$x + 4 = 38$$
The pieces are 34 inches and 38 inches in length.

Chapter 5 Test

1.   Factoring the polynomial: $5x - 10 = 5(x-2)$

2.   Factoring the polynomial: $18x^2y - 9xy - 36xy^2 = 9xy(2x - 1 - 4y)$

3.   Factoring the polynomial: $x^2 + 2ax - 3bx - 6ab = x(x+2a) - 3b(x+2a) = (x+2a)(x-3b)$

4.   Factoring the polynomial: $xy + 4x - 7y - 28 = x(y+4) - 7(y+4) = (y+4)(x-7)$

5.   Factoring the polynomial: $x^2 - 5x + 6 = (x-2)(x-3)$ **6.**   Factoring the polynomial: $x^2 - x - 6 = (x-3)(x+2)$

7.   Factoring the polynomial: $a^2 - 16 = (a+4)(a-4)$ **8.**   The polynomial $x^2 + 25$ cannot be factored.

9.   Factoring the polynomial: $x^4 - 81 = (x^2 + 9)(x^2 - 9) = (x^2 + 9)(x+3)(x-3)$

10.   Factoring the polynomial: $27x^2 - 75y^2 = 3(9x^2 - 25y^2) = 3(3x + 5y)(3x - 5y)$

11.   Factoring the polynomial: $x^3 + 5x^2 - 9x - 45 = x^2(x+5) - 9(x+5) = (x+5)(x^2 - 9) = (x+5)(x+3)(x-3)$

12.   Factoring the polynomial: $x^2 - bx + 5x - 5b = x(x-b) + 5(x-b) = (x-b)(x+5)$

13.   Factoring the polynomial: $4a^2 + 22a + 10 = 2(2a^2 + 11a + 5) = 2(2a+1)(a+5)$

14.   Factoring the polynomial: $3m^2 - 3m - 18 = 3(m^2 - m - 6) = 3(m-3)(m+2)$

15.   Factoring the polynomial: $6y^2 + 7y - 5 = (3y+5)(2y-1)$

16.   Factoring the polynomial: $12x^3 - 14x^2 - 10x = 2x(6x^2 - 7x - 5) = 2x(3x-5)(2x+1)$

17. Solving the equation by factoring:
$$x^2 + 7x + 12 = 0$$
$$(x+4)(x+3) = 0$$
$$x = -4, -3$$

18. Solving the equation by factoring:
$$x^2 - 4x + 4 = 0$$
$$(x-2)^2 = 0$$
$$x - 2 = 0$$
$$x = 2$$

19. Solving the equation by factoring:
$$x^2 - 36 = 0$$
$$(x+6)(x-6) = 0$$
$$x = -6, 6$$

20. Solving the equation by factoring:
$$x^2 = x + 20$$
$$x^2 - x - 20 = 0$$
$$(x+4)(x-5) = 0$$
$$x = -4, 5$$

21. Solving the equation by factoring:
$$x^2 - 11x = -30$$
$$x^2 - 11x + 30 = 0$$
$$(x-6)(x-5) = 0$$
$$x = 5, 6$$

22. Solving the equation by factoring:
$$y^3 = 16y$$
$$y^3 - 16y = 0$$
$$y(y^2 - 16) = 0$$
$$y(y+4)(y-4) = 0$$
$$y = 0, -4, 4$$

23. Solving the equation by factoring:
$$2a^2 = a + 15$$
$$2a^2 - a - 15 = 0$$
$$(2a+5)(a-3) = 0$$
$$a = -\frac{5}{2}, 3$$

24. Solving the equation by factoring:
$$30x^3 - 20x^2 = 10x$$
$$30x^3 - 20x^2 - 10x = 0$$
$$10x(3x^2 - 2x - 1) = 0$$
$$10x(3x+1)(x-1) = 0$$
$$x = 0, -\frac{1}{3}, 1$$

25. Let x and $20 - x$ represent the numbers. The equation is:
$$x(20 - x) = 64$$
$$20x - x^2 = 64$$
$$0 = x^2 - 20x + 64$$
$$0 = (x-16)(x-4)$$
$$x = 4, 16$$
$$20 - x = 16, 4$$
The numbers are 4 and 16.

26. Let x and $x + 2$ represent the two integers. The equation is:
$$x(x+2) = x + x + 2 + 7$$
$$x^2 + 2x = 2x + 9$$
$$x^2 - 9 = 0$$
$$(x+3)(x-3) = 0$$
$$x = -3, 3$$
$$x + 2 = -1, 5$$
The integers are either –3 and –1, or 3 and 5.

27. Let w represent the width and $3w + 5$ represent the length. The equation is:
$$w(3w+5) = 42$$
$$3w^2 + 5w = 42$$
$$3w^2 + 5w - 42 = 0$$
$$(3w+14)(w-3) = 0$$
$$w = 3 \qquad \left(w = -\frac{14}{3} \text{ is impossible} \right)$$
$$3w + 5 = 14$$
The width is 3 feet and the length is 14 feet.

28. Let x and $2x + 2$ represent the two legs. The equation is:

$$x^2 + (2x+2)^2 = 13^2$$
$$x^2 + 4x^2 + 8x + 4 = 169$$
$$5x^2 + 8x - 165 = 0$$
$$(5x + 33)(x - 5) = 0$$
$$x = 5 \qquad \left(x = -\tfrac{33}{5} \text{ is impossible}\right)$$
$$2x + 2 = 12$$

The two legs are 5 meters and 12 meters in length.

29. Setting $C = \$800$:

$$800 = 200 + 500x - 100x^2$$
$$100x^2 - 500x + 600 = 0$$
$$100\left(x^2 - 5x + 6\right) = 0$$
$$100(x - 2)(x - 3) = 0$$
$$x = 2, 3$$

The company can manufacture either 200 items or 300 items.

30. The revenue is given by: $R = xp = (900 - 100p)p$. Setting $R = \$1,800$:

$$1800 = (900 - 100p)p$$
$$1800 = 900p - 100p^2$$
$$100p^2 - 900p + 1800 = 0$$
$$100\left(p^2 - 9p + 18\right) = 0$$
$$100(p - 6)(p - 3) = 0$$
$$p = 3, 6$$

The manufacturer should sell the items at either \$3 or \$6.

Chapter 6
Rational Expressions

6.1 Reducing Rational Expressions to Lowest Terms

1. Reducing the rational expression: $\dfrac{5}{5x-10} = \dfrac{5}{5(x-2)} = \dfrac{1}{x-2}$. The variable restriction is $x \neq 2$.

3. Reducing the rational expression: $\dfrac{a-3}{a^2-9} = \dfrac{1(a-3)}{(a+3)(a-3)} = \dfrac{1}{a+3}$. The variable restriction is $a \neq -3, 3$.

5. Reducing the rational expression: $\dfrac{x+5}{x^2-25} = \dfrac{1(x+5)}{(x+5)(x-5)} = \dfrac{1}{x-5}$. The variable restriction is $x \neq -5, 5$.

7. Reducing the rational expression: $\dfrac{2x^2-8}{4} = \dfrac{2(x^2-4)}{4} = \dfrac{2(x+2)(x-2)}{4} = \dfrac{(x+2)(x-2)}{2}$

    There are no variable restrictions.

9. Reducing the rational expression: $\dfrac{2x-10}{3x-6} = \dfrac{2(x-5)}{3(x-2)}$. The variable restriction is $x \neq 2$.

11. Reducing the rational expression: $\dfrac{10a+20}{5a+10} = \dfrac{10(a+2)}{5(a+2)} = \dfrac{2}{1} = 2$

13. Reducing the rational expression: $\dfrac{5x^2-5}{4x+4} = \dfrac{5(x^2-1)}{4(x+1)} = \dfrac{5(x+1)(x-1)}{4(x+1)} = \dfrac{5(x-1)}{4}$

15. Reducing the rational expression: $\dfrac{x-3}{x^2-6x+9} = \dfrac{1(x-3)}{(x-3)^2} = \dfrac{1}{x-3}$

17. Reducing the rational expression: $\dfrac{3x+15}{3x^2+24x+45} = \dfrac{3(x+5)}{3(x^2+8x+15)} = \dfrac{3(x+5)}{3(x+5)(x+3)} = \dfrac{1}{x+3}$

19. Reducing the rational expression: $\dfrac{a^2-3a}{a^3-8a^2+15a} = \dfrac{a(a-3)}{a(a^2-8a+15)} = \dfrac{a(a-3)}{a(a-3)(a-5)} = \dfrac{1}{a-5}$

21. Reducing the rational expression: $\dfrac{3x-2}{9x^2-4} = \dfrac{1(3x-2)}{(3x+2)(3x-2)} = \dfrac{1}{3x+2}$

23. Reducing the rational expression: $\dfrac{x^2+8x+15}{x^2+5x+6}=\dfrac{(x+5)(x+3)}{(x+2)(x+3)}=\dfrac{x+5}{x+2}$

25. Reducing the rational expression: $\dfrac{2m^3-2m^2-12m}{m^2-5m+6}=\dfrac{2m(m^2-m-6)}{(m-2)(m-3)}=\dfrac{2m(m-3)(m+2)}{(m-2)(m-3)}=\dfrac{2m(m+2)}{m-2}$

27. Reducing the rational expression: $\dfrac{x^3+3x^2-4x}{x^3-16x}=\dfrac{x(x^2+3x-4)}{x(x^2-16)}=\dfrac{x(x+4)(x-1)}{x(x+4)(x-4)}=\dfrac{x-1}{x-4}$

29. Reducing the rational expression: $\dfrac{4x^3-10x^2+6x}{2x^3+x^2-3x}=\dfrac{2x(2x^2-5x+3)}{x(2x^2+x-3)}=\dfrac{2x(2x-3)(x-1)}{x(2x+3)(x-1)}=\dfrac{2(2x-3)}{2x+3}$

31. Reducing the rational expression: $\dfrac{4x^2-12x+9}{4x^2-9}=\dfrac{(2x-3)^2}{(2x+3)(2x-3)}=\dfrac{2x-3}{2x+3}$

33. Reducing the rational expression: $\dfrac{x+3}{x^4-81}=\dfrac{x+3}{(x^2+9)(x^2-9)}=\dfrac{x+3}{(x^2+9)(x+3)(x-3)}=\dfrac{1}{(x^2+9)(x-3)}$

35. Reducing the rational expression: $\dfrac{3x^2+x-10}{x^4-16}=\dfrac{(3x-5)(x+2)}{(x^2+4)(x^2-4)}=\dfrac{(3x-5)(x+2)}{(x^2+4)(x+2)(x-2)}=\dfrac{3x-5}{(x^2+4)(x-2)}$

37. Reducing the rational expression: $\dfrac{42x^3-20x^2-48x}{6x^2-5x-4}=\dfrac{2x(21x^2-10x-24)}{(3x-4)(2x+1)}=\dfrac{2x(7x+6)(3x-4)}{(3x-4)(2x+1)}=\dfrac{2x(7x+6)}{2x+1}$

39. Reducing the rational expression: $\dfrac{xy+3x+2y+6}{xy+3x+5y+15}=\dfrac{x(y+3)+2(y+3)}{x(y+3)+5(y+3)}=\dfrac{(y+3)(x+2)}{(y+3)(x+5)}=\dfrac{x+2}{x+5}$

41. Reducing the rational expression: $\dfrac{x^2-3x+ax-3a}{x^2-3x+bx-3b}=\dfrac{x(x-3)+a(x-3)}{x(x-3)+b(x-3)}=\dfrac{(x-3)(x+a)}{(x-3)(x+b)}=\dfrac{x+a}{x+b}$

43. Reducing the rational expression: $\dfrac{xy+bx+ay+ab}{xy+bx+3y+3b}=\dfrac{x(y+b)+a(y+b)}{x(y+b)+3(y+b)}=\dfrac{(y+b)(x+a)}{(y+b)(x+3)}=\dfrac{x+a}{x+3}$

45. Writing as a ratio: $\frac{8}{6}=\frac{4}{3}$ 47. Writing as a ratio: $\frac{200}{250}=\frac{4}{5}$

49. Writing as a ratio: $\frac{32}{4}=\frac{8}{1}$

51. Completing the table:

Checks Written x	Total Cost $2.00+0.15x$	Cost per Check $\dfrac{2.00+0.15x}{x}$
0	$2.00	undefined
5	$2.75	$0.55
10	$3.50	$0.35
15	$4.25	$0.28
20	$5.00	$0.25

53. The average speed is: $\dfrac{122 \text{ miles}}{3 \text{ hours}}\approx 40.7 \text{ miles / hour}$

55. The average speed is: $\dfrac{785 \text{ feet}}{20 \text{ minutes}}=39.25 \text{ feet / minute}$

57. The average speed is: $\dfrac{518 \text{ feet}}{40 \text{ seconds}}=12.95 \text{ feet / second}$

59. Her average speed on level ground is: $\dfrac{20 \text{ minutes}}{2 \text{ miles}} = 10 \text{ minutes}/\text{mile}$, or $\dfrac{2 \text{ miles}}{20 \text{ minutes}} = 0.1 \text{ miles}/\text{minute}$

Her average speed downhill is: $\dfrac{40 \text{ minutes}}{6 \text{ miles}} = \dfrac{20}{3} \text{ minutes}/\text{mile}$, or $\dfrac{6 \text{ miles}}{40 \text{ minutes}} = \dfrac{3}{20} \text{ miles}/\text{minute}$

61. The average fuel consumption is: $\dfrac{168 \text{ miles}}{3.5 \text{ gallons}} = 48 \text{ miles}/\text{gallon}$

63. Substituting $x = 5$ and $y = 4$: $\dfrac{x^2 - y^2}{x - y} = \dfrac{5^2 - 4^2}{5 - 4} = \dfrac{25 - 16}{5 - 4} = \dfrac{9}{1} = 9$. The result is equal to $5 + 4 = 9$.

65. Completing the table:

x	$\dfrac{x-3}{3-x}$
-2	-1
-1	-1
0	-1
1	-1
2	-1

67. Completing the table:

x	$\dfrac{x-5}{x^2-25}$	$\dfrac{1}{x+5}$
0	$\frac{1}{5}$	$\frac{1}{5}$
-2	$\frac{1}{3}$	$\frac{1}{3}$
2	$\frac{1}{7}$	$\frac{1}{7}$
-5	undefined	undefined
5	undefined	$\frac{1}{10}$

Simplifying: $\dfrac{x-3}{3-x} = \dfrac{-(3-x)}{3-x} = -1$

69. Simplifying the expression: $\dfrac{27x^5}{9x^2} - \dfrac{45x^8}{15x^5} = 3x^3 - 3x^3 = 0$

71. Simplifying the expression: $\dfrac{72a^3b^7}{9ab^5} + \dfrac{64a^5b^3}{8a^3b} = 8a^2b^2 + 8a^2b^2 = 16a^2b^2$

73. Dividing by the monomial: $\dfrac{38x^7 + 42x^5 - 84x^3}{2x^3} = \dfrac{38x^7}{2x^3} + \dfrac{42x^5}{2x^3} - \dfrac{84x^3}{2x^3} = 19x^4 + 21x^2 - 42$

75. Dividing by the monomial: $\dfrac{28a^5b^5 + 36ab^4 - 44a^4b}{4ab} = \dfrac{28a^5b^5}{4ab} + \dfrac{36ab^4}{4ab} - \dfrac{44a^4b}{4ab} = 7a^4b^4 + 9b^3 - 11a^3$

6.2 Multiplication and Division of Rational Expressions

1. Simplifying the expression: $\dfrac{x+y}{3} \cdot \dfrac{6}{x+y} = \dfrac{6(x+y)}{3(x+y)} = 2$

3. Simplifying the expression: $\dfrac{2x+10}{x^2} \cdot \dfrac{x^3}{4x+20} = \dfrac{2(x+5)}{x^2} \cdot \dfrac{x^3}{4(x+5)} = \dfrac{2x^3(x+5)}{4x^2(x+5)} = \dfrac{x}{2}$

5. Simplifying the expression: $\dfrac{9}{2a-8} \div \dfrac{3}{a-4} = \dfrac{9}{2a-8} \cdot \dfrac{a-4}{3} = \dfrac{9}{2(a-4)} \cdot \dfrac{a-4}{3} = \dfrac{9(a-4)}{6(a-4)} = \dfrac{3}{2}$

7. Simplifying the expression:

$\dfrac{x+1}{x^2-9} \div \dfrac{2x+2}{x+3} = \dfrac{x+1}{x^2-9} \cdot \dfrac{x+3}{2x+2} = \dfrac{x+1}{(x+3)(x-3)} \cdot \dfrac{x+3}{2(x+1)} = \dfrac{(x+1)(x+3)}{2(x+3)(x-3)(x+1)} = \dfrac{1}{2(x-3)}$

9. Simplifying the expression: $\dfrac{a^2+5a}{7a} \cdot \dfrac{4a^2}{a^2+4a} = \dfrac{a(a+5)}{7a} \cdot \dfrac{4a^2}{a(a+4)} = \dfrac{4a^3(a+5)}{7a^2(a+4)} = \dfrac{4a(a+5)}{7(a+4)}$

11. Simplifying the expression:

$\dfrac{y^2-5y+6}{2y+4} \div \dfrac{2y-6}{y+2} = \dfrac{y^2-5y+6}{2y+4} \cdot \dfrac{y+2}{2y-6} = \dfrac{(y-2)(y-3)}{2(y+2)} \cdot \dfrac{y+2}{2(y-3)} = \dfrac{(y-2)(y-3)(y+2)}{4(y+2)(y-3)} = \dfrac{y-2}{4}$

13.   Simplifying the expression:

$$\frac{2x-8}{x^2-4}\cdot\frac{x^2+6x+8}{x-4}=\frac{2(x-4)}{(x+2)(x-2)}\cdot\frac{(x+4)(x+2)}{x-4}=\frac{2(x-4)(x+4)(x+2)}{(x+2)(x-2)(x-4)}=\frac{2(x+4)}{x-2}$$

15.   Simplifying the expression:

$$\frac{x-1}{x^2-x-6}\cdot\frac{x^2+5x+6}{x^2-1}=\frac{x-1}{(x-3)(x+2)}\cdot\frac{(x+2)(x+3)}{(x+1)(x-1)}=\frac{(x-1)(x+2)(x+3)}{(x-3)(x+2)(x+1)(x-1)}=\frac{x+3}{(x-3)(x+1)}$$

17.   Simplifying the expression:

$$\frac{a^2+10a+25}{a+5}\div\frac{a^2-25}{a-5}=\frac{a^2+10a+25}{a+5}\cdot\frac{a-5}{a^2-25}=\frac{(a+5)^2}{a+5}\cdot\frac{a-5}{(a+5)(a-5)}=\frac{(a+5)^2(a-5)}{(a+5)^2(a-5)}=1$$

19.   Simplifying the expression:

$$\frac{y^3-5y^2}{y^4+3y^3+2y^2}\div\frac{y^2-5y+6}{y^2-2y-3}=\frac{y^3-5y^2}{y^4+3y^3+2y^2}\cdot\frac{y^2-2y-3}{y^2-5y+6}$$

$$=\frac{y^2(y-5)}{y^2(y+2)(y+1)}\cdot\frac{(y-3)(y+1)}{(y-2)(y-3)}$$

$$=\frac{y^2(y-5)(y-3)(y+1)}{y^2(y+2)(y+1)(y-2)(y-3)}$$

$$=\frac{y-5}{(y+2)(y-2)}$$

21.   Simplifying the expression:

$$\frac{2x^2+17x+21}{x^2+2x-35}\cdot\frac{x^2-25}{2x^2-7x-15}=\frac{(2x+3)(x+7)}{(x+7)(x-5)}\cdot\frac{(x+5)(x-5)}{(2x+3)(x-5)}=\frac{(2x+3)(x+7)(x+5)(x-5)}{(x+7)(x-5)^2(2x+3)}=\frac{x+5}{x-5}$$

23.   Simplifying the expression:

$$\frac{2x^2+10x+12}{4x^2+24x+32}\cdot\frac{2x^2+18x+40}{x^2+8x+15}=\frac{2(x^2+5x+6)}{4(x^2+6x+8)}\cdot\frac{2(x^2+9x+20)}{x^2+8x+15}$$

$$=\frac{2(x+2)(x+3)}{4(x+4)(x+2)}\cdot\frac{2(x+5)(x+4)}{(x+5)(x+3)}$$

$$=\frac{4(x+2)(x+3)(x+4)(x+5)}{4(x+2)(x+3)(x+4)(x+5)}$$

$$=1$$

25.   Simplifying the expression:

$$\frac{2a^2+7a+3}{a^2-16}\div\frac{4a^2+8a+3}{2a^2-5a-12}=\frac{2a^2+7a+3}{a^2-16}\cdot\frac{2a^2-5a-12}{4a^2+8a+3}$$

$$=\frac{(2a+1)(a+3)}{(a+4)(a-4)}\cdot\frac{(2a+3)(a-4)}{(2a+1)(2a+3)}$$

$$=\frac{(2a+1)(a+3)(2a+3)(a-4)}{(a+4)(a-4)(2a+1)(2a+3)}$$

$$=\frac{a+3}{a+4}$$

27. Simplifying the expression:

$$\frac{4y^2-12y+9}{y^2-36}\div\frac{2y^2-5y+3}{y^2+5y-6}=\frac{4y^2-12y+9}{y^2-36}\cdot\frac{y^2+5y-6}{2y^2-5y+3}$$

$$=\frac{(2y-3)^2}{(y+6)(y-6)}\cdot\frac{(y+6)(y-1)}{(2y-3)(y-1)}$$

$$=\frac{(2y-3)^2(y+6)(y-1)}{(y+6)(y-6)(2y-3)(y-1)}$$

$$=\frac{2y-3}{y-6}$$

29. Simplifying the expression:

$$\frac{x^2-1}{6x^2+42x+60}\cdot\frac{7x^2+17x+6}{x+1}\cdot\frac{6x+30}{7x^2-11x-6}=\frac{(x+1)(x-1)}{6(x+5)(x+2)}\cdot\frac{(7x+3)(x+2)}{x+1}\cdot\frac{6(x+5)}{(7x+3)(x-2)}$$

$$=\frac{6(x+1)(x-1)(7x+3)(x+2)(x+5)}{6(x+5)(x+2)(x+1)(7x+3)(x-2)}$$

$$=\frac{x-1}{x-2}$$

31. Simplifying the expression:

$$\frac{18x^3+21x^2-60x}{21x^2-25x-4}\cdot\frac{28x^2-17x-3}{16x^3+28x^2-30x}=\frac{3x\left(6x^2+7x-20\right)}{21x^2-25x-4}\cdot\frac{28x^2-17x-3}{2x\left(8x^2+14x-15\right)}$$

$$=\frac{3x(3x-4)(2x+5)}{(7x+1)(3x-4)}\cdot\frac{(7x+1)(4x-3)}{2x(4x-3)(2x+5)}$$

$$=\frac{3x(3x-4)(2x+5)(7x+1)(4x-3)}{2x(7x+1)(3x-4)(4x-3)(2x+5)}$$

$$=\tfrac{3}{2}$$

33. Simplifying the expression: $\left(x^2-9\right)\left(\dfrac{2}{x+3}\right)=\dfrac{(x+3)(x-3)}{1}\cdot\dfrac{2}{x+3}=\dfrac{2(x+3)(x-3)}{x+3}=2(x-3)$

35. Simplifying the expression: $a(a+5)(a-5)\left(\dfrac{2}{a^2-25}\right)=\dfrac{a(a+5)(a-5)}{1}\cdot\dfrac{2}{(a+5)(a-5)}=\dfrac{2a(a+5)(a-5)}{(a+5)(a-5)}=2a$

37. Simplifying the expression: $\left(x^2-x-6\right)\left(\dfrac{x+1}{x-3}\right)=\dfrac{(x-3)(x+2)}{1}\cdot\dfrac{x+1}{x-3}=\dfrac{(x-3)(x+2)(x+1)}{x-3}=(x+2)(x+1)$

39. Simplifying the expression: $\left(x^2-4x-5\right)\left(\dfrac{-2x}{x+1}\right)=\dfrac{(x-5)(x+1)}{1}\cdot\dfrac{-2x}{x+1}=\dfrac{-2x(x-5)(x+1)}{x+1}=-2x(x-5)$

41. Simplifying the expression:

$$\frac{x^2-9}{x^2-3x}\cdot\frac{2x+10}{xy+5x+3y+15}=\frac{(x+3)(x-3)}{x(x-3)}\cdot\frac{2(x+5)}{x(y+5)+3(y+5)}=\frac{2(x+3)(x-3)(x+5)}{x(x-3)(y+5)(x+3)}=\frac{2(x+5)}{x(y+5)}$$

43. Simplifying the expression:

$$\frac{2x^2+4x}{x^2-y^2}\cdot\frac{x^2+3x+xy+3y}{x^2+5x+6}=\frac{2x(x+2)}{(x+y)(x-y)}\cdot\frac{x(x+3)+y(x+3)}{(x+2)(x+3)}=\frac{2x(x+2)(x+3)(x+y)}{(x+y)(x-y)(x+2)(x+3)}=\frac{2x}{x-y}$$

45. Simplifying the expression:

$$\frac{x^3-3x^2+4x-12}{x^4-16}\cdot\frac{3x^2+5x-2}{3x^2-10x+3}=\frac{x^2(x-3)+4(x-3)}{\left(x^2+4\right)\left(x^2-4\right)}\cdot\frac{(3x-1)(x+2)}{(3x-1)(x-3)}$$

$$=\frac{(x-3)\left(x^2+4\right)}{\left(x^2+4\right)(x+2)(x-2)}\cdot\frac{(3x-1)(x+2)}{(3x-1)(x-3)}$$

$$=\frac{(x-3)\left(x^2+4\right)(3x-1)(x+2)}{\left(x^2+4\right)(x+2)(x-2)(3x-1)(x-3)}$$

$$=\frac{1}{x-2}$$

47. Simplifying the expression: $\left(1-\frac{1}{2}\right)\left(1-\frac{1}{3}\right)\left(1-\frac{1}{4}\right)\left(1-\frac{1}{5}\right)=\left(\frac{2}{2}-\frac{1}{2}\right)\left(\frac{3}{3}-\frac{1}{3}\right)\left(\frac{4}{4}-\frac{1}{4}\right)\left(\frac{5}{5}-\frac{1}{5}\right)=\frac{1}{2}\cdot\frac{2}{3}\cdot\frac{3}{4}\cdot\frac{4}{5}=\frac{1}{5}$

49. Simplifying the expression: $\left(1-\frac{1}{2}\right)\left(1-\frac{1}{3}\right)\left(1-\frac{1}{4}\right)\bullet\bullet\bullet\left(1-\frac{1}{99}\right)\left(1-\frac{1}{100}\right)=\frac{1}{2}\cdot\frac{2}{3}\cdot\frac{3}{4}\bullet\bullet\bullet\frac{98}{99}\cdot\frac{99}{100}=\frac{1}{100}$

51. Since 5,280 feet = 1 mile, the height is: $\dfrac{14,494\text{ feet}}{5,280\text{ feet / mile}}\approx2.7$ miles

53. Converting to miles per hour: $\dfrac{1088\text{ feet}}{1\text{ second}}\bullet\dfrac{1\text{ mile}}{5280\text{ feet}}\bullet\dfrac{60\text{ seconds}}{1\text{ minute}}\bullet\dfrac{60\text{ minutes}}{1\text{ hour}}\approx742$ miles / hour

55. Converting to miles per hour: $\dfrac{785\text{ feet}}{20\text{ minutes}}\bullet\dfrac{60\text{ minutes}}{1\text{ hour}}\bullet\dfrac{1\text{ mile}}{5280\text{ feet}}\approx0.45$ miles / hour

57. Converting to miles per hour: $\dfrac{518\text{ feet}}{40\text{ seconds}}\bullet\dfrac{60\text{ seconds}}{1\text{ minute}}\bullet\dfrac{60\text{ minutes}}{1\text{ hour}}\bullet\dfrac{1\text{ mile}}{5280\text{ feet}}\approx8.8$ miles / hour

59. Her average speed on level ground is: $\dfrac{2\text{ miles}}{1/3\text{ hour}}=6$ miles / hour

Her average speed downhill is: $\dfrac{6\text{ miles}}{2/3\text{ hour}}=9$ miles / hour

61. Adding the fractions: $\frac{1}{2}+\frac{5}{2}=\frac{6}{2}=3$

63. Adding the fractions: $2+\frac{3}{4}=\frac{2\bullet4}{1\bullet4}+\frac{3}{4}=\frac{8}{4}+\frac{3}{4}=\frac{11}{4}$

65. Adding the fractions: $\frac{1}{10}+\frac{3}{14}=\frac{1\bullet7}{10\bullet7}+\frac{3\bullet5}{14\bullet5}=\frac{7}{70}+\frac{15}{70}=\frac{22}{70}=\frac{11}{35}$

67. Simplifying the expression: $\dfrac{10x^4}{2x^2}+\dfrac{12x^6}{3x^4}=5x^2+4x^2=9x^2$

69. Simplifying the expression: $\dfrac{12a^2b^5}{3ab^3}+\dfrac{14a^4b^7}{7a^3b^5}=4ab^2+2ab^2=6ab^2$

6.3 Addition and Subtraction of Rational Expressions

1. Combining the fractions: $\dfrac{3}{x}+\dfrac{4}{x}=\dfrac{7}{x}$

3. Combining the fractions: $\dfrac{9}{a}-\dfrac{5}{a}=\dfrac{4}{a}$

5. Combining the fractions: $\dfrac{1}{x+1}+\dfrac{x}{x+1}=\dfrac{1+x}{x+1}=1$

7. Combining the fractions: $\dfrac{y^2}{y-1}-\dfrac{1}{y-1}=\dfrac{y^2-1}{y-1}=\dfrac{(y+1)(y-1)}{y-1}=y+1$

9. Combining the fractions: $\dfrac{x^2}{x+2}+\dfrac{4x+4}{x+2}=\dfrac{x^2+4x+4}{x+2}=\dfrac{(x+2)^2}{x+2}=x+2$

11. Combining the fractions: $\dfrac{x^2}{x-2} - \dfrac{4x-4}{x-2} = \dfrac{x^2 - 4x + 4}{x-2} = \dfrac{(x-2)^2}{x-2} = x-2$

13. Combining the fractions: $\dfrac{x+2}{x+6} - \dfrac{x-4}{x+6} = \dfrac{x+2-x+4}{x+6} = \dfrac{6}{x+6}$

15. Combining the fractions: $\dfrac{y}{2} - \dfrac{2}{y} = \dfrac{y \bullet y}{2 \bullet y} - \dfrac{2 \bullet 2}{y \bullet 2} = \dfrac{y^2}{2y} - \dfrac{4}{2y} = \dfrac{y^2 - 4}{2y} = \dfrac{(y+2)(y-2)}{2y}$

17. Combining the fractions: $\dfrac{1}{2} + \dfrac{a}{3} = \dfrac{1 \bullet 3}{2 \bullet 3} + \dfrac{a \bullet 2}{3 \bullet 2} = \dfrac{3}{6} + \dfrac{2a}{6} = \dfrac{2a+3}{6}$

19. Combining the fractions: $\dfrac{x}{x+1} + \dfrac{3}{4} = \dfrac{x \bullet 4}{(x+1) \bullet 4} + \dfrac{3 \bullet (x+1)}{4 \bullet (x+1)} = \dfrac{4x}{4(x+1)} + \dfrac{3x+3}{4(x+1)} = \dfrac{4x+3x+3}{4(x+1)} = \dfrac{7x+3}{4(x+1)}$

21. Combining the fractions:

$\dfrac{x+1}{x-2} - \dfrac{4x+7}{5x-10} = \dfrac{(x+1) \bullet 5}{(x-2) \bullet 5} - \dfrac{4x+7}{5(x-2)} = \dfrac{5x+5}{5(x-2)} - \dfrac{4x+7}{5(x-2)} = \dfrac{5x+5-4x-7}{5(x-2)} = \dfrac{x-2}{5(x-2)} = \dfrac{1}{5}$

23. Combining the fractions:

$\dfrac{4x-2}{3x+12} - \dfrac{x-2}{x+4} = \dfrac{4x-2}{3(x+4)} - \dfrac{(x-2) \bullet 3}{(x+4) \bullet 3} = \dfrac{4x-2}{3(x+4)} - \dfrac{3x-6}{3(x+4)} = \dfrac{4x-2-3x+6}{3(x+4)} = \dfrac{x+4}{3(x+4)} = \dfrac{1}{3}$

25. Combining the fractions:

$\dfrac{6}{x(x-2)} + \dfrac{3}{x} = \dfrac{6}{x(x-2)} + \dfrac{3 \bullet (x-2)}{x \bullet (x-2)} = \dfrac{6}{x(x-2)} + \dfrac{3x-6}{x(x-2)} = \dfrac{6+3x-6}{x(x-2)} = \dfrac{3x}{x(x-2)} = \dfrac{3}{x-2}$

27. Combining the fractions:

$\dfrac{4}{a} - \dfrac{12}{a^2+3a} = \dfrac{4 \bullet (a+3)}{a \bullet (a+3)} - \dfrac{12}{a(a+3)} = \dfrac{4a+12}{a(a+3)} - \dfrac{12}{a(a+3)} = \dfrac{4a+12-12}{a(a+3)} = \dfrac{4a}{a(a+3)} = \dfrac{4}{a+3}$

29. Combining the fractions:

$\dfrac{2}{x+5} - \dfrac{10}{x^2-25} = \dfrac{2 \bullet (x-5)}{(x+5) \bullet (x-5)} - \dfrac{10}{(x+5)(x-5)}$

$= \dfrac{2x-10}{(x+5)(x-5)} - \dfrac{10}{(x+5)(x-5)}$

$= \dfrac{2x-10-10}{(x+5)(x-5)}$

$= \dfrac{2x-20}{(x+5)(x-5)}$

$= \dfrac{2(x-10)}{(x+5)(x-5)}$

31. Combining the fractions:

$\dfrac{x-4}{x-3} + \dfrac{6}{x^2-9} = \dfrac{(x-4) \bullet (x+3)}{(x-3) \bullet (x+3)} + \dfrac{6}{(x+3)(x-3)}$

$= \dfrac{x^2-x-12}{(x+3)(x-3)} + \dfrac{6}{(x+3)(x-3)}$

$= \dfrac{x^2-x-12+6}{(x+3)(x-3)}$

$= \dfrac{x^2-x-6}{(x+3)(x-3)}$

$= \dfrac{(x-3)(x+2)}{(x+3)(x-3)}$

$= \dfrac{x+2}{x+3}$

33. Combining the fractions:

$$\frac{a-4}{a-3}+\frac{5}{a^2-a-6}=\frac{(a-4)\cdot(a+2)}{(a-3)\cdot(a+2)}+\frac{5}{(a-3)(a+2)}$$

$$=\frac{a^2-2a-8}{(a-3)(a+2)}+\frac{5}{(a-3)(a+2)}$$

$$=\frac{a^2-2a-8+5}{(a-3)(a+2)}$$

$$=\frac{a^2-2a-3}{(a-3)(a+2)}$$

$$=\frac{(a-3)(a+1)}{(a-3)(a+2)}$$

$$=\frac{a+1}{a+2}$$

35. Combining the fractions:

$$\frac{8}{x^2-16}-\frac{7}{x^2-x-12}=\frac{8}{(x+4)(x-4)}-\frac{7}{(x-4)(x+3)}$$

$$=\frac{8(x+3)}{(x+4)(x-4)(x+3)}-\frac{7(x+4)}{(x+4)(x-4)(x+3)}$$

$$=\frac{8x+24}{(x+4)(x-4)(x+3)}-\frac{7x+28}{(x+4)(x-4)(x+3)}$$

$$=\frac{8x+24-7x-28}{(x+4)(x-4)(x+3)}$$

$$=\frac{x-4}{(x+4)(x-4)(x+3)}$$

$$=\frac{1}{(x+4)(x+3)}$$

37. Combining the fractions:

$$\frac{4y}{y^2+6y+5}-\frac{3y}{y^2+5y+4}=\frac{4y}{(y+5)(y+1)}-\frac{3y}{(y+4)(y+1)}$$

$$=\frac{4y(y+4)}{(y+5)(y+1)(y+4)}-\frac{3y(y+5)}{(y+5)(y+1)(y+4)}$$

$$=\frac{4y^2+16y}{(y+5)(y+1)(y+4)}-\frac{3y^2+15y}{(y+5)(y+1)(y+4)}$$

$$=\frac{4y^2+16y-3y^2-15y}{(y+5)(y+1)(y+4)}$$

$$=\frac{y^2+y}{(y+5)(y+1)(y+4)}$$

$$=\frac{y(y+1)}{(y+5)(y+1)(y+4)}$$

$$=\frac{y}{(y+5)(y+4)}$$

39. Combining the fractions:

$$\frac{4x+1}{x^2+5x+4} - \frac{x+3}{x^2+4x+3} = \frac{4x+1}{(x+4)(x+1)} - \frac{x+3}{(x+3)(x+1)}$$

$$= \frac{(4x+1)(x+3)}{(x+4)(x+1)(x+3)} - \frac{(x+3)(x+4)}{(x+4)(x+1)(x+3)}$$

$$= \frac{4x^2+13x+3}{(x+4)(x+1)(x+3)} - \frac{x^2+7x+12}{(x+4)(x+1)(x+3)}$$

$$= \frac{4x^2+13x+3-x^2-7x-12}{(x+4)(x+1)(x+3)}$$

$$= \frac{3x^2+6x-9}{(x+4)(x+1)(x+3)}$$

$$= \frac{3(x+3)(x-1)}{(x+4)(x+1)(x+3)}$$

$$= \frac{3(x-1)}{(x+4)(x+1)}$$

41. Combining the fractions:

$$\frac{1}{x} + \frac{x}{3x+9} - \frac{3}{x^2+3x} = \frac{1}{x} + \frac{x}{3(x+3)} - \frac{3}{x(x+3)}$$

$$= \frac{1 \cdot 3(x+3)}{x \cdot 3(x+3)} + \frac{x \cdot x}{3(x+3) \cdot x} - \frac{3 \cdot 3}{x(x+3) \cdot 3}$$

$$= \frac{3x+9}{3x(x+3)} + \frac{x^2}{3x(x+3)} - \frac{9}{3x(x+3)}$$

$$= \frac{3x+9+x^2-9}{3x(x+3)}$$

$$= \frac{x^2+3x}{3x(x+3)}$$

$$= \frac{x(x+3)}{3x(x+3)}$$

$$= \frac{1}{3}$$

43. Completing the table:

Number x	Reciprocal $\frac{1}{x}$	Sum $1+\frac{1}{x}$	Sum $\frac{x+1}{x}$
1	1	2	2
2	$\frac{1}{2}$	$\frac{3}{2}$	$\frac{3}{2}$
3	$\frac{1}{3}$	$\frac{4}{3}$	$\frac{4}{3}$
4	$\frac{1}{4}$	$\frac{5}{4}$	$\frac{5}{4}$

45. Completing the table:

x	$x+\frac{4}{x}$	$\frac{x^2+4}{x}$	$x+4$
1	5	5	5
2	4	4	6
3	$\frac{13}{3}$	$\frac{13}{3}$	7
4	5	5	8

47. Combining the fractions: $1 + \dfrac{1}{x+2} = \dfrac{1 \cdot (x+2)}{1 \cdot (x+2)} + \dfrac{1}{x+2} = \dfrac{x+2}{x+2} + \dfrac{1}{x+2} = \dfrac{x+2+1}{x+2} = \dfrac{x+3}{x+2}$

49. Combining the fractions: $1 - \dfrac{1}{x+3} = \dfrac{1 \cdot (x+3)}{1 \cdot (x+3)} - \dfrac{1}{x+3} = \dfrac{x+3}{x+3} - \dfrac{1}{x+3} = \dfrac{x+3-1}{x+3} = \dfrac{x+2}{x+3}$

51. The expression is: $x + 2\left(\dfrac{1}{x}\right) = x + \dfrac{2}{x} = \dfrac{x \cdot x}{1 \cdot x} + \dfrac{2}{x} = \dfrac{x^2}{x} + \dfrac{2}{x} = \dfrac{x^2+2}{x}$

53. Represent the two numbers as x and $2x$. The expression is: $\dfrac{1}{x} + \dfrac{1}{2x} = \dfrac{1 \cdot 2}{x \cdot 2} + \dfrac{1}{2x} = \dfrac{2}{2x} + \dfrac{1}{2x} = \dfrac{3}{2x}$

55.   Solving the equation:
$$2x + 3(x - 3) = 6$$
$$2x + 3x - 9 = 6$$
$$5x - 9 = 6$$
$$5x = 15$$
$$x = 3$$

57.   Solving the equation:
$$x - 3(x + 3) = x - 3$$
$$x - 3x - 9 = x - 3$$
$$-2x - 9 = x - 3$$
$$-3x - 9 = -3$$
$$-3x = 6$$
$$x = -2$$

59.   Solving the equation:
$$7 - 2(3x + 1) = 4x + 3$$
$$7 - 6x - 2 = 4x + 3$$
$$-6x + 5 = 4x + 3$$
$$-10x + 5 = 3$$
$$-10x = -2$$
$$x = \tfrac{1}{5}$$

61.   Solving the equation:

$$x^2 + 5x + 6 = 0$$
$$(x + 2)(x + 3) = 0$$
$$x = -2, -3$$

63.   Solving the quadratic equation:
$$x^2 - x = 6$$
$$x^2 - x - 6 = 0$$
$$(x - 3)(x + 2) = 0$$
$$x = -2, 3$$

65.   Solving the quadratic equation:

$$x^2 - 5x = 0$$
$$x(x - 5) = 0$$
$$x = 0, 5$$

6.4 Equations Involving Rational Expressions

1.   Multiplying both sides of the equation by 6:
$$6\left(\frac{x}{3} + \frac{1}{2}\right) = 6\left(-\frac{1}{2}\right)$$
$$2x + 3 = -3$$
$$2x = -6$$
$$x = -3$$
Since $x = -3$ checks in the original equation, the solution is $x = -3$.

3.   Multiplying both sides of the equation by $5a$:
$$5a\left(\frac{4}{a}\right) = 5a\left(\frac{1}{5}\right)$$
$$20 = a$$
Since $a = 20$ checks in the original equation, the solution is $a = 20$.

5.   Multiplying both sides of the equation by x:
$$x\left(\frac{3}{x} + 1\right) = x\left(\frac{2}{x}\right)$$
$$3 + x = 2$$
$$x = -1$$
Since $x = -1$ checks in the original equation, the solution is $x = -1$.

7.   Multiplying both sides of the equation by $5a$:
$$5a\left(\frac{3}{a} - \frac{2}{a}\right) = 5a\left(\frac{1}{5}\right)$$
$$15 - 10 = a$$
$$a = 5$$
Since $a = 5$ checks in the original equation, the solution is $a = 5$.

9. Multiplying both sides of the equation by $2x$:
$$2x\left(\frac{3}{x}+2\right)=2x\left(\tfrac{1}{2}\right)$$
$$6+4x=x$$
$$6=-3x$$
$$x=-2$$
Since $x=-2$ checks in the original equation, the solution is $x=-2$.

11. Multiplying both sides of the equation by $4y$:
$$4y\left(\frac{1}{y}-\tfrac{1}{2}\right)=4y\left(-\tfrac{1}{4}\right)$$
$$4-2y=-y$$
$$4=y$$
Since $y=4$ checks in the original equation, the solution is $y=4$.

13. Multiplying both sides of the equation by x^2:
$$x^2\left(1-\frac{8}{x}\right)=x^2\left(-\frac{15}{x^2}\right)$$
$$x^2-8x=-15$$
$$x^2-8x+15=0$$
$$(x-3)(x-5)=0$$
$$x=3,5$$
Both $x=3$ and $x=5$ check in the original equation.

15. Multiplying both sides of the equation by $2x$:
$$2x\left(\frac{x}{2}-\frac{4}{x}\right)=2x\left(-\tfrac{7}{2}\right)$$
$$x^2-8=-7x$$
$$x^2+7x-8=0$$
$$(x+8)(x-1)=0$$
$$x=-8,1$$
Both $x=-8$ and $x=1$ check in the original equation.

17. Multiplying both sides of the equation by 6:
$$6\left(\frac{x-3}{2}+\frac{2x}{3}\right)=6\left(\tfrac{5}{6}\right)$$
$$3(x-3)+2(2x)=5$$
$$3x-9+4x=5$$
$$7x-9=5$$
$$7x=14$$
$$x=2$$
Since $x=2$ checks in the original equation, the solution is $x=2$.

19. Multiplying both sides of the equation by 12:
$$12\left(\frac{x+1}{3}+\frac{x-3}{4}\right)=12\left(\tfrac{1}{6}\right)$$
$$4(x+1)+3(x-3)=2$$
$$4x+4+3x-9=2$$
$$7x-5=2$$
$$7x=7$$
$$x=1$$
Since $x=1$ checks in the original equation, the solution is $x=1$.

21. Multiplying both sides of the equation by $5(x+2)$:
$$5(x+2)\cdot\frac{6}{x+2}=5(x+2)\cdot\tfrac{3}{5}$$
$$30=3x+6$$
$$24=3x$$
$$x=8$$
Since $x=8$ checks in the original equation, the solution is $x=8$.

23. Multiplying both sides of the equation by $(y-2)(y-3)$:

$$(y-2)(y-3) \cdot \frac{3}{y-2} = (y-2)(y-3) \cdot \frac{2}{y-3}$$
$$3(y-3) = 2(y-2)$$
$$3y-9 = 2y-4$$
$$y = 5$$

Since $y = 5$ checks in the original equation, the solution is $y = 5$.

25. Multiplying both sides of the equation by $3(x-2)$:

$$3(x-2)\left(\frac{x}{x-2} + \frac{2}{3}\right) = 3(x-2)\left(\frac{2}{x-2}\right)$$
$$3x + 2(x-2) = 6$$
$$3x + 2x - 4 = 6$$
$$5x - 4 = 6$$
$$5x = 10$$
$$x = 2$$

Since $x = 2$ does not check in the original equation, there is no solution ($\varnothing$).

27. Multiplying both sides of the equation by $2(x-2)$:

$$2(x-2)\left(\frac{x}{x-2} + \frac{3}{2}\right) = 2(x-2) \cdot \frac{9}{2(x-2)}$$
$$2x + 3(x-2) = 9$$
$$2x + 3x - 6 = 9$$
$$5x - 6 = 9$$
$$5x = 15$$
$$x = 3$$

Since $x = 3$ checks in the original equation, the solution is $x = 3$.

29. Multiplying both sides of the equation by $x^2 + 5x + 6 = (x+2)(x+3)$:

$$(x+2)(x+3)\left(\frac{5}{x+2} + \frac{1}{x+3}\right) = (x+2)(x+3) \cdot \frac{-1}{(x+2)(x+3)}$$
$$5(x+3) + 1(x+2) = -1$$
$$5x + 15 + x + 2 = -1$$
$$6x + 17 = -1$$
$$6x = -18$$
$$x = -3$$

Since $x = -3$ does not check in the original equation, there is no solution ($\varnothing$).

31. Multiplying both sides of the equation by $x^2 - 4 = (x+2)(x-2)$:

$$(x+2)(x-2)\left(\frac{8}{x^2-4} + \frac{3}{x+2}\right) = (x+2)(x-2) \cdot \frac{1}{x-2}$$
$$8 + 3(x-2) = 1(x+2)$$
$$8 + 3x - 6 = x + 2$$
$$3x + 2 = x + 2$$
$$2x = 0$$
$$x = 0$$

Since $x = 0$ checks in the original equation, the solution is $x = 0$.

33. Multiplying both sides of the equation by $2(a-3)$:

$$2(a-3)\left(\frac{a}{2}+\frac{3}{a-3}\right)=2(a-3)\cdot\frac{a}{a-3}$$
$$a(a-3)+6=2a$$
$$a^2-3a+6=2a$$
$$a^2-5a+6=0$$
$$(a-2)(a-3)=0$$
$$a=2,3$$

Since $a=3$ does not check in the original equation, the solution is $a=2$.

35. Since $y^2-4=(y+2)(y-2)$ and $y^2+2y=y(y+2)$, the LCD is $y(y+2)(y-2)$. Multiplying by the LCD:

$$y(y+2)(y-2)\cdot\frac{6}{(y+2)(y-2)}=y(y+2)(y-2)\cdot\frac{4}{y(y+2)}$$
$$6y=4(y-2)$$
$$6y=4y-8$$
$$2y=-8$$
$$y=-4$$

Since $y=-4$ checks in the original equation, the solution is $y=-4$.

37. Since $a^2-9=(a+3)(a-3)$ and $a^2+a-12=(a+4)(a-3)$, the LCD is $(a+3)(a-3)(a+4)$. Multiplying by the LCD:

$$(a+3)(a-3)(a+4)\cdot\frac{2}{(a+3)(a-3)}=(a+3)(a-3)(a+4)\cdot\frac{3}{(a+4)(a-3)}$$
$$2(a+4)=3(a+3)$$
$$2a+8=3a+9$$
$$-a+8=9$$
$$-a=1$$
$$a=-1$$

Since $a=-1$ checks in the original equation, the solution is $a=-1$.

39. Multiplying both sides of the equation by $x^2-4x-5=(x-5)(x+1)$:

$$(x-5)(x+1)\left(\frac{3x}{x-5}-\frac{2x}{x+1}\right)=(x-5)(x+1)\cdot\frac{-42}{(x-5)(x+1)}$$
$$3x(x+1)-2x(x-5)=-42$$
$$3x^2+3x-2x^2+10x=-42$$
$$x^2+13x+42=0$$
$$(x+7)(x+6)=0$$
$$x=-7,-6$$

Both $x=-7$ and $x=-6$ check in the original equation.

41. Multiplying both sides of the equation by $x^2+5x+6=(x+2)(x+3)$:

$$(x+2)(x+3)\left(\frac{2x}{x+2}\right)=(x+2)(x+3)\left(\frac{x}{x+3}-\frac{3}{x^2+5x+6}\right)$$
$$2x(x+3)=x(x+2)-3$$
$$2x^2+6x=x^2+2x-3$$
$$x^2+4x+3=0$$
$$(x+3)(x+1)=0$$
$$x=-3,-1$$

Since $x=-3$ does not check in the original equation, the solution is $x=-1$.

43.   Solving the equation:
$$x + \frac{4}{x} = 5$$
$$x\left(x + \frac{4}{x}\right) = x(5)$$
$$x^2 + 4 = 5x$$
$$x^2 - 5x + 4 = 0$$
$$(x-1)(x-4) = 0$$
$$x = 1, 4$$
The solution is consistent with the table.

45.   Let x represent the number. The equation is:
$$2(x-3) - 5 = 3$$
$$2x - 6 - 5 = 3$$
$$2x - 11 = 3$$
$$2x = 14$$
$$x = 7$$
The number is 7.

47.   Let w represent the width, and $2w + 5$ represent the length. Using the perimeter formula:
$$2w + 2(2w+5) = 34$$
$$2w + 4w + 10 = 34$$
$$6w + 10 = 34$$
$$6w = 24$$
$$w = 4$$
$$2w + 5 = 13$$
The length is 13 inches and the width is 4 inches.

49.   Let x and $x + 2$ represent the two integers. The equation is:
$$x(x+2) = 48$$
$$x^2 + 2x = 48$$
$$x^2 + 2x - 48 = 0$$
$$(x+8)(x-6) = 0$$
$$x = -8, 6$$
$$x + 2 = -6, 8$$
The two integers are either –8 and –6, or 6 and 8.

51.   Let x and $x + 2$ represent the two legs. The equation is:
$$x^2 + (x+2)^2 = 10^2$$
$$x^2 + x^2 + 4x + 4 = 100$$
$$2x^2 + 4x - 96 = 0$$
$$x^2 + 2x - 48 = 0$$
$$(x+8)(x-6) = 0$$
$$x = 6 \quad (x = -8 \text{ is impossible})$$
$$x + 2 = 8$$
The legs are 6 inches and 8 inches.

6.5 Applications

1.   Let x and $3x$ represent the two numbers. The equation is:
$$\frac{1}{x} + \frac{1}{3x} = \frac{16}{3}$$
$$3x\left(\frac{1}{x} + \frac{1}{3x}\right) = 3x\left(\frac{16}{3}\right)$$
$$3 + 1 = 16x$$
$$16x = 4$$
$$x = \frac{1}{4}$$
$$3x = \frac{3}{4}$$
The numbers are $\frac{1}{4}$ and $\frac{3}{4}$.

3. Let x represent the number. The equation is:

$$x + \frac{1}{x} = \frac{13}{6}$$
$$6x\left(x + \frac{1}{x}\right) = 6x\left(\frac{13}{6}\right)$$
$$6x^2 + 6 = 13x$$
$$6x^2 - 13x + 6 = 0$$
$$(3x - 2)(2x - 3) = 0$$
$$x = \frac{2}{3}, \frac{3}{2}$$

The number is either $\frac{2}{3}$ or $\frac{3}{2}$.

5. Let x represent the number. The equation is:

$$\frac{7 + x}{9 + x} = \frac{5}{7}$$
$$7(9 + x) \cdot \frac{7 + x}{9 + x} = 7(9 + x) \cdot \frac{5}{7}$$
$$7(7 + x) = 5(9 + x)$$
$$49 + 7x = 45 + 5x$$
$$49 + 2x = 45$$
$$2x = -4$$
$$x = -2$$

The number is -2.

7. Let x and $x + 2$ represent the two integers. The equation is:

$$\frac{1}{x} + \frac{1}{x + 2} = \frac{5}{12}$$
$$12x(x + 2)\left(\frac{1}{x} + \frac{1}{x + 2}\right) = 12x(x + 2)\left(\frac{5}{12}\right)$$
$$12(x + 2) + 12x = 5x(x + 2)$$
$$12x + 24 + 12x = 5x^2 + 10x$$
$$0 = 5x^2 - 14x - 24$$
$$(5x + 6)(x - 4) = 0$$
$$x = 4 \quad \left(x = -\frac{6}{5} \text{ is impossible}\right)$$
$$x + 2 = 6$$

The integers are 4 and 6.

9. Let x represent the rate of the boat in still water:

	d	r	t
Upstream	26	$x-3$	$\frac{26}{x-3}$
Downstream	38	$x+3$	$\frac{38}{x+3}$

The equation is:

$$\frac{26}{x - 3} = \frac{38}{x + 3}$$
$$(x + 3)(x - 3) \cdot \frac{26}{x - 3} = (x + 3)(x - 3) \cdot \frac{38}{x + 3}$$
$$26(x + 3) = 38(x - 3)$$
$$26x + 78 = 38x - 114$$
$$-12x + 78 = -114$$
$$-12x = -192$$
$$x = 16$$

The speed of the boat in still water is 16 mph.

11.   Let x represent the plane speed in still air:

	d	r	t
Against Wind	140	$x-20$	$\dfrac{140}{x-20}$
With Wind	160	$x+20$	$\dfrac{160}{x+20}$

The equation is:

$$\frac{140}{x-20}=\frac{160}{x+20}$$

$$(x+20)(x-20)\cdot\frac{140}{x-20}=(x+20)(x-20)\cdot\frac{160}{x+20}$$

$$140(x+20)=160(x-20)$$

$$140x+2800=160x-3200$$

$$-20x+2800=-3200$$

$$-20x=-6000$$

$$x=300$$

The plane speed in still air is 300 mph.

15.   Let x represent her rate downhill:

	d	r	t
Level Ground	2	$x-3$	$\dfrac{2}{x-3}$
Downhill	6	x	$\dfrac{6}{x}$

The equation is:

$$\frac{2}{x-3}+\frac{6}{x}=1$$

$$x(x-3)\left(\frac{2}{x-3}+\frac{6}{x}\right)=x(x-3)\cdot1$$

$$2x+6(x-3)=x(x-3)$$

$$2x+6x-18=x^2-3x$$

$$8x-18=x^2-3x$$

$$0=x^2-11x+18$$

$$0=(x-2)(x-9)$$

$$x=9 \qquad (x=2 \text{ is impossible})$$

Tina runs 9 mph on the downhill part of the course.

13.   Let x and $x+20$ represent the rates of each plane:

	d	r	t
Plane 1	285	$x+20$	$\dfrac{285}{x+20}$
Plane 2	255	x	$\dfrac{255}{x}$

The equation is:

$$\frac{285}{x+20}=\frac{255}{x}$$

$$x(x+20)\cdot\frac{285}{x+20}=x(x+20)\cdot\frac{255}{x}$$

$$285x=255(x+20)$$

$$285x=255x+5100$$

$$30x=5100$$

$$x=170$$

$$x+20=190$$

The plane speeds are 170 mph and 190 mph.

17. Let x represent her rate on level ground:

	d	r	t
Level Ground	4	x	$\dfrac{4}{x}$
Downhill	5	$x+2$	$\dfrac{5}{x+2}$

The equation is:

$$\frac{4}{x}+\frac{5}{x+2}=1$$

$$x(x+2)\left(\frac{4}{x}+\frac{5}{x+2}\right)=x(x+2)\cdot 1$$

$$4(x+2)+5x=x(x+2)$$

$$4x+8+5x=x^2+2x$$

$$9x+8=x^2+2x$$

$$0=x^2-7x-8$$

$$0=(x-8)(x+1)$$

$$x=8 \qquad (x=-1 \text{ is impossible})$$

Jerri jogs 8 mph on level ground.

19. Let t represent the time to fill the pool with both pipes left open. The equation is:

$$\frac{1}{12}-\frac{1}{15}=\frac{1}{t}$$

$$60t\left(\frac{1}{12}-\frac{1}{15}\right)=60t\cdot\frac{1}{t}$$

$$5t-4t=60$$

$$t=60$$

It will take 60 hours to fill the pool with both pipes left open.

21. Let t represent the time to fill the bathtub with both faucets open. The equation is:

$$\frac{1}{10}+\frac{1}{12}=\frac{1}{t}$$

$$60t\left(\frac{1}{10}+\frac{1}{12}\right)=60t\cdot\frac{1}{t}$$

$$6t+5t=60$$

$$11t=60$$

$$t=\frac{60}{11}=5\frac{5}{11}$$

It will take $5\frac{5}{11}$ minutes to fill the tub with both faucets open.

23. Let t represent the time to fill the sink with both the faucet and the drain left open. The equation is:

$$\frac{1}{3}-\frac{1}{4}=\frac{1}{t}$$

$$12t\left(\frac{1}{3}-\frac{1}{4}\right)=12t\cdot\frac{1}{t}$$

$$4t-3t=12$$

$$t=12$$

It will take 12 minutes for the sink to overflow with both the faucet and drain left open.

25.   Sketching the line graph:

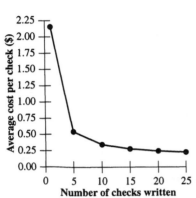

27.   Sketching the graph:

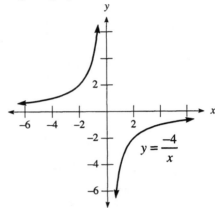

29.   Sketching the graph:

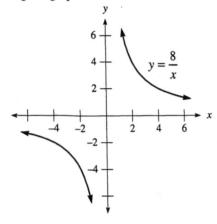

31.   Sketching the graph:

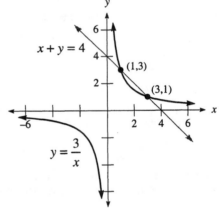

The intersection points are (1,3) and (3,1).

33.   Factoring the polynomial: $15a^3b^3 - 20a^2b - 35ab^2 = 5ab\left(3a^2b^2 - 4a - 7b\right)$

35.   Factoring the polynomial: $x^2 - 4x - 12 = (x - 6)(x + 2)$

37.   Factoring the polynomial: $x^4 - 16 = \left(x^2 + 4\right)\left(x^2 - 4\right) = \left(x^2 + 4\right)(x + 2)(x - 2)$

39.   Factoring the polynomial: $5x^3 - 25x^2 - 30x = 5x\left(x^2 - 5x - 6\right) = 5x(x - 6)(x + 1)$

41.   Solving the equation by factoring:

$$x^2 - 6x = 0$$
$$x(x - 6) = 0$$
$$x = 0, 6$$

43.   Solving the equation by factoring:
$$x(x + 2) = 80$$
$$x^2 + 2x = 80$$
$$x^2 + 2x - 80 = 0$$
$$(x + 10)(x - 8) = 0$$
$$x = -10, 8$$

45. Let x and $x + 3$ represent the two integers. The equation is:

$$x^2 + (x+3)^2 = 15^2$$
$$x^2 + x^2 + 6x + 9 = 225$$
$$2x^2 + 6x - 216 = 0$$
$$x^2 + 3x - 108 = 0$$
$$(x+12)(x-9) = 0$$
$$x = 9 \quad (x = -12 \text{ is impossible})$$
$$x + 3 = 12$$

The two legs are 9 inches and 12 inches.

6.6 Complex Fractions

1. Simplifying the complex fraction: $\dfrac{\frac{3}{4}}{\frac{7}{8}} = \dfrac{\frac{3}{4} \cdot 8}{\frac{7}{8} \cdot 8} = \dfrac{6}{1} = 6$

3. Simplifying the complex fraction: $\dfrac{\frac{2}{3}}{4} = \dfrac{\frac{2}{3} \cdot 3}{4 \cdot 3} = \dfrac{2}{12} = \dfrac{1}{6}$

5. Simplifying the complex fraction: $\dfrac{\frac{x^2}{y}}{\frac{x}{y^3}} = \dfrac{\frac{x^2}{y} \cdot y^3}{\frac{x}{y^3} \cdot y^3} = \dfrac{x^2 y^2}{x} = xy^2$

7. Simplifying the complex fraction: $\dfrac{\frac{4x^3}{y^6}}{\frac{8x^2}{y^7}} = \dfrac{\frac{4x^3}{y^6} \cdot y^7}{\frac{8x^2}{y^7} \cdot y^7} = \dfrac{4x^3 y}{8x^2} = \dfrac{xy}{2}$

9. Simplifying the complex fraction: $\dfrac{y + \frac{1}{x}}{x + \frac{1}{y}} = \dfrac{\left(y + \frac{1}{x}\right) \cdot xy}{\left(x + \frac{1}{y}\right) \cdot xy} = \dfrac{xy^2 + y}{x^2 y + x} = \dfrac{y(xy+1)}{x(xy+1)} = \dfrac{y}{x}$

11. Simplifying the complex fraction: $\dfrac{1 + \frac{1}{a}}{1 - \frac{1}{a}} = \dfrac{\left(1 + \frac{1}{a}\right) \cdot a}{\left(1 - \frac{1}{a}\right) \cdot a} = \dfrac{a+1}{a-1}$

13. Simplifying the complex fraction: $\dfrac{\frac{x+1}{x^2-9}}{\frac{2}{x+3}} = \dfrac{\frac{x+1}{(x+3)(x-3)} \cdot (x+3)(x-3)}{\frac{2}{x+3} \cdot (x+3)(x-3)} = \dfrac{x+1}{2(x-3)}$

15. Simplifying the complex fraction: $\dfrac{\frac{1}{a+2}}{\frac{1}{a^2-a-6}} = \dfrac{\frac{1}{a+2} \cdot (a-3)(a+2)}{\frac{1}{(a-3)(a+2)} \cdot (a-3)(a+2)} = \dfrac{a-3}{1} = a-3$

17. Simplifying the complex fraction: $\dfrac{1 - \frac{9}{y^2}}{1 - \frac{1}{y} - \frac{6}{y^2}} = \dfrac{\left(1 - \frac{9}{y^2}\right) \cdot y^2}{\left(1 - \frac{1}{y} - \frac{6}{y^2}\right) \cdot y^2} = \dfrac{y^2 - 9}{y^2 - y - 6} = \dfrac{(y+3)(y-3)}{(y+2)(y-3)} = \dfrac{y+3}{y+2}$

19. Simplifying the complex fraction: $\dfrac{\dfrac{1}{y}+\dfrac{1}{x}}{\dfrac{1}{xy}} = \dfrac{\left(\dfrac{1}{y}+\dfrac{1}{x}\right)\bullet xy}{\left(\dfrac{1}{xy}\right)\bullet xy} = \dfrac{x+y}{1} = x+y$

21. Simplifying the complex fraction: $\dfrac{1-\dfrac{1}{a^2}}{1-\dfrac{1}{a}} = \dfrac{\left(1-\dfrac{1}{a^2}\right)\bullet a^2}{\left(1-\dfrac{1}{a}\right)\bullet a^2} = \dfrac{a^2-1}{a^2-a} = \dfrac{(a+1)(a-1)}{a(a-1)} = \dfrac{a+1}{a}$

23. Simplifying the complex fraction: $\dfrac{\dfrac{1}{10x}-\dfrac{y}{10x^2}}{\dfrac{1}{10}-\dfrac{y}{10x}} = \dfrac{\left(\dfrac{1}{10x}-\dfrac{y}{10x^2}\right)\bullet 10x^2}{\left(\dfrac{1}{10}-\dfrac{y}{10x}\right)\bullet 10x^2} = \dfrac{x-y}{x^2-xy} = \dfrac{1(x-y)}{x(x-y)} = \dfrac{1}{x}$

25. Simplifying the complex fraction: $\dfrac{\dfrac{1}{a+1}+2}{\dfrac{1}{a+1}+3} = \dfrac{\left(\dfrac{1}{a+1}+2\right)\bullet(a+1)}{\left(\dfrac{1}{a+1}+3\right)\bullet(a+1)} = \dfrac{1+2(a+1)}{1+3(a+1)} = \dfrac{1+2a+2}{1+3a+3} = \dfrac{2a+3}{3a+4}$

27. Simplifying each parenthesis first:

$$1-\dfrac{1}{x} = \dfrac{x}{x}-\dfrac{1}{x} = \dfrac{x-1}{x}$$

$$1-\dfrac{1}{x+1} = \dfrac{x+1}{x+1}-\dfrac{1}{x+1} = \dfrac{x}{x+1}$$

$$1-\dfrac{1}{x+2} = \dfrac{x+2}{x+2}-\dfrac{1}{x+2} = \dfrac{x+1}{x+2}$$

Now performing the multiplication: $\left(1-\dfrac{1}{x}\right)\left(1-\dfrac{1}{x+1}\right)\left(1-\dfrac{1}{x+2}\right) = \dfrac{x-1}{x}\bullet\dfrac{x}{x+1}\bullet\dfrac{x+1}{x+2} = \dfrac{x-1}{x+2}$

29. Simplifying each parenthesis first:

$$1+\dfrac{1}{x+3} = \dfrac{x+3}{x+3}+\dfrac{1}{x+3} = \dfrac{x+4}{x+3}$$

$$1+\dfrac{1}{x+2} = \dfrac{x+2}{x+2}+\dfrac{1}{x+2} = \dfrac{x+3}{x+2}$$

$$1+\dfrac{1}{x+1} = \dfrac{x+1}{x+1}+\dfrac{1}{x+1} = \dfrac{x+2}{x+1}$$

Now performing the multiplication: $\left(1+\dfrac{1}{x+3}\right)\left(1+\dfrac{1}{x+2}\right)\left(1+\dfrac{1}{x+1}\right) = \dfrac{x+4}{x+3}\bullet\dfrac{x+3}{x+2}\bullet\dfrac{x+2}{x+1} = \dfrac{x+4}{x+1}$

31. Simplifying each term in the sequence:

$$2+\dfrac{1}{2+1} = 2+\dfrac{1}{3} = \dfrac{6}{3}+\dfrac{1}{3} = \dfrac{7}{3}$$

$$2+\dfrac{1}{2+\dfrac{1}{2+1}} = 2+\dfrac{1}{\dfrac{7}{3}} = 2+\dfrac{3}{7} = \dfrac{14}{7}+\dfrac{3}{7} = \dfrac{17}{7}$$

$$2+\dfrac{2}{2+\dfrac{1}{2+\dfrac{1}{2+1}}} = 2+\dfrac{1}{\dfrac{17}{7}} = 2+\dfrac{7}{17} = \dfrac{34}{17}+\dfrac{7}{17} = \dfrac{41}{17}$$

33. Completing the table:

Number x	Reciprocal $\frac{1}{x}$	Quotient $\frac{x}{1/x}$	Square x^2
1	1	1	1
2	$\frac{1}{2}$	4	4
3	$\frac{1}{3}$	9	9
4	$\frac{1}{4}$	16	16

35. Completing the table:

Number x	Reciprocal $\frac{1}{x}$	Sum $1+\frac{1}{x}$	Quotient $\dfrac{1+\frac{1}{x}}{\frac{1}{x}}$
1	1	2	2
2	$\frac{1}{2}$	$\frac{3}{2}$	3
3	$\frac{1}{3}$	$\frac{4}{3}$	4
4	$\frac{1}{4}$	$\frac{5}{4}$	5

37. Solving the inequality:
$$2x+3<5$$
$$2x+3-3<5-3$$
$$2x<2$$
$$\tfrac{1}{2}(2x)<\tfrac{1}{2}(2)$$
$$x<1$$

39. Solving the inequality:
$$-3x\le 21$$
$$-\tfrac{1}{3}(-3x)\ge -\tfrac{1}{3}(21)$$
$$x\ge -7$$

41. Solving the inequality:
$$-2x+8>-4$$
$$-2x+8-8>-4-8$$
$$-2x>-12$$
$$-\tfrac{1}{2}(-2x)<-\tfrac{1}{2}(-12)$$
$$x<6$$

43. Solving the inequality:
$$4-2(x+1)\ge -2$$
$$4-2x-2\ge -2$$
$$-2x+2\ge -2$$
$$-2x+2-2\ge -2-2$$
$$-2x\ge -4$$
$$-\tfrac{1}{2}(-2x)\le -\tfrac{1}{2}(-4)$$
$$x\le 2$$

6.7 Proportions

1. Solving the proportion:
$$\frac{x}{2}=\frac{6}{12}$$
$$12x=12$$
$$x=1$$

3. Solving the proportion:
$$\frac{2}{5}=\frac{4}{x}$$
$$2x=20$$
$$x=10$$

5. Solving the proportion:
$$\frac{10}{20}=\frac{20}{x}$$
$$10x=400$$
$$x=40$$

7. Solving the proportion:
$$\frac{a}{3}=\frac{5}{12}$$
$$12a=15$$
$$a=\frac{15}{12}=\frac{5}{4}$$

9. Solving the proportion:
$$\frac{2}{x}=\frac{6}{7}$$
$$6x=14$$
$$x=\frac{14}{6}=\frac{7}{3}$$

11. Solving the proportion:
$$\frac{x+1}{3}=\frac{4}{x}$$
$$x^2+x=12$$
$$x^2+x-12=0$$
$$(x+4)(x-3)=0$$
$$x=-4,3$$

13.   Solving the proportion:
$$\frac{x}{2} = \frac{8}{x}$$
$$x^2 = 16$$
$$x^2 - 16 = 0$$
$$(x+4)(x-4) = 0$$
$$x = -4, 4$$

15.   Solving the proportion:
$$\frac{4}{a+2} = \frac{a}{2}$$
$$a^2 + 2a = 8$$
$$a^2 + 2a - 8 = 0$$
$$(a+4)(a-2) = 0$$
$$a = -4, 2$$

17.   Solving the proportion:
$$\frac{1}{x} = \frac{x-5}{6}$$
$$x^2 - 5x = 6$$
$$x^2 - 5x - 6 = 0$$
$$(x-6)(x+1) = 0$$
$$x = -1, 6$$

19.   Comparing hits to games, the proportion is:
$$\frac{6}{18} = \frac{x}{45}$$
$$18x = 270$$
$$x = 15$$
He will get 15 hits in 45 games.

21.   Comparing ml alcohol to ml water, the proportion is:
$$\frac{12}{16} = \frac{x}{28}$$
$$16x = 336$$
$$x = 21$$
The solution will have 21 ml of alcohol.

23.   Comparing grams of fat to total grams, the proportion is:
$$\frac{13}{100} = \frac{x}{350}$$
$$100x = 4550$$
$$x = 45.5$$
There are 45.5 grams of fat in 350 grams of ice cream.

25.   Comparing inches on the map to actual miles, the proportion is:
$$\frac{3.5}{100} = \frac{x}{420}$$
$$100x = 1470$$
$$x = 14.7$$
They are 14.7 inches apart on the map.

27.   Comparing miles to hours, the proportion is:
$$\frac{245}{5} = \frac{x}{7}$$
$$5x = 1715$$
$$x = 343$$
He will travel 343 miles.

29.   Reducing the fraction: $\dfrac{x^2 - x - 6}{x^2 - 9} = \dfrac{(x-3)(x+2)}{(x+3)(x-3)} = \dfrac{x+2}{x+3}$

31.   Multiplying the fractions:
$$\frac{x^2 - 25}{x+4} \cdot \frac{2x+8}{x^2 - 9x + 20} = \frac{(x+5)(x-5)}{x+4} \cdot \frac{2(x+4)}{(x-5)(x-4)} = \frac{2(x+5)(x-5)(x+4)}{(x+4)(x-5)(x-4)} = \frac{2(x+5)}{x-4}$$

33.   Adding the fractions: $\dfrac{x}{x^2-16} + \dfrac{4}{x^2-16} = \dfrac{x+4}{x^2-16} = \dfrac{1(x+4)}{(x+4)(x-4)} = \dfrac{1}{x-4}$

Chapter 6 Review

1.   Reducing the rational expression: $\dfrac{7}{14x-28} = \dfrac{7}{14(x-2)} = \dfrac{1}{2(x-2)}$. The variable restriction is $x \neq 2$.

3.   Reducing the rational expression: $\dfrac{8x-4}{4x+12} = \dfrac{4(2x-1)}{4(x+3)} = \dfrac{2x-1}{x+3}$. The variable restriction is $x \neq -3$.

5.   Reducing the rational expression: $\dfrac{3x^3+16x^2-12x}{2x^3+9x^2-18x} = \dfrac{x\left(3x^2+16x-12\right)}{x\left(2x^2+9x-18\right)} = \dfrac{x(3x-2)(x+6)}{x(2x-3)(x+6)} = \dfrac{3x-2}{2x-3}$

   The variable restriction is $x \neq -6, \frac{3}{2}$.

7.   Reducing the rational expression: $\dfrac{x^2+5x-14}{x+7} = \dfrac{(x+7)(x-2)}{x+7} = x-2$. The variable restriction is $x \neq -7$.

9.   Reducing the rational expression: $\dfrac{xy+bx+ay+ab}{xy+5x+ay+5a} = \dfrac{x(y+b)+a(y+b)}{x(y+5)+a(y+5)} = \dfrac{(y+b)(x+a)}{(y+5)(x+a)} = \dfrac{y+b}{y+5}$

   The variable restriction is $y \neq -5, x \neq -a$.

11. Performing the operations:

$$\dfrac{x^2+8x+16}{x^2+x-12} \div \dfrac{x^2-16}{x^2-x-6} = \dfrac{x^2+8x+16}{x^2+x-12} \cdot \dfrac{x^2-x-6}{x^2-16}$$

$$= \dfrac{(x+4)^2}{(x+4)(x-3)} \cdot \dfrac{(x+2)(x-3)}{(x+4)(x-4)}$$

$$= \dfrac{(x+4)^2(x+2)(x-3)}{(x+4)^2(x-3)(x-4)}$$

$$= \dfrac{x+2}{x-4}$$

13. Performing the operations:

$$\dfrac{3x^2-2x-1}{x^2+6x+8} \div \dfrac{3x^2+13x+4}{x^2+8x+16} = \dfrac{3x^2-2x-1}{x^2+6x+8} \cdot \dfrac{x^2+8x+16}{3x^2+13x+4}$$

$$= \dfrac{(3x+1)(x-1)}{(x+4)(x+2)} \cdot \dfrac{(x+4)^2}{(3x+1)(x+4)}$$

$$= \dfrac{(x+4)^2(3x+1)(x-1)}{(x+4)^2(x+2)(3x+1)}$$

$$= \dfrac{x-1}{x+2}$$

15. Performing the operations: $\dfrac{x^2}{x-9} - \dfrac{18x-81}{x-9} = \dfrac{x^2-18x+81}{x-9} = \dfrac{(x-9)^2}{x-9} = x-9$

17. Performing the operations: $\dfrac{x}{x+9} + \dfrac{5}{x} = \dfrac{x \cdot x}{(x+9)\cdot x} + \dfrac{5\cdot(x+9)}{x\cdot(x+9)} = \dfrac{x^2}{x(x+9)} + \dfrac{5x+45}{x(x+9)} = \dfrac{x^2+5x+45}{x(x+9)}$

19. Performing the operations:

$$\frac{3}{x^2-36}-\frac{2}{x^2-4x-12}=\frac{3}{(x+6)(x-6)}-\frac{2}{(x-6)(x+2)}$$

$$=\frac{3(x+2)}{(x+6)(x-6)(x+2)}-\frac{2(x+6)}{(x+6)(x-6)(x+2)}$$

$$=\frac{3x+6}{(x+6)(x-6)(x+2)}-\frac{2x+12}{(x+6)(x-6)(x+2)}$$

$$=\frac{3x+6-2x-12}{(x+6)(x-6)(x+2)}$$

$$=\frac{x-6}{(x+6)(x-6)(x+2)}$$

$$=\frac{1}{(x+6)(x+2)}$$

21. Multiplying both sides of the equation by $2x$:

$$2x\left(\frac{3}{x}+\frac{1}{2}\right)=2x\left(\frac{5}{x}\right)$$

$$6+x=10$$

$$x=4$$

Since $x=4$ checks in the original equation, the solution is $x=4$.

23. Multiplying both sides of the equation by x^2:

$$x^2\left(1-\frac{7}{x}\right)=x^2\left(\frac{-6}{x^2}\right)$$

$$x^2-7x=-6$$

$$x^2-7x+6=0$$

$$(x-6)(x-1)=0$$

$$x=1,6$$

Both $x=1$ and $x=6$ check in the original equation.

25. Since $y^2-16=(y+4)(y-4)$ and $y^2+4y=y(y+4)$, multiply each side of the equation by $y(y+4)(y-4)$:

$$y(y+4)(y-4)\cdot\frac{2}{(y+4)(y-4)}=y(y+4)(y-4)\cdot\frac{10}{y(y+4)}$$

$$2y=10(y-4)$$

$$2y=10y-40$$

$$-8y=-40$$

$$y=5$$

Since $y=5$ checks in the original equation, the solution is $y=5$.

27. Let x represent the speed of the boat in still water. Completing the table:

	d	r	t
Upstream	48	$x-3$	$\dfrac{48}{x-3}$
Downstream	72	$x+3$	$\dfrac{72}{x+3}$

The equation is:
$$\frac{48}{x-3} = \frac{72}{x+3}$$
$$(x+3)(x-3) \cdot \frac{48}{x-3} = (x+3)(x-3) \cdot \frac{72}{x+3}$$
$$48(x+3) = 72(x-3)$$
$$48x+144 = 72x-216$$
$$-24x+144 = -216$$
$$-24x = -360$$
$$x = 15$$
The speed of the boat in still water is 15 mph.

29. Simplifying the complex fraction: $\dfrac{\dfrac{x+4}{x^2-16}}{\dfrac{2}{x-4}} = \dfrac{\dfrac{x+4}{(x+4)(x-4)}}{\dfrac{2}{x-4}} = \dfrac{\dfrac{1}{x-4} \cdot (x-4)}{\dfrac{2}{x-4} \cdot (x-4)} = \dfrac{1}{2}$

31. Simplifying the complex fraction: $\dfrac{\dfrac{1}{a-2}+4}{\dfrac{1}{a-2}+1} = \dfrac{\left(\dfrac{1}{a-2}+4\right)(a-2)}{\left(\dfrac{1}{a-2}+1\right)(a-2)} = \dfrac{1+4(a-2)}{1+1(a-2)} = \dfrac{1+4a-8}{1+a-2} = \dfrac{4a-7}{a-1}$

33. Writing as a fraction: $\dfrac{40 \text{ seconds}}{3 \text{ minutes}} = \dfrac{40 \text{ seconds}}{180 \text{ seconds}} = \dfrac{2}{9}$

35. Solving the proportion:
$$\frac{a}{3} = \frac{12}{a}$$
$$a^2 = 36$$
$$a^2 - 36 = 0$$
$$(a+6)(a-6) = 0$$
$$a = -6, 6$$

Chapters 1-6 Cumulative Review

1. Simplifying the expression: $8-11 = 8+(-11) = -3$

3. Simplifying the expression: $\dfrac{-48}{12} = -4$

5. Simplifying the expression: $5x-4-9x = 5x-9x-4 = -4x-4$

7. Simplifying the expression: $9^{-2} = \dfrac{1}{9^2} = \dfrac{1}{81}$

9. Simplifying the expression: $4^1+9^0+(-7)^0 = 4+1+1 = 6$

11. Simplifying the expression:
$$\left(4a^3-10a^2+6\right)-\left(6a^3+5a-7\right) = 4a^3-10a^2+6-6a^3-5a+7 = -2a^3-10a^2-5a+13$$

13. Simplifying the expression: $\dfrac{x^2}{x-7} - \dfrac{14x-49}{x-7} = \dfrac{x^2-14x+49}{x-7} = \dfrac{(x-7)^2}{x-7} = x-7$

15.   Simplifying the expression: $\dfrac{\dfrac{x-2}{x^2+6x+8}}{\dfrac{4}{x+4}} = \dfrac{\dfrac{x-2}{(x+4)(x+2)}\cdot(x+4)(x+2)}{\dfrac{4}{x+4}\cdot(x+4)(x+2)} = \dfrac{x-2}{4(x+2)}$

17.   Solving the equation:

$x - \frac{3}{4} = \frac{5}{6}$

$\qquad x = \frac{3}{4} + \frac{5}{6}$

$\qquad x = \frac{9}{12} + \frac{10}{12}$

$\qquad x = \frac{19}{12}$

19.   Solving the equation:

$98r^2 - 18 = 0$

$2\left(49r^2 - 9\right) = 0$

$2(7r+3)(7r-3) = 0$

$\qquad r = -\frac{3}{7}, \frac{3}{7}$

21.   Multiplying each side of the equation by $3x$:

$3x\left(\dfrac{5}{x} - \dfrac{1}{3}\right) = 3x\left(\dfrac{3}{x}\right)$

$\qquad 15 - x = 9$

$\qquad\quad -x = -6$

$\qquad\qquad x = 6$

Since $x = 6$ checks in the original equation, the solution is $x = 6$.

23.   Multiplying each side of the equation by $3(x-3)$:

$3(x-3)\cdot\dfrac{x}{3} = 3(x-3)\cdot\dfrac{6}{x-3}$

$\qquad x(x-3) = 18$

$\qquad x^2 - 3x = 18$

$\qquad x^2 - 3x - 18 = 0$

$\qquad (x-6)(x+3) = 0$

$\qquad\qquad x = 6, -3$

Both $x = -3$ and $x = 6$ check in the original equation.

25.   Multiplying the first equation by -5 and the second equation by 3:

$-45x - 70y = 20$

$\;\;\;45x - 24y = 27$

Adding the two equations:

$-94y = 47$

$\qquad y = -\frac{1}{2}$

Substituting into the first equation:

$9x + 14\left(-\frac{1}{2}\right) = -4$

$\qquad 9x - 7 = -4$

$\qquad\quad 9x = 3$

$\qquad\qquad x = \frac{1}{3}$

The solution is $\left(\frac{1}{3}, -\frac{1}{2}\right)$.

27.   To clear each equation of fractions, multiply the first equation by 6 and the second equation by 12:

$6\left(\frac{1}{2}x + \frac{1}{3}y\right) = 6(-1)$ $12\left(\frac{1}{3}x\right) = 12\left(\frac{1}{4}y + 5\right)$

$\qquad 3x + 2y = -6$ $\qquad 4x = 3y + 60$

$\qquad\qquad\qquad\qquad\qquad\qquad 4x - 3y = 60$

The system of equations is:

$3x + 2y = -6$

$4x - 3y = 60$

Multiplying the first equation by 3 and the second equation by 2:

$9x + 6y = -18$

$8x - 6y = 120$

Adding the two equations:
$$17x = 102$$
$$x = 6$$
Substituting into $3x + 2y = -6$:
$$3(6) + 2y = -6$$
$$18 + 2y = -6$$
$$2y = -24$$
$$y = -12$$
The solution is $(6, -12)$.

29. Graphing the equation:

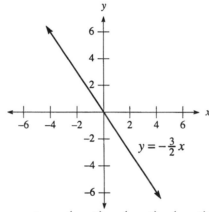

$$y = -\frac{3}{2}x$$

31. Factoring the polynomial: $xy + 5x + ay + 5a = x(y + 5) + a(y + 5) = (y + 5)(x + a)$

33. Factoring the polynomial: $20y^2 - 27y + 9 = (5y - 3)(4y - 3)$

35. Factoring the polynomial: $16x^2 + 72xy + 81y^2 = (4x + 9y)(4x + 9y) = (4x + 9y)^2$

37. Checking each ordered pair:
$$(3, -1): \quad 2(3) - 5 = 6 - 5 = 1 \neq -1$$
$$(1, -3): \quad 2(1) - 5 = 2 - 5 = -3$$
$$(-2, 9): \quad 2(-2) - 5 = -4 - 5 = -9 \neq 9$$

The ordered pair $(1, -3)$ is a solution to the equation.

39. The expression is: $(-3)(4) + (-2) = -12 + (-2) = -14$

41. The rational numbers are: -3, 0, $\frac{3}{4}$, 1.5

43. Using long division:

$$\begin{array}{r} x - 7 \\ x + 4 \overline{) x^2 - 3x - 28} \\ \underline{x^2 + 4x} \\ -7x - 28 \\ \underline{-7x - 28} \\ 0 \end{array}$$

The quotient is $x - 7$.

45. Solving for l:
$$2l + 2w = P$$
$$2l = P - 2w$$
$$l = \frac{P - 2w}{2}$$

47.   Let x represent the amount invested at 6% and y represent the amount invested at 8%. The system of equations is:
$$x + y = 18000$$
$$0.06x + 0.08y = 1290$$
Multiplying the first equation by -0.06:
$$-0.06x - 0.06y = -1080$$
$$0.06x + 0.08y = 1290$$
Adding the two equations:
$$0.02y = 210$$
$$y = 10500$$
Substituting into the first equation:
$$x + 10500 = 18000$$
$$x = 7500$$
Ms. Jones invested $7,500 at 6% and $10,500 at 8%.

Chapter 6 Test

1.   Reducing the rational expression: $\dfrac{x^2 - 16}{x^2 - 8x + 16} = \dfrac{(x+4)(x-4)}{(x-4)^2} = \dfrac{x+4}{x-4}$

2.   Reducing the rational expression: $\dfrac{10a + 20}{5a^2 + 20a + 20} = \dfrac{10(a+2)}{5(a^2 + 4a + 4)} = \dfrac{10(a+2)}{5(a+2)^2} = \dfrac{2}{a+2}$

3.   Reducing the rational expression: $\dfrac{xy + 7x + 5y + 35}{x^2 + ax + 5x + 5a} = \dfrac{x(y+7) + 5(y+7)}{x(x+a) + 5(x+a)} = \dfrac{(y+7)(x+5)}{(x+a)(x+5)} = \dfrac{y+7}{x+a}$

4.   Performing the operations: $\dfrac{3x - 12}{4} \cdot \dfrac{8}{2x - 8} = \dfrac{3(x-4)}{4} \cdot \dfrac{8}{2(x-4)} = \dfrac{24(x-4)}{8(x-4)} = 3$

5.   Performing the operations:
$$\dfrac{x^2 - 49}{x+1} \div \dfrac{x+7}{x^2 - 1} = \dfrac{x^2 - 49}{x+1} \cdot \dfrac{x^2 - 1}{x+7}$$
$$= \dfrac{(x+7)(x-7)}{x+1} \cdot \dfrac{(x+1)(x-1)}{x+7}$$
$$= \dfrac{(x+7)(x-7)(x+1)(x-1)}{(x+1)(x+7)}$$
$$= (x-7)(x-1)$$

6.   Performing the operations:
$$\dfrac{x^2 - 3x - 10}{x^2 - 8x + 15} \div \dfrac{3x^2 + 2x - 8}{x^2 + x - 12} = \dfrac{x^2 - 3x - 10}{x^2 - 8x + 15} \cdot \dfrac{x^2 + x - 12}{3x^2 + 2x - 8}$$
$$= \dfrac{(x-5)(x+2)}{(x-5)(x-3)} \cdot \dfrac{(x+4)(x-3)}{(3x-4)(x+2)}$$
$$= \dfrac{(x-5)(x+2)(x+4)(x-3)}{(x-5)(x-3)(3x-4)(x+2)}$$
$$= \dfrac{x+4}{3x-4}$$

7.   Performing the operations: $\left(x^2 - 9\right)\left(\dfrac{x+2}{x+3}\right) = \dfrac{(x+3)(x-3)}{1} \cdot \dfrac{x+2}{x+3} = \dfrac{(x+3)(x-3)(x+2)}{x+3} = (x-3)(x+2)$

8.   Performing the operations: $\dfrac{3}{x-2} - \dfrac{6}{x-2} = \dfrac{3-6}{x-2} = \dfrac{-3}{x-2}$

9.   Performing the operations:
$$\frac{x}{x^2-9}+\frac{4}{4x-12}=\frac{x}{(x+3)(x-3)}+\frac{4}{4(x-3)}$$
$$=\frac{x}{(x+3)(x-3)}+\frac{1\cdot(x+3)}{(x+3)(x-3)}$$
$$=\frac{x}{(x+3)(x-3)}+\frac{x+3}{(x+3)(x-3)}$$
$$=\frac{2x+3}{(x+3)(x-3)}$$

10.   Performing the operations:
$$\frac{2x}{x^2-1}+\frac{x}{x^2-3x+2}=\frac{2x}{(x+1)(x-1)}+\frac{x}{(x-1)(x-2)}$$
$$=\frac{2x\cdot(x-2)}{(x+1)(x-1)(x-2)}+\frac{x\cdot(x+1)}{(x+1)(x-1)(x-2)}$$
$$=\frac{2x^2-4x}{(x+1)(x-1)(x-2)}+\frac{x^2+x}{(x+1)(x-1)(x-2)}$$
$$=\frac{3x^2-3x}{(x+1)(x-1)(x-2)}$$
$$=\frac{3x(x-1)}{(x+1)(x-1)(x-2)}$$
$$=\frac{3x}{(x+1)(x-2)}$$

11.   Multiplying both sides of the equation by 15:
$$15\cdot\frac{7}{5}=15\cdot\frac{x+2}{3}$$
$$21=5(x+2)$$
$$21=5x+10$$
$$11=5x$$
$$x=\frac{11}{5}$$
Since $x=\frac{11}{5}$ checks in the original equation, the solution is $x=\frac{11}{5}$.

12.   Multiplying both sides of the equation by $x(x+4)$:
$$x(x+4)\cdot\frac{10}{x+4}=x(x+4)\cdot\left(\frac{6}{x}-\frac{4}{x}\right)$$
$$10x=6(x+4)-4(x+4)$$
$$10x=6x+24-4x-16$$
$$10x=2x+8$$
$$8x=8$$
$$x=1$$
Since $x=1$ checks in the original equation, the solution is $x=1$.

13.   Multiplying both sides of the equation by $x^2-x-2=(x-2)(x+1)$:
$$(x-2)(x+1)\left(\frac{3}{x-2}-\frac{4}{x+1}\right)=(x-2)(x+1)\cdot\frac{5}{(x-2)(x+1)}$$
$$3(x+1)-4(x-2)=5$$
$$3x+3-4x+8=5$$
$$-x+11=5$$
$$-x=-6$$
$$x=6$$
Since $x=6$ checks in the original equation, the solution is $x=6$.

14. Let x represent the speed of the boat in still water. Completing the table:

	d	r	t
Upstream	26	$x-2$	$\dfrac{26}{x-2}$
Downstream	34	$x+2$	$\dfrac{34}{x+2}$

The equation is:
$$\frac{26}{x-2} = \frac{34}{x+2}$$
$$(x+2)(x-2)\cdot\frac{26}{x-2} = (x+2)(x-2)\cdot\frac{34}{x+2}$$
$$26(x+2) = 34(x-2)$$
$$26x+52 = 34x-68$$
$$-8x+52 = -68$$
$$-8x = -120$$
$$x = 15$$
The speed of the boat in still water is 15 mph.

15. Let t represent the time to empty the pool with both pipes open. The equation is:
$$\tfrac{1}{12} - \tfrac{1}{15} = \frac{1}{t}$$
$$60t\left(\tfrac{1}{12} - \tfrac{1}{15}\right) = 60t\cdot\frac{1}{t}$$
$$5t - 4t = 60$$
$$t = 60$$
It will take 60 hours to empty the pool with both pipes open.

16. The ratio of alcohol to water is given by: $\dfrac{27\text{ ml}}{54\text{ ml}} = \tfrac{1}{2}$

The ratio of alcohol to total volume is given by: $\dfrac{27\text{ ml}}{81\text{ ml}} = \tfrac{1}{3}$

17. Comparing defective parts to total parts, the proportion is:
$$\frac{8}{100} = \frac{x}{1650}$$
$$100x = 13200$$
$$x = 132$$
The machine can be expected to produce 132 defective parts.

18. Simplifying the complex fraction: $\dfrac{\frac{3}{4}}{\frac{7}{8}} = \dfrac{3}{4}\cdot\dfrac{8}{7} = \dfrac{6}{7}$

19. Simplifying the complex fraction: $\dfrac{1+\frac{1}{x}}{1-\frac{1}{x}} = \dfrac{\left(1+\frac{1}{x}\right)\cdot x}{\left(1-\frac{1}{x}\right)\cdot x} = \dfrac{x+1}{x-1}$

20. Simplifying the complex fraction: $\dfrac{1-\frac{16}{x^2}}{1-\frac{2}{x}-\frac{8}{x^2}} = \dfrac{\left(1-\frac{16}{x^2}\right)\cdot x^2}{\left(1-\frac{2}{x}-\frac{8}{x^2}\right)\cdot x^2} = \dfrac{x^2-16}{x^2-2x-8} = \dfrac{(x+4)(x-4)}{(x-4)(x+2)} = \dfrac{x+4}{x+2}$

Chapter 7
Transitions

7.1   Review of Solving Equations

1.   Solving the equation:
$$2x - 4 = 6$$
$$2x = 10$$
$$x = \frac{10}{2} = 5$$

3.   Solving the equation:
$$-300y + 100 = 500$$
$$-300y = 400$$
$$y = -\frac{4}{3}$$

5.   Solving the equation:
$$-x = 2$$
$$x = -1 \cdot 2 = -2$$

7.   Solving the equation:
$$-a = -\frac{3}{4}$$
$$a = -1 \cdot \left(-\frac{3}{4}\right) = \frac{3}{4}$$

9.   Solving the equation:
$$-\frac{3}{5}a + 2 = 8$$
$$-\frac{3}{5}a = 6$$
$$a = -\frac{5}{3} \cdot 6 = -10$$

11.   Solving the equation:
$$2x - 5 = 3x + 2$$
$$-x - 5 = 2$$
$$-x = 7$$
$$x = -7$$

13.   Solving the equation:
$$5 - 2x = 3x + 1$$
$$5 - 5x = 1$$
$$-5x = -4$$
$$x = \frac{4}{5}$$

15.   Solving the equation:
$$5(y + 2) - 4(y + 1) = 3$$
$$5y + 10 - 4y - 4 = 3$$
$$y + 6 = 3$$
$$y = -3$$

17.   Solving the equation:
$$6 - 7(m - 3) = -1$$
$$6 - 7m + 21 = -1$$
$$-7m + 27 = -1$$
$$-7m = -28$$
$$m = 4$$

19.   Solving the equation:
$$7 + 3(x + 2) = 4(x - 1)$$
$$7 + 3x + 6 = 4x - 4$$
$$3x + 13 = 4x - 4$$
$$-x + 13 = -4$$
$$-x = -17$$
$$x = 17$$

21. Solving the equation:
$$\tfrac{1}{2}x + \tfrac{1}{4} = \tfrac{1}{3}x + \tfrac{5}{4}$$
$$12\left(\tfrac{1}{2}x + \tfrac{1}{4}\right) = 12\left(\tfrac{1}{3}x + \tfrac{5}{4}\right)$$
$$6x + 3 = 4x + 15$$
$$2x + 3 = 15$$
$$2x = 12$$
$$x = 6$$

23. Solving the equation:
$$\tfrac{1}{2}x + \tfrac{1}{3}x + \tfrac{1}{4}x = 13$$
$$12\left(\tfrac{1}{2}x + \tfrac{1}{3}x + \tfrac{1}{4}x\right) = 12(13)$$
$$6x + 4x + 3x = 156$$
$$13x = 156$$
$$x = 12$$

25. Solving the equation:
$$0.08x + 0.09(9,000 - x) = 750$$
$$0.08x + 810 - 0.09x = 750$$
$$-0.01x + 810 = 750$$
$$-0.01x = -60$$
$$x = 6,000$$

27. Solving the equation:
$$0.35x - 0.2 = 0.15x + 0.1$$
$$0.2x - 0.2 = 0.1$$
$$0.2x = 0.3$$
$$2x = 3$$
$$x = \tfrac{3}{2}$$

29. Solving the equation:
$$x^2 - 5x - 6 = 0$$
$$(x + 1)(x - 6) = 0$$
$$x = -1, 6$$

31. Solving the equation:
$$x^3 - 5x^2 + 6x = 0$$
$$x\left(x^2 - 5x + 6\right) = 0$$
$$x(x - 2)(x - 3) = 0$$
$$x = 0, 2, 3$$

33. Solving the equation:
$$3y^2 + 11y - 4 = 0$$
$$(3y - 1)(y + 4) = 0$$
$$y = -4, \tfrac{1}{3}$$

35. Solving the equation:
$$\tfrac{1}{10}t^2 - \tfrac{5}{2} = 0$$
$$10\left(\tfrac{1}{10}t^2 - \tfrac{5}{2}\right) = 10(0)$$
$$t^2 - 25 = 0$$
$$(t + 5)(t - 5) = 0$$
$$t = -5, 5$$

37. Solving the equation:
$$\tfrac{1}{5}y^2 - 2 = -\tfrac{3}{10}y$$
$$10\left(\tfrac{1}{5}y^2 - 2\right) = 10\left(-\tfrac{3}{10}y\right)$$
$$2y^2 - 20 = -3y$$
$$2y^2 + 3y - 20 = 0$$
$$(y + 4)(2y - 5) = 0$$
$$y = -4, \tfrac{5}{2}$$

39. Solving the equation:
$$9x^2 - 12x = 0$$
$$3x(3x - 4) = 0$$
$$x = 0, \tfrac{4}{3}$$

41. Solving the equation:
$$0.02r + 0.01 = 0.15r^2$$
$$2r + 1 = 15r^2$$
$$15r^2 - 2r - 1 = 0$$
$$(5r + 1)(3r - 1) = 0$$
$$r = -\tfrac{1}{5}, \tfrac{1}{3}$$

43. Solving the equation:
$$9a^3 = 16a$$
$$9a^3 - 16a = 0$$
$$a\left(9a^2 - 16\right) = 0$$
$$a(3a + 4)(3a - 4) = 0$$
$$a = -\tfrac{4}{3}, 0, \tfrac{4}{3}$$

45. Solving the equation:
$$-100x = 10x^2$$
$$0 = 10x^2 + 100x$$
$$0 = 10x(x + 10)$$
$$x = -10, 0$$

47. Solving the equation:
$$(x + 6)(x - 2) = -7$$
$$x^2 + 4x - 12 = -7$$
$$x^2 + 4x - 5 = 0$$
$$(x + 5)(x - 1) = 0$$
$$x = -5, 1$$

49. Solving the equation:
$$(x+1)^2 = 3x+7$$
$$x^2 + 2x + 1 = 3x + 7$$
$$x^2 - x - 6 = 0$$
$$(x+2)(x-3) = 0$$
$$x = -2, 3$$

51. Solving the equation:
$$x^3 + 3x^2 - 4x - 12 = 0$$
$$x^2(x+3) - 4(x+3) = 0$$
$$(x+3)(x^2 - 4) = 0$$
$$(x+3)(x+2)(x-2) = 0$$
$$x = -3, -2, 2$$

53. Solving the equation:
$$2x^3 + 3x^2 - 8x - 12 = 0$$
$$x^2(2x+3) - 4(2x+3) = 0$$
$$(2x+3)(x^2 - 4) = 0$$
$$(2x+3)(x+2)(x-2) = 0$$
$$x = -2, -\tfrac{3}{2}, 2$$

55. Solving the equation:
$$3x - 6 = 3(x+4)$$
$$3x - 6 = 3x + 12$$
$$-6 = 12$$
Since this statement is false, there is no solution ($\emptyset$).

57. Solving the equation:
$$4y + 2 - 3y + 5 = 3 + y + 4$$
$$y + 7 = y + 7$$
$$7 = 7$$
Since this statement is true, the solution is all real numbers.

59. Solving the equation:
$$2(4t - 1) + 3 = 5t + 4 + 3t$$
$$8t - 2 + 3 = 8t + 4$$
$$8t + 1 = 8t + 4$$
$$1 = 4$$
Since this statement is false, there is no solution ($\emptyset$).

61. Let w represent the width and $2w$ represent the length. Using the perimeter formula:
$$2w + 2(2w) = 60$$
$$2w + 4w = 60$$
$$6w = 60$$
$$w = 10$$
The dimensions are 10 feet by 20 feet.

63. Let w represent the width and $2w$ represent the length. Using the perimeter formula:
$$2w + 2(2w) = 48$$
$$2w + 4w = 48$$
$$6w = 48$$
$$w = 8$$
The width is 8 feet and the length is 16 feet. Finding the cost: $C = 1.75(32) + 2.25(16) = 56 + 36 = 92$
The cost to build the pen is $92.00.

65. The total money collected is: $1204 - \$250 = \$954$
Let x represent the amount of her sales (not including tax). Since this amount includes the tax collected, the equation is:
$$x + 0.06x = 954$$
$$1.06x = 954$$
$$x = 900$$
Her sales were $900, so the sales tax is: $0.06(900) = \$54$

67. Completing the table:

Age (years)	Maximum Heart Rate (beats per minute)
18	202
19	201
20	200
21	199
22	198
23	197

69. Completing the table:

Resting Heart Rate (beats per minute)	Training Heart Rate (beats per minute)
60	144
62	145
64	146
68	147
70	148
72	149

71. Let x and $x + 2$ represent the two odd integers. The equation is:

$$x^2 + (x+2)^2 = 34$$
$$x^2 + x^2 + 4x + 4 = 34$$
$$2x^2 + 4x - 30 = 0$$
$$x^2 + 2x - 15 = 0$$
$$(x+5)(x-3) = 0$$
$$x = -5, 3$$
$$x + 2 = -3, 5$$

The integers are either –5 and –3, or 3 and 5.

73. Let h represent the height the ladder makes with the building. Using the Pythagorean theorem:

$$7^2 + h^2 = 25^2$$
$$49 + h^2 = 625$$
$$h^2 = 576$$
$$h = 24$$

The ladder reaches a height of 24 feet along the building.

75. Let w represent the width and $3w + 2$ represent the length. Using the area formula:

$$w(3w + 2) = 16$$
$$3w^2 + 2w = 16$$
$$3w^2 + 2w - 16 = 0$$
$$(3w+8)(w-2) = 0$$
$$w = 2 \quad (w = -\tfrac{8}{3} \text{ is impossible})$$

The dimensions are 2 feet by 8 feet.

77. Multiplying each side of the equation by 6:

$$6\left(\frac{x}{3} - \frac{1}{2}\right) = 6\left(\frac{5}{2}\right)$$
$$2x - 3 = 15$$
$$2x = 18$$
$$x = 9$$

Since $x = 9$ checks in the original equation, the solution is $x = 9$.

79. Multiplying each side of the equation by x^2:

$$x^2\left(1 - \frac{5}{x}\right) = x^2\left(\frac{-6}{x^2}\right)$$
$$x^2 - 5x = -6$$
$$x^2 - 5x + 6 = 0$$
$$(x-2)(x-3) = 0$$
$$x = 2, 3$$

Both $x = 2$ and $x = 3$ check in the original equation.

81.   Multiplying each side of the equation by $2(a-4)$:
$$2(a-4)\left(\frac{a}{a-4}-\frac{a}{2}\right)=2(a-4)\cdot\frac{4}{a-4}$$
$$2a-a(a-4)=8$$
$$2a-a^2+4a=8$$
$$-a^2+6a-8=0$$
$$a^2-6a+8=0$$
$$(a-4)(a-2)=0$$
$$a=2,4$$
Since $a=4$ does not check in the original equation, the solution is $a=2$.

7.2   Equations with Absolute Value

1.   Solving the equation:
$$|x|=4$$
$$x=-4,4$$

3.   Solving the equation:
$$2=|a|$$
$$a=-2,2$$

5.   The equation $|x|=-3$ has no solution, or $\varnothing$.

7.   Solving the equation:
$$|a|+2=3$$
$$|a|=1$$
$$a=-1,1$$

9.   Solving the equation:
$$|y|+4=3$$
$$|y|=-1$$
The equation $|y|=-1$ has no solution, or $\varnothing$.

11.   Solving the equation:
$$4=|x|-2$$
$$|x|=6$$
$$x=-6,6$$

13.   Solving the equation:
$$|x-2|=5$$
$$x-2=-5,5$$
$$x=-3,7$$

15.   Solving the equation:
$$|a-4|=\tfrac{5}{3}$$
$$a-4=-\tfrac{5}{3},\tfrac{5}{3}$$
$$a=\tfrac{7}{3},\tfrac{17}{3}$$

17.   Solving the equation:
$$1=|3-x|$$
$$3-x=-1,1$$
$$-x=-4,-2$$
$$x=2,4$$

19.   Solving the equation:
$$\left|\tfrac{3}{5}a+\tfrac{1}{2}\right|=1$$
$$\tfrac{3}{5}a+\tfrac{1}{2}=-1,1$$
$$\tfrac{3}{5}a=-\tfrac{3}{2},\tfrac{1}{2}$$
$$a=-\tfrac{5}{2},\tfrac{5}{6}$$

21.   Solving the equation:
$$60=|20x-40|$$
$$20x-40=-60,60$$
$$20x=-20,100$$
$$x=-1,5$$

23.   Since $|2x+1|=-3$ is impossible, there is no solution, or $\varnothing$.

25. Solving the equation:
$$\left|\tfrac{3}{4}x - 6\right| = 9$$
$$\tfrac{3}{4}x - 6 = -9, 9$$
$$\tfrac{3}{4}x = -3, 15$$
$$3x = -12, 60$$
$$x = -4, 20$$

27. Solving the equation:
$$\left|1 - \tfrac{1}{2}a\right| = 3$$
$$1 - \tfrac{1}{2}a = -3, 3$$
$$-\tfrac{1}{2}a = -4, 2$$
$$a = -4, 8$$

29. Solving the equation:
$$\left|3x + 4\right| + 1 = 7$$
$$\left|3x + 4\right| = 6$$
$$3x + 4 = -6, 6$$
$$3x = -10, 2$$
$$x = -\tfrac{10}{3}, \tfrac{2}{3}$$

31. Solving the equation:
$$\left|3 - 2y\right| + 4 = 3$$
$$\left|3 - 2y\right| = -1$$

Since this equation is impossible, there is no solution, or $\varnothing$.

33. Solving the equation:

$$3 + \left|4t - 1\right| = 8$$
$$\left|4t - 1\right| = 5$$
$$4t - 1 = -5, 5$$
$$4t = -4, 6$$
$$t = -1, \tfrac{3}{2}$$

35. Solving the equation:
$$\left|9 - \tfrac{3}{5}x\right| + 6 = 12$$
$$\left|9 - \tfrac{3}{5}x\right| = 6$$
$$9 - \tfrac{3}{5}x = -6, 6$$
$$-\tfrac{3}{5}x = -15, -3$$
$$-3x = -75, -15$$
$$x = 5, 25$$

37. Solving the equation:

$$5 = \left|\tfrac{2x}{7} + \tfrac{4}{7}\right| - 3$$
$$\left|\tfrac{2x}{7} + \tfrac{4}{7}\right| = 8$$
$$\tfrac{2x}{7} + \tfrac{4}{7} = -8, 8$$
$$2x + 4 = -56, 56$$
$$2x = -60, 52$$
$$x = -30, 26$$

39. Solving the equation:

$$2 = -8 + \left|4 - \tfrac{1}{2}y\right|$$
$$\left|4 - \tfrac{1}{2}y\right| = 10$$
$$4 - \tfrac{1}{2}y = -10, 10$$
$$-\tfrac{1}{2}y = -14, 6$$
$$y = -12, 28$$

41. Solving the equation:
$$\left|3a + 1\right| = \left|2a - 4\right|$$

$$3a + 1 = 2a - 4 \qquad \text{or} \qquad 3a + 1 = -2a + 4$$
$$a + 1 = -4 \qquad\qquad\qquad 5a = 3$$
$$a = -5 \qquad\qquad\qquad\quad a = \tfrac{3}{5}$$

43. Solving the equation:
$$\left|x - \tfrac{1}{3}\right| = \left|\tfrac{1}{2}x + \tfrac{1}{6}\right|$$

$$x - \tfrac{1}{3} = \tfrac{1}{2}x + \tfrac{1}{6} \qquad \text{or} \qquad x - \tfrac{1}{3} = -\tfrac{1}{2}x - \tfrac{1}{6}$$
$$6x - 2 = 3x + 1 \qquad\qquad\qquad 6x - 2 = -3x - 1$$
$$3x - 2 = 1 \qquad\qquad\qquad\quad 9x - 2 = -1$$
$$3x = 3 \qquad\qquad\qquad\qquad 9x = 1$$
$$x = 1 \qquad\qquad\qquad\qquad x = \tfrac{1}{9}$$

45. Solving the equation:

$$|y-2| = |y+3|$$

$y-2 = y+3$	or	$y-2 = -y-3$
$-2 = -3$		$2y = -1$
$y =$ impossible		$y = -\frac{1}{2}$

47. Solving the equation:

$$|3x-1| = |3x+1|$$

$3x-1 = 3x+1$	or	$3x-1 = -3x-1$
$-1 = 1$		$6x = 0$
$x =$ impossible		$x = 0$

49. Solving the equation:

$$|3-m| = |m+4|$$

$3-m = m+4$	or	$3-m = -m-4$
$-2m = 1$		$3 = -4$
$m = -\frac{1}{2}$		$m =$ impossible

51. Solving the equation:

$$|0.03 - 0.01x| = |0.04 + 0.05x|$$

$0.03 - 0.01x = 0.04 + 0.05x$	or	$0.03 - 0.01x = -0.04 - 0.05x$
$-0.06x = 0.01$		$0.04x = -0.07$
$x = -\frac{1}{6}$		$x = -\frac{7}{4}$

53. Since $|x-2| = |2-x|$ is always true, the solution set is all real numbers.

55. Since $\left|\frac{x}{5} - 1\right| = \left|1 - \frac{x}{5}\right|$ is always true, the solution set is all real numbers.

57. Setting $R = 722$:

$$-60|x-11| + 962 = 722$$
$$-60|x-11| = -240$$
$$|x-11| = 4$$
$$x - 11 = -4, 4$$
$$x = 7, 15$$

The revenue was 722 million dollars in the years 1987 and 1995.

59. Solving for t:

$$d = rt$$
$$\frac{d}{r} = \frac{rt}{r}$$
$$t = \frac{d}{r}$$

61. Finding the percent:

$$p \cdot 60 = 15$$
$$p = \frac{15}{60} = 0.25$$

So 25% of 60 is 15.

63. Solving the inequality:

$$x - 5 > 8$$
$$x > 13$$

65. Solving the inequality:

$$\frac{1}{4}x \geq 1$$
$$x \geq 4$$

67. Solving the inequality:

$$4 - 2x < 12$$
$$-2x < 8$$
$$x > -4$$

7.3   Compound Inequalities and Interval Notation

1.   Graphing the solution set:

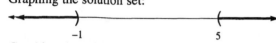

3.   Graphing the solution set:

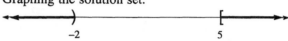

5.   Graphing the solution set:

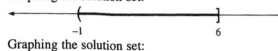

7.   Graphing the solution set:

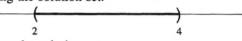

9.   Graphing the solution set:

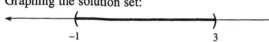

11.   Graphing the solution set:

13.   Graphing the solution set:

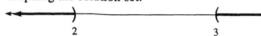

15.   Graphing the solution set:

17.   Solving the compound inequality:

$$3x-1<5 \qquad \text{or} \qquad 5x-5>10$$
$$3x<6 \qquad\qquad\qquad 5x>15$$
$$x<2 \qquad\qquad\qquad x>3$$

The solution set is $(-\infty,2)\cup(3,\infty)$.

Graphing the solution set:

19.   Solving the compound inequality:

$$x-2>-5 \qquad \text{and} \qquad x+7<13$$
$$x>-3 \qquad\qquad\qquad x<6$$

The solution set is $(-3,6)$.

Graphing the solution set:

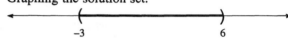

21.   Solving the compound inequality:

$$11x\le 22 \qquad \text{or} \qquad 12x\ge 36$$
$$x\le 2 \qquad\qquad\qquad x\ge 3$$

The solution set is $(-\infty,2]\cup[3,\infty)$.

Graphing the solution set:

23.   Solving the compound inequality:

$$3x-5\le 10 \qquad \text{and} \qquad 2x+1\ge -5$$
$$3x\le 15 \qquad\qquad\qquad 2x\ge -6$$
$$x\le 5 \qquad\qquad\qquad x\ge -3$$

The solution set is $[-3,5]$.

Graphing the solution set:

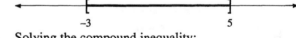

25.   Solving the compound inequality:

$$2x-3<8 \qquad \text{and} \qquad 3x+1>-10$$
$$2x<11 \qquad\qquad\qquad 3x>-11$$
$$x<\frac{11}{2} \qquad\qquad\qquad x>-\frac{11}{3}$$

The solution set is $\left(-\frac{11}{3},\frac{11}{2}\right)$.

Graphing the solution set:

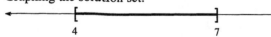

27.   Solving the compound inequality:

$$2x-1<3 \qquad \text{and} \qquad 3x-2>1$$
$$2x<4 \qquad\qquad\qquad 3x>3$$
$$x<2 \qquad\qquad\qquad x>1$$

The solution set is $(1,2)$.

Graphing the solution set:

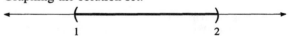

29.   Solving the compound inequality:

$$-1\le x-5\le 2$$
$$4\le x\le 7$$

The solution set is $[4,7]$.

Graphing the solution set:

31.   Solving the compound inequality:

$$-4\le 2x\le 6$$
$$-2\le x\le 3$$

The solution set is $[-2,3]$.

Graphing the solution set:

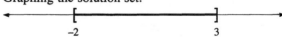

33. Solving the compound inequality:
$$-3 < 2x + 1 < 5$$
$$-4 < 2x < 4$$
$$-2 < x < 2$$

The solution set is $(-2, 2)$.

Graphing the solution set:

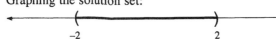

35. Solving the compound inequality:
$$0 \le 3x + 2 \le 7$$
$$-2 \le 3x \le 5$$
$$-\frac{2}{3} \le x \le \frac{5}{3}$$

The solution set is $\left[-\frac{2}{3}, \frac{5}{3}\right]$.

Graphing the solution set:

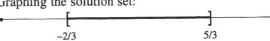

37. Solving the compound inequality:
$$-7 < 2x + 3 < 11$$
$$-10 < 2x < 8$$
$$-5 < x < 4$$
The solution set is $(-5, 4)$.
Graphing the solution set:

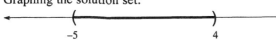

39. Solving the compound inequality:
$$-60 < 40a + 20 < 60$$
$$-80 < 40a < 40$$
$$-2 < a < 1$$
The solution set is $(-2, 1)$.
Graphing the solution set:

41. The inequality is $-2 < x < 3$.

43. The inequality is $-2 \le x \le 3$.

45. **a.** The three inequalities are $2x + x > 10$, $2x + 10 > x$, and $x + 10 > 2x$.

   b. For $x + 10 > 2x$ we have $x < 10$. For $3x > 10$ we have $x > \frac{10}{3}$. The compound inequality is $\frac{10}{3} < x < 10$.

47. Graphing the inequality:

49. Let x represent the number. Solving the inequality:
$$5 < 2x - 3 < 7$$
$$8 < 2x < 10$$
$$4 < x < 5$$
The number is between 4 and 5.

51. Let w represent the width and $w + 4$ represent the length. Using the perimeter formula:
$$20 < 2w + 2(w + 4) < 30$$
$$20 < 2w + 2w + 8 < 30$$
$$20 < 4w + 8 < 30$$
$$12 < 4w < 22$$
$$3 < w < \frac{11}{2}$$

The width is between 3 inches and $\frac{11}{2} = 5\frac{1}{2}$ inches.

53. Simplifying the expression: $-|-5| = -(5) = -5$

55. Simplifying the expression: $-3 - 4(-2) = -3 + 8 = 5$

57. Simplifying the expression: $5|3 - 8| - 6|2 - 5| = 5|-5| - 6|-3| = 5(5) - 6(3) = 25 - 18 = 7$

59. Simplifying the expression: $5 - 2[-3(5 - 7) - 8] = 5 - 2[-3(-2) - 8] = 5 - 2(6 - 8) = 5 - 2(-2) = 5 + 4 = 9$

61. The expression is: $-3 - (-9) = -3 + 9 = 6$

63. Applying the distributive property: $\frac{1}{2}(4x - 6) = \frac{1}{2} \cdot 4x - \frac{1}{2} \cdot 6 = 2x - 3$

65. The integers are: $-3, 0, 2$

7.4   Inequalities Involving Absolute Value

1. Solving the inequality:
    $|x| < 3$
    $-3 < x < 3$
    Graphing the solution set:
    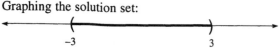
    -3 3

3. Solving the inequality:
    $|x| \geq 2$
    $x \leq -2 \text{ or } x \geq 2$
    Graphing the solution set:
    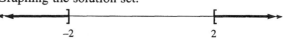
    -2 2

5. Solving the inequality:
    $|x| + 2 < 5$
    $|x| < 3$
    $-3 < x < 3$
    Graphing the solution set:
    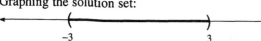
    -3 3

7. Solving the inequality:
    $|t| - 3 > 4$
    $|t| > 7$
    $t < -7 \text{ or } t > 7$
    Graphing the solution set:

    -7 7

9. Since the inequality $|y| < -5$ is never true, there is no solution, or $\varnothing$. Graphing the solution set:
    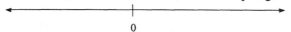
    0

11. Since the inequality $|x| \geq -2$ is always true, the solution set is all real numbers.
    Graphing the solution set:
    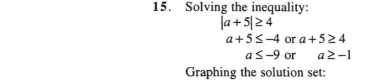
    0

13. Solving the inequality:
    $|x - 3| < 7$
    $-7 < x - 3 < 7$
    $-4 < x < 10$
    Graphing the solution set:
    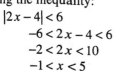
    -4 10

15. Solving the inequality:
    $|a + 5| \geq 4$
    $a + 5 \leq -4 \text{ or } a + 5 \geq 4$
    $a \leq -9 \text{ or } \quad a \geq -1$
    Graphing the solution set:

    -9 -1

17. Since the inequality $|a - 1| < -3$ is never true, there is no solution, or $\varnothing$. Graphing the solution set:
    0

19. Solving the inequality:
    $|2x - 4| < 6$
    $-6 < 2x - 4 < 6$
    $-2 < 2x < 10$
    $-1 < x < 5$
    Graphing the solution set:
    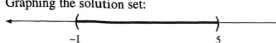
    -1 5

21. Solving the inequality:
    $|3y + 9| \geq 6$
    $3y + 9 \leq -6 \qquad \text{or} \qquad 3y + 9 \geq 6$
    $3y \leq -15 \qquad\qquad\qquad 3y \geq -3$
    $y \leq -5 \qquad\qquad\qquad y \geq -1$
    Graphing the solution set:
    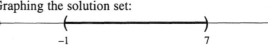
    -5 -1

23. Solving the inequality:
    $|2k + 3| \geq 7$
    $2k + 3 \leq -7 \qquad \text{or} \qquad 2k + 3 \geq 7$
    $2k \leq -10 \qquad\qquad\qquad 2k \geq 4$
    $k \leq -5 \qquad\qquad\qquad k \geq 2$
    Graphing the solution set:
    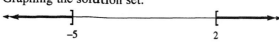
    -5 2

25. Solving the inequality:
    $|x - 3| + 2 < 6$
    $|x - 3| < 4$
    $-4 < x - 3 < 4$
    $-1 < x < 7$
    Graphing the solution set:
    -1 7

27. Solving the inequality:
$$|2a+1|+4 \geq 7$$
$$|2a+1| \geq 3$$

$$2a+1 \leq -3 \quad \text{or} \quad 2a+1 \geq 3$$
$$2a \leq -4 \qquad\qquad 2a \geq 2$$
$$a \leq -2 \qquad\qquad a \geq 1$$

Graphing the solution set:

29. Solving the inequality:
$$|3x+5|-8 < 5$$
$$|3x+5| < 13$$
$$-13 < 3x+5 < 13$$
$$-18 < 3x < 8$$
$$-6 < x < \frac{8}{3}$$

Graphing the solution set:

31. Solving the inequality:

$$|5-x| > 3$$
$$5-x < -3 \quad \text{or} \quad 5-x > 3$$
$$-x < -8 \qquad\qquad -x > -2$$
$$x > 8 \qquad\qquad x < 2$$

Graphing the solution set:

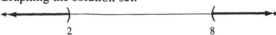

33. Solving the inequality:
$$\left|3 - \frac{2}{3}x\right| \geq 5$$
$$3 - \frac{2}{3}x \leq -5 \quad \text{or} \quad 3 - \frac{2}{3}x \geq 5$$
$$-\frac{2}{3}x \leq -8 \qquad\qquad -\frac{2}{3}x \geq 2$$
$$-2x \leq -24 \qquad\qquad -2x \geq 6$$
$$x \geq 12 \qquad\qquad x \leq -3$$

Graphing the solution set:

35. Solving the inequality:
$$\left|2 - \frac{1}{2}x\right| > 1$$
$$2 - \frac{1}{2}x < -1 \quad \text{or} \quad 2 - \frac{1}{2}x > 1$$
$$-\frac{1}{2}x < -3 \qquad\qquad -\frac{1}{2}x > -1$$
$$x > 6 \qquad\qquad x < 2$$

Graphing the solution set:

37. Solving the inequality:

$$|x-1| < 0.01$$
$$-0.01 < x-1 < 0.01$$
$$0.99 < x < 1.01$$

39. Solving the inequality:
$$|2x+1| \geq \frac{1}{5}$$
$$2x+1 \leq -\frac{1}{5} \quad \text{or} \quad 2x+1 \geq \frac{1}{5}$$
$$2x \leq -\frac{6}{5} \qquad\qquad 2x \geq -\frac{4}{5}$$
$$x \leq -\frac{3}{5} \qquad\qquad x \geq -\frac{2}{5}$$

41. Solving the inequality:
$$\left|\frac{3x-2}{5}\right| \leq \frac{1}{2}$$
$$-\frac{1}{2} \leq \frac{3x-2}{5} \leq \frac{1}{2}$$
$$-\frac{5}{2} \leq 3x-2 \leq \frac{5}{2}$$
$$-\frac{1}{2} \leq 3x \leq \frac{9}{2}$$
$$-\frac{1}{6} \leq x \leq \frac{3}{2}$$

43. Solving the inequality:

$$\left|2x - \frac{1}{5}\right| < 0.3$$
$$-0.3 < 2x - 0.2 < 0.3$$
$$-0.1 < 2x < 0.5$$
$$-0.05 < x < 0.25$$

45. Writing as an absolute value inequality: $|x| \leq 4$

47. Writing as an absolute value inequality: $|x-5| \leq 1$

49. The absolute value inequality is: $|x-65| \leq 10$

51. Solving the inequality:
$$|v - 455| < 23$$
$$-23 < v - 455 < 23$$
$$432 < v < 478$$
The color of the fireworks display is blue.

53. Factoring by grouping: $2ax - 3a + 8x - 12 = a(2x - 3) + 4(2x - 3) = (a + 4)(2x - 3)$

55. Factoring: $x^2 - 2x - 35 = (x - 7)(x + 5)$ **57.** Factoring: $x^2 - xy - 6y^2 = (x + 2y)(x - 3y)$

59. Factoring: $x^2 - 9 = (x + 3)(x - 3)$

61. Solving the equation:
$$9x^2 = 12x - 4$$
$$9x^2 - 12x + 4 = 0$$
$$(3x - 2)^2 = 0$$
$$3x = 2$$
$$x = \tfrac{2}{3}$$

7.5   Factoring the Sum and Difference of Two Cubes

1. Factoring: $x^3 - y^3 = (x - y)(x^2 + xy + y^2)$ **3.** Factoring: $a^3 + 8 = (a + 2)(a^2 - 2a + 4)$

5. Factoring: $27 + x^3 = (3 + x)(9 - 3x + x^2)$ **7.** Factoring: $y^3 - 1 = (y - 1)(y^2 + y + 1)$

9. Factoring: $10r^3 - 1250 = 10(r^3 - 125) = 10(r - 5)(r^2 + 5r + 25)$

11. Factoring: $64 + 27a^3 = (4 + 3a)(16 - 12a + 9a^2)$ **13.** Factoring: $8x^3 - 27y^3 = (2x - 3y)(4x^2 + 6xy + 9y^2)$

15. Factoring: $t^3 + \tfrac{1}{27} = \left(t + \tfrac{1}{3}\right)\left(t^2 - \tfrac{1}{3}t + \tfrac{1}{9}\right)$ **17.** Factoring: $27x^3 - \tfrac{1}{27} = \left(3x - \tfrac{1}{3}\right)\left(9x^2 + x + \tfrac{1}{9}\right)$

19. Factoring: $64a^3 + 125b^3 = (4a + 5b)(16a^2 - 20ab + 25b^2)$

21. Factoring: $x^2 - 81 = (x + 9)(x - 9)$ **23.** Factoring: $x^2 + 2x - 15 = (x - 3)(x + 5)$

25. Factoring: $64 - t^2 = (8 + t)(8 - t)$

27. Factoring: $x^2y^2 + 2y^2 + x^2 + 2 = y^2(x^2 + 2) + 1(x^2 + 2) = (x^2 + 2)(y^2 + 1)$

29. Factoring: $2a^3b + 6a^2b + 2ab = 2ab(a^2 + 3a + 1)$ **31.** The polynomial $x^2 + x + 1$ does not factor (prime).

33. Factoring: $12a^2 - 75 = 3(4a^2 - 25) = 3(2a + 5)(2a - 5)$

35. Factoring: $9x^2 - 12xy + 4y^2 = (3x - 2y)^2$ **37.** Factoring: $25 - 10t + t^2 = (5 - t)^2$

39. Factoring: $4x^3 + 16xy^2 = 4x(x^2 + 4y^2)$

41. Factoring: $2y^3 + 20y^2 + 50y = 2y(y^2 + 10y + 25) = 2y(y + 5)^2$

43. Factoring: $a^7 + 8a^4b^3 = a^4(a^3 + 8b^3) = a^4(a + 2b)(a^2 - 2ab + 4b^2)$

45. Factoring: $x^3 + 5x^2 - 9x - 45 = x^2(x + 5) - 9(x + 5) = (x + 5)(x^2 - 9) = (x + 5)(x + 3)(x - 3)$

47. Factoring: $5a^2 + 10ab + 5b^2 = 5(a^2 + 2ab + b^2) = 5(a + b)^2$

49. The polynomial $x^2 + 49$ does not factor (prime).

51. Factoring: $3x^2 + 15xy + 18y^2 = 3(x^2 + 5xy + 6y^2) = 3(x + 2y)(x + 3y)$

53. Factoring: $4y^4 - 12y^2 + 9 = (2y^2 - 3)(2y^2 - 3) = (2y^2 - 3)^2$

55. Factoring: $x^2(x - 3) - 14x(x - 3) + 49(x - 3) = (x - 3)(x^2 - 14x + 49) = (x - 3)(x - 7)^2$

57. Factoring: $x^2 - 64 = (x+8)(x-8)$ **59.** Factoring: $8 - 14x - 15x^2 = (2-5x)(4+3x)$

61. Factoring: $49a^7 - 9a^5 = a^5\left(49a^2 - 9\right) = a^5(7a+3)(7a-3)$

63. Factoring: $r^2 - \frac{1}{25} = \left(r + \frac{1}{5}\right)\left(r - \frac{1}{5}\right)$ **65.** The polynomial $49x^2 + 9y^2$ does not factor (prime).

67. Factoring: $100x^2 - 100x - 600 = 100\left(x^2 - x - 6\right) = 100(x-3)(x+2)$

69. Factoring: $25a^3 + 20a^2 + 3a = a\left(25a^2 + 20a + 3\right) = a(5a+3)(5a+1)$

71. Factoring: $3x^4 - 14x^2 - 5 = \left(3x^2 + 1\right)\left(x^2 - 5\right)$

73. Factoring: $24a^5b - 3a^2b = 3a^2b\left(8a^3 - 1\right) = 3a^2b(2a-1)\left(4a^2 + 2a + 1\right)$

75. Factoring: $64 - r^3 = (4-r)\left(16 + 4r + r^2\right)$

77. Factoring: $20x^4 - 45x^2 = 5x^2\left(4x^2 - 9\right) = 5x^2(2x+3)(2x-3)$

79. Factoring: $400t^2 - 900 = 100\left(4t^2 - 9\right) = 100(2t+3)(2t-3)$

81. Graphing the two equations:

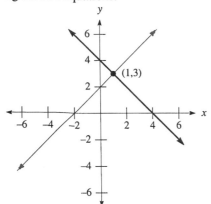

The intersection point is $(1,3)$.

83. Adding the two equations yields:
$$2x = 2$$
$$x = 1$$
Substituting into the first equation:
$$1 + y = 4$$
$$y = 3$$
The solution is $(1,3)$.

85. Substituting into the first equation:
$$x + 2x - 1 = 2$$
$$3x - 1 = 2$$
$$3x = 3$$
$$x = 1$$
Substituting into the first equation:
$$1 + y = 2$$
$$y = 1$$
The solution is $(1,1)$.

7.6   Review of Systems of Linear Equations in Two Variables

1.   The intersection point is (4,3).

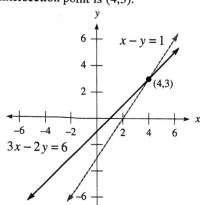

3.   The intersection point is (−5,−6).

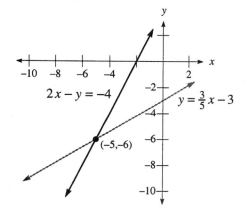

5.   The intersection point is (4,2).

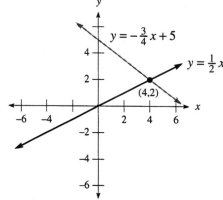

7.   The lines are parallel. There is no solution to the system.

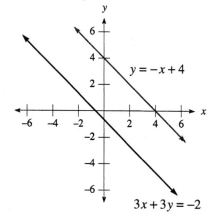

9. The lines coincide. Any solution to one of the equations is a solution to the other.

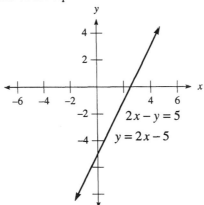

11. Solving the two equations:
$$x + y = 5$$
$$3x - y = 3$$
Adding yields:
$$4x = 8$$
$$x = 2$$
The solution is $(2,3)$.

13. Multiply the first equation by -1:
$$-3x - y = 4$$
$$4x + y = 5$$
Adding yields: $x = 1$. The solution is $(1,1)$.

15. Multiply the first equation by -2:
$$-6x + 4y = -12$$
$$6x - 4y = 12$$
Adding yields $0 = 0$, so the lines coincide. The solution is $\{(x, y) \mid 3x - 2y = 6\}$.

17. Multiply the first equation by 3:
$$3x + 6y = 0$$
$$2x - 6y = 5$$
Adding yields:
$$5x = 5$$
$$x = 1$$
The solution is $\left(1, -\frac{1}{2}\right)$.

19. Multiply the first equation by -2:
$$-4x + 10y = -32$$
$$4x - 3y = 11$$
Adding yields:
$$7y = -21$$
$$y = -3$$
The solution is $\left(\frac{1}{2}, -3\right)$.

21. Multiply the first equation by 3 and the second equation by -2:
$$18x + 9y = -3$$
$$-18x - 10y = -2$$
Adding yields:
$$-y = -5$$
$$y = 5$$
The solution is $\left(-\frac{8}{3}, 5\right)$.

23. Multiply the first equation by 2 and the second equation by 3:
$$8x + 6y = 28$$
$$27x - 6y = 42$$
Adding yields:
$$35x = 70$$
$$x = 2$$
The solution is $(2,2)$.

25.   Multiply the first equation by 2:
$$4x - 10y = 6$$
$$-4x + 10y = 3$$
Adding yields $0 = 9$, which is false (parallel lines). There is no solution ($\varnothing$).

27.   To clear each equation of fractions, multiply the first equation by 12 and the second equation by 30:
$$3x - 2y = -24$$
$$-5x + 6y = 120$$
Multiply the first equation by 3:
$$9x - 6y = -72$$
$$-5x + 6y = 120$$
Adding yields $4x = 48$, so $x = 12$. Substituting into the first equation:
$$36 - 2y = -24$$
$$-2y = -60$$
$$y = 30$$
The solution is (12,30).

29.   To clear each equation of fractions, multiply the first equation by 6 and the second equation by 20:
$$3x + 2y = 78$$
$$8x + 5y = 200$$
Multiply the first equation by 5 and the second equation by –2:
$$15x + 10y = 390$$
$$-16x - 10y = -400$$
Adding yields:
$$-x = -10$$
$$x = 10$$
The solution is (10,24).

31.   Substituting into the first equation:

$$7(2y + 9) - y = 24$$
$$14y + 63 - y = 24$$
$$13y = -39$$
$$y = -3$$

The solution is (3,–3).

33.   Substituting into the first equation:
$$6x - \left(-\tfrac{3}{4}x - 1\right) = 10$$
$$6x + \tfrac{3}{4}x + 1 = 10$$
$$\tfrac{27}{4}x = 9$$
$$27x = 36$$
$$x = \tfrac{4}{3}$$
The solution is $\left(\tfrac{4}{3}, -2\right)$.

35.   Substituting $x = y + 4$ into the second equation:
$$2(y + 4) - 3y = 6$$
$$2y + 8 - 3y = 6$$
$$-y + 8 = 6$$
$$-y = -2$$
$$y = 2$$
The solution is (6,2).

37.   Substituting into the first equation:

$$4x - 4 = 3x - 2$$
$$x - 4 = -2$$
$$x = 2$$

The solution is (2,4).

39.   Solving the first equation for y yields $y = 2x - 5$. Substituting into the second equation:
$$4x - 2(2x - 5) = 10$$
$$4x - 4x + 10 = 10$$
$$10 = 10$$
Since this statement is true, the two lines coincide. The solution is $\{(x,y) \mid 2x - y = 5\}$.

41. Substituting into the first equation:
$$\frac{1}{3}\left(\frac{3}{2}y\right) - \frac{1}{2}y = 0$$
$$\frac{1}{2}y - \frac{1}{2}y = 0$$
$$0 = 0$$

Since this statement is true, the two lines coincide. The solution is $\left\{(x,y)\mid x = \frac{3}{2}y\right\}$.

43. Multiply the first equation by 2 and the second equation by 7:
$$8x - 14y = 6$$
$$35x + 14y = -21$$

Adding yields:
$$43x = -15$$
$$x = -\frac{15}{43}$$

Substituting into the original second equation:
$$5\left(-\frac{15}{43}\right) + 2y = -3$$
$$-\frac{75}{43} + 2y = -3$$
$$2y = -\frac{54}{43}$$
$$y = -\frac{27}{43}$$

The solution is $\left(-\frac{15}{43}, -\frac{27}{43}\right)$.

45. Multiply the first equation by 3 and the second equation by 8:
$$27x - 24y = 12$$
$$16x + 24y = 48$$

Adding yields:
$$43x = 60$$
$$x = \frac{60}{43}$$

Substituting into the original second equation:
$$2\left(\frac{60}{43}\right) + 3y = 6$$
$$\frac{120}{43} + 3y = 6$$
$$3y = \frac{138}{43}$$
$$y = \frac{46}{43}$$

The solution is $\left(\frac{60}{43}, \frac{46}{43}\right)$.

47. Multiply the first equation by 2 and the second equation by 5:
$$6x - 10y = 4$$
$$35x + 10y = 5$$

Adding yields:
$$41x = 9$$
$$x = \frac{9}{41}$$

Substituting into the original second equation:
$$7\left(\frac{9}{41}\right) + 2y = 1$$
$$\frac{63}{41} + 2y = 1$$
$$2y = -\frac{22}{41}$$
$$y = -\frac{11}{41}$$

The solution is $\left(\frac{9}{41}, -\frac{11}{41}\right)$.

49. Multiply the second equation by 100:
$$x + y = 10000$$
$$6x + 5y = 56000$$
Multiply the first equation by –5:
$$-5x - 5y = -50000$$
$$6x + 5y = 56000$$
Adding yields $x = 6000$. The solution is (6000,4000).

51. Multiplying the first equation by $\frac{2}{3}$ yields the equation $4x - 6y = 2$. For the lines to coincide, the value is $c = 2$.

53. Let x and y represent the two numbers. The system of equations is:
$$y = 2x + 3$$
$$x + y = 18$$
Substituting into the second equation:
$$x + 2x + 3 = 18$$
$$3x = 15$$
$$x = 5$$
$$y = 2(5) + 3 = 13$$
The two numbers are 5 and 13.

55. Let a represent the number of adult tickets and c represent the number of children's tickets. The system of equations is:
$$a + c = 925$$
$$2a + c = 1150$$
Multiply the first equation by –1:
$$-a - c = -925$$
$$2a + c = 1150$$
Adding yields:
$$a = 225$$
$$c = 700$$
There were 225 adult tickets and 700 children's tickets sold.

57. Let x represent the amount of 20% disinfectant and y represent the amount of 14% disinfectant.
The system of equations is:
$$x + y = 15$$
$$0.20x + 0.14y = 0.16(15)$$
Multiplying the first equation by –0.14:
$$-0.14x - 0.14y = -2.1$$
$$0.20x + 0.14y = 2.4$$
Adding yields:
$$0.06x = 0.3$$
$$x = 5$$
$$y = 10$$
The mixture contains 5 gallons of 20% disinfectant and 10 gallons of 14% disinfectant.

59. Let b represent the rate of the boat and c represent the rate of the current. The system of equations is:
$$2(b + c) = 24$$
$$3(b - c) = 18$$
The system of equations simplifies to:
$$b + c = 12$$
$$b - c = 6$$
Adding yields:
$$2b = 18$$
$$b = 9$$
$$c = 3$$
The rate of the boat is 9 mph and the rate of the current is 3 mph.

61. Let a represent the rate of the airplane and w represent the rate of the wind. The system of equations is:
$$2(a+w)=600$$
$$\tfrac{5}{2}(a-w)=600$$
The system of equations simplifies to:
$$a+w=300$$
$$a-w=240$$
Adding yields:
$$2a=540$$
$$a=270$$
$$w=30$$
The rate of the airplane is 270 mph and the rate of the wind is 30 mph.

63. Simplifying: $\left(3x^2\right)^3\left(5y^4\right)^2=\left(27x^6\right)\left(25y^8\right)=675x^6y^8$

65. Simplifying: $\dfrac{x^5}{x^{-2}}=x^{5-(-2)}=x^7$

67. Simplifying: $\left(5\times10^{-8}\right)\left(4\times10^3\right)=20\times10^{-5}=2\times10^{-4}$

69. Subtracting: $\left(7a^2+8a+9\right)-\left(5a^2-2a+1\right)=7a^2+8a+9-5a^2+2a-1=2a^2+10a+8$

71. Multiplying: $(4a+9)(4a-9)=16a^2+36a-36a-81=16a^2-81$

73. Multiplying: $6xy^2\left(4x^2-3xy+2y^2\right)=24x^3y^2-18x^2y^3+12xy^4$

75. Using long division:

$$
\begin{array}{r}
x-2 \\
x-3\,\overline{)\,x^2-5x+8} \\
\underline{x^2-3x} \\
-2x+8 \\
\underline{-2x+6} \\
2
\end{array}
$$

The quotient is $x-2+\dfrac{2}{x-3}$.

7.7 Systems of Linear Equations in Three Variables

1. Adding the first two equations and the first and third equations results in the system:
$$2x+3z=5$$
$$2x-2z=0$$
Solving the second equation yields $x=z$, now substituting:
$$2z+3z=5$$
$$5z=5$$
$$z=1$$
So $x=1$, now substituting into the original first equation:
$$1+y+1=4$$
$$y+2=4$$
$$y=2$$
The solution is $(1,2,1)$.

3. Adding the first two equations and the first and third equations results in the system:
$$2x + 3z = 13$$
$$3x - 3z = -3$$
Adding yields:
$$5x = 10$$
$$x = 2$$
Substituting to find z:
$$2(2) + 3z = 13$$
$$4 + 3z = 13$$
$$3z = 9$$
$$z = 3$$
Substituting into the original first equation:
$$2 + y + 3 = 6$$
$$y + 5 = 6$$
$$y = 1$$
The solution is (2,1,3).

5. Adding the second and third equations:
$$5x + z = 11$$
Multiplying the second equation by 2:
$$x + 2y + z = 3$$
$$4x - 2y + 4z = 12$$
Adding yields:
$$5x + 5z = 15$$
$$x + z = 3$$
So the system becomes:
$$5x + z = 11$$
$$x + z = 3$$
Multiply the second equation by -1:
$$5x + z = 11$$
$$-x - z = -3$$
Adding yields:
$$4x = 8$$
$$x = 2$$
Substituting to find z:
$$5(2) + z = 11$$
$$z + 10 = 11$$
$$z = 1$$
Substituting into the original first equation:
$$2 + 2y + 1 = 3$$
$$2y + 3 = 3$$
$$2y = 0$$
$$y = 0$$
The solution is (2,0,1).

7. Multiply the second equation by -1 and add it to the first equation:
$$2x + 3y - 2z = 4$$
$$-x - 3y + 3z = -4$$
Adding results in the equation $x + z = 0$. Multiply the second equation by 2 and add it to the third equation:
$$2x + 6y - 6z = 8$$
$$3x - 6y + z = -3$$
Adding results in the equation:
$$5x - 5z = 5$$
$$x - z = 1$$

So the system becomes:
$$x - z = 1$$
$$x + z = 0$$
Adding yields:
$$2x = 1$$
$$x = \tfrac{1}{2}$$
Substituting to find z:
$$\tfrac{1}{2} + z = 0$$
$$z = -\tfrac{1}{2}$$
Substituting into the original first equation:
$$2\left(\tfrac{1}{2}\right) + 3y - 2\left(-\tfrac{1}{2}\right) = 4$$
$$1 + 3y + 1 = 4$$
$$3y + 2 = 4$$
$$3y = 2$$
$$y = \tfrac{2}{3}$$
The solution is $\left(\tfrac{1}{2}, \tfrac{2}{3}, -\tfrac{1}{2}\right)$.

9. Multiply the first equation by 2 and add it to the second equation:
$$-2x + 8y - 6z = 4$$
$$2x - 8y + 6z = 1$$
Adding yields $0 = 5$, which is false. There is no solution (inconsistent system).

11. To clear the system of fractions, multiply the first equation by 2 and the second equation by 3:
$$x - 2y + 2z = 0$$
$$6x + y + 3z = 6$$
$$x + y + z = -4$$
Multiply the third equation by 2 and add it to the first equation:
$$x - 2y + 2z = 0$$
$$2x + 2y + 2z = -8$$
Adding yields the equation $3x + 4z = -8$. Multiply the third equation by -1 and add it to the second equation:
$$6x + y + 3z = 6$$
$$-x - y - z = 4$$
Adding yields the equation $5x + 2z = 10$. So the system becomes:
$$3x + 4z = -8$$
$$5x + 2z = 10$$
Multiply the second equation by -2:
$$3x + 4z = -8$$
$$-10x - 4z = -20$$
Adding yields:
$$-7x = -28$$
$$x = 4$$
Substituting to find z:
$$3(4) + 4z = -8$$
$$12 + 4z = -8$$
$$4z = -20$$
$$z = -5$$
Substituting into the original third equation:
$$4 + y - 5 = -4$$
$$y - 1 = -4$$
$$y = -3$$
The solution is $(4, -3, -5)$.

13.   Multiply the first equation by –2 and add it to the third equation:
$$-4x + 2y + 6z = -2$$
$$4x - 2y - 6z = 2$$
Adding yields $0 = 0$, which is true. Since there are now less equations than unknowns, there is no unique solution (dependent system).

15.   Multiply the second equation by 3 and add it to the first equation:
$$2x - y + 3z = 4$$
$$3x + 6y - 3z = -9$$
Adding yields the equation $5x + 5y = -5$, or $x + y = -1$.
Multiply the second equation by 2 and add it to the third equation:
$$2x + 4y - 2z = -6$$
$$4x + 3y + 2z = -5$$
Adding yields the equation $6x + 7y = -11$. So the system becomes:
$$6x + 7y = -11$$
$$x + y = -1$$
Multiply the second equation by –6:
$$6x + 7y = -11$$
$$-6x - 6y = 6$$
Adding yields $y = -5$. Substituting to find x:
$$6x + 7(-5) = -11$$
$$6x - 35 = -11$$
$$6x = 24$$
$$x = 4$$
Substituting into the original first equation:
$$2(4) - (-5) + 3z = 4$$
$$13 + 3z = 4$$
$$3z = -9$$
$$z = -3$$
The solution is $(4, -5, -3)$.

17.   Adding the second and third equations results in the equation $x + y = 9$. Since this is the same as the first equation, there are less equations than unknowns. There is no unique solution (dependent system).

19.   Adding the second and third equations results in the equation $4x + y = 3$. So the system becomes:
$$4x + y = 3$$
$$2x + y = 2$$
Multiplying the second equation by –1:
$$4x + y = 3$$
$$-2x - y = -2$$
Adding yields:
$$2x = 1$$
$$x = \frac{1}{2}$$
Substituting to find y:
$$2\left(\frac{1}{2}\right) + y = 2$$
$$1 + y = 2$$
$$y = 1$$
Substituting into the original second equation:
$$1 + z = 3$$
$$z = 2$$
The solution is $\left(\frac{1}{2}, 1, 2\right)$.

21.   Multiply the third equation by 2 and add it to the second equation:

$$6y - 4z = 1$$
$$2x + 4z = 2$$

Adding yields the equation $2x + 6y = 3$. So the system becomes:

$$2x - 3y = 0$$
$$2x + 6y = 3$$

Multiply the first equation by 2:

$$4x - 6y = 0$$
$$2x + 6y = 3$$

Adding yields:

$$6x = 3$$
$$x = \tfrac{1}{2}$$

Substituting to find y:

$$2\left(\tfrac{1}{2}\right) + 6y = 3$$
$$1 + 6y = 3$$
$$6y = 2$$
$$y = \tfrac{1}{3}$$

Substituting into the original third equation to find z:

$$\tfrac{1}{2} + 2z = 1$$
$$2z = \tfrac{1}{2}$$
$$z = \tfrac{1}{4}$$

The solution is $\left(\tfrac{1}{2}, \tfrac{1}{3}, \tfrac{1}{4}\right)$.

23.   To clear each equation of fractions, multiply the first equation by 6, the second equation by 10, and the third equation by 12:

$$3x + 4y = 15$$
$$2x - 5z = -3$$
$$4y - 3z = 9$$

Multiply the third equation by -1 and add it to the first equation:

$$3x + 4y = 15$$
$$-4y + 3z = -9$$

Adding yields:

$$3x + 3z = 6$$
$$x + z = 2$$

So the system becomes:

$$x + z = 2$$
$$2x - 5z = -3$$

Multiply the first equation by 5:

$$5x + 5z = 10$$
$$2x - 5z = -3$$

Adding yields:

$$7x = 7$$
$$x = 1$$

Substituting yields $z = 1$. Substituting to find y:

$$3 + 4y = 15$$
$$4y = 12$$
$$y = 3$$

The solution is $(1, 3, 1)$.

25.   To clear each equation of fractions, multiply the first equation by 4, the second equation by 12, and the third equation by 6:

$$2x - y + 2z = -8$$
$$3x - y - 4z = 3$$
$$x + 2y - 3z = 9$$

Multiply the first equation by -1 and add it to the first equation:

$$-2x + y - 2z = 8$$
$$3x - y - 4z = 3$$

Adding yields the equation $x - 6z = 11$. Multiply the first equation by 2 and add it to the third equation:

$$4x - 2y + 4z = -16$$
$$x + 2y - 3z = 9$$

Adding yields the equation $5x + z = -7$. So the system becomes:

$$5x + z = -7$$
$$x - 6z = 11$$

Multiply the first equation by 6:

$$30x + 6z = -42$$
$$x - 6z = 11$$

Adding yields:

$$31x = -31$$
$$x = -1$$

Substituting to find z:

$$-5 + z = -7$$
$$z = -2$$

Substituting to find y:

$$-1 + 2y + 6 = 9$$
$$2y + 5 = 9$$
$$2y = 4$$
$$y = 2$$

The solution is $(-1, 2, -2)$.

27.   Divide the second equation by 5 and the third equation by 10 to produce the system:

$$x - y - z = 0$$
$$x + 4y = 16$$
$$2y - z = 5$$

Multiply the third equation by -1 and add it to the first equation:

$$x - y - z = 0$$
$$-2y + z = -5$$

Adding yields the equation $x - 3y = -5$. So the system becomes:

$$x + 4y = 16$$
$$x - 3y = -5$$

Multiply the second equation by -1:

$$x + 4y = 16$$
$$-x + 3y = 5$$

Adding yields:

$$7y = 21$$
$$y = 3$$

Substituting to find x:

$$x + 12 = 16$$
$$x = 4$$

Substituting to find z:

$$6 - z = 5$$
$$z = 1$$

The currents are 4 amps, 3 amps, and 1 amp.

29. Let x, y and z represent the amounts invested in the three accounts. The system of equations is:
$$x + y + z = 2200$$
$$z = 3x$$
$$0.06x + 0.08y + 0.09z = 178$$
Substituting into the first equation:
$$x + y + 3x = 2200$$
$$4x + y = 2200$$
Substituting into the third equation:
$$0.06x + 0.08y + 0.09(3x) = 178$$
$$0.33x + 0.08y = 178$$
The system of equations becomes:
$$4x + y = 2200$$
$$0.33x + 0.08y = 178$$
Multiply the first equation by -0.08:
$$-0.32x - 0.08y = -176$$
$$0.33x + 0.08y = 178$$
Adding yields:
$$0.01x = 2$$
$$x = 200$$
$$z = 3(200) = 600$$
$$y = 2200 - 4(200) = 1400$$
He invested \$200 at 6%, \$1,400 at 8%, and \$600 at 9%.

31. Let n, d, and q represent the number of nickels, dimes, and quarters. The system of equations is:
$$n + d + q = 9$$
$$0.05n + 0.10d + 0.25q = 1.20$$
$$d = n$$
Substituting into the first equation:
$$n + n + q = 9$$
$$2n + q = 9$$
Substituting into the second equation:
$$0.05n + 0.10n + 0.25q = 1.20$$
$$0.15n + 0.25q = 1.20$$
The system of equations becomes:
$$2n + q = 9$$
$$0.15n + 0.25q = 1.20$$
Multiplying the first equation by -0.25:
$$-0.50n - 0.25q = -2.25$$
$$0.15n + 0.25q = 1.20$$
Adding yields:
$$-0.35n = -1.05$$
$$n = 3$$
$$d = 3$$
$$q = 9 - 2(3) = 3$$
The collection contains 3 nickels, 3 dimes, and 3 quarters.

33. Reducing the rational expression: $\dfrac{x^2 - x - 6}{x^2 - 9} = \dfrac{(x-3)(x+2)}{(x+3)(x-3)} = \dfrac{x+2}{x+3}$

35. Performing the operations:
$$\frac{x^2 - 25}{x + 4} \cdot \frac{2x + 8}{x^2 - 9x + 20} = \frac{(x+5)(x-5)}{x+4} \cdot \frac{2(x+4)}{(x-4)(x-5)} = \frac{2(x+5)(x-5)(x+4)}{(x+4)(x-4)(x-5)} = \frac{2(x+5)}{x-4}$$

37. Performing the operations: $\dfrac{x}{x^2 - 16} + \dfrac{4}{x^2 - 16} = \dfrac{x+4}{x^2 - 16} = \dfrac{x+4}{(x+4)(x-4)} = \dfrac{1}{x-4}$

39. Simplifying the complex fraction: $\dfrac{1-\dfrac{25}{x^2}}{1-\dfrac{8}{x}+\dfrac{15}{x^2}}\cdot\dfrac{x^2}{x^2}=\dfrac{x^2-25}{x^2-8x+15}=\dfrac{(x+5)(x-5)}{(x-5)(x-3)}=\dfrac{x+5}{x-3}$

41. Multiplying each side of the equation by $x^2-9=(x+3)(x-3)$:

$$(x+3)(x-3)\left(\dfrac{x}{x^2-9}-\dfrac{3}{x-3}\right)=(x+3)(x-3)\bullet\dfrac{1}{x+3}$$
$$x-3(x+3)=x-3$$
$$x-3x-9=x-3$$
$$-2x-9=x-3$$
$$-3x=6$$
$$x=-2$$

Since $x=-2$ checks in the original equation, the solution is $x=-2$.

43. Let t represent the time to fill the pool with both pipes open. The equation is:

$$\tfrac{1}{8}-\tfrac{1}{12}=\dfrac{1}{t}$$
$$24t\left(\tfrac{1}{8}-\tfrac{1}{12}\right)=24t\bullet\dfrac{1}{t}$$
$$3t-2t=24$$
$$t=24$$

It will take 24 hours to fill the pool with both pipes left open.

45. The variation equation is $y=Kx$. Finding K:

$$8=K\bullet 12$$
$$K=\tfrac{2}{3}$$

So $y=\tfrac{2}{3}x$. Substituting $x=36$: $y=\tfrac{2}{3}(36)=24$

Chapter 7 Review

1. Solving the equation:

$$4x-2=7x+7$$
$$-3x-2=7$$
$$-3x=9$$
$$x=-3$$

3. Solving the equation:

$$8-3(2t+1)=5(t+2)$$
$$8-6t-3=5t+10$$
$$-6t+5=5t+10$$
$$-11t+5=10$$
$$-11t=5$$
$$t=-\tfrac{5}{11}$$

5. Solving the equation:

$$2x^2-5x=12$$
$$2x^2-5x-12=0$$
$$(2x+3)(x-4)=0$$
$$x=-\tfrac{3}{2},4$$

7. Solving the equation:

$$(x-2)(x-3)=2$$
$$x^2-5x+6=2$$
$$x^2-5x+4=0$$
$$(x-1)(x-4)=0$$
$$x=1,4$$

9. Let w represent the width and $3w$ represent the length. Using the perimeter formula:

$$2w+2(3w)=32$$
$$2w+6w=32$$
$$8w=32$$
$$w=4$$

The dimensions are 4 feet by 12 feet.

11. Solving the equation:
$$|x-3|=1$$
$$x-3=-1,1$$
$$x=2,4$$

13. Solving the equation:
$$|2y-3|=5$$
$$2y-3=-5,5$$
$$2y=-2,8$$
$$y=-1,4$$

15. Solving the equation:
$$|4x-3|+2=11$$
$$|4x-3|=9$$
$$4x-3=-9,9$$
$$4x=-6,12$$
$$x=-\tfrac{3}{2},3$$

17. Solving the inequality:
$$\tfrac{3}{4}x+1\le10$$
$$\tfrac{3}{4}x\le9$$
$$3x\le36$$
$$x\le12$$

The solution set is $(-\infty,12]$.

19. Solving the inequality:
$$\tfrac{1}{3}\le\tfrac{1}{6}x\le1$$
$$2\le x\le6$$

The solution set is $[2,6]$.

21. Solving the inequality:
$$5t+1\le3t-2 \quad\text{or}\quad -7t\le-21$$
$$2t\le-3 \qquad\qquad\qquad t\ge3$$
$$t\le-\tfrac{3}{2} \qquad\qquad\qquad t\ge3$$

The solution set is $\left(-\infty,-\tfrac{3}{2}\right]\cup[3,\infty)$.

23. Solving the inequality:

$$|y-2|<3$$
$$-3<y-2<3$$
$$-1<y<5$$

Graphing the solution set:

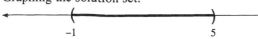

25. Solving the inequality:
$$|2t+1|-3<2$$
$$|2t+1|<5$$
$$-5<2t+1<5$$
$$-6<2t<4$$
$$-3<t<2$$

Graphing the solution set:

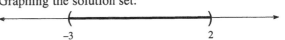

27. Solving the inequality:
$$|5+8t|+4\le1$$
$$|5+8t|\le-3$$
Since this statement is never true, there is no solution, or $\varnothing$.

29. Factoring: $x^4-16=\left(x^2+4\right)\left(x^2-4\right)=\left(x^2+4\right)(x+2)(x-2)$

31. Factoring: $a^3-8=(a-2)\left(a^2+2a+4\right)$

33. Factoring: $3a^3b-27ab^3=3ab\left(a^2-9b^2\right)=3ab(a-3b)(a+3b)$

35. Factoring: $36-25a^2=(6+5a)(6-5a)$

37. Multiply the second equation by –2:
$$6x-5y=-5$$
$$-6x-2y=-2$$
Adding yields:
$$-7y=-7$$
$$y=1$$
Substituting to find x:
$$3x+1=1$$
$$3x=0$$
$$x=0$$
The solution is $(0,1)$.

39. Multiply the first equation by 3 and the second equation by 4:
$$-21x + 12y = -3$$
$$20x - 12y = 0$$
Adding yields:
$$-x = -3$$
$$x = 3$$
Substitute to find y:
$$-21 + 4y = -1$$
$$4y = 20$$
$$y = 5$$
The solution is $(3,5)$.

41. Substitute into the first equation:
$$x + x - 1 = 2$$
$$2x - 1 = 2$$
$$2x = 3$$
$$x = \frac{3}{2}$$
$$y = \frac{3}{2} - 1 = \frac{1}{2}$$
The solution is $\left(\frac{3}{2}, \frac{1}{2}\right)$.

43. Substitute into the first equation:
$$3(-3y + 4) + 7y = 6$$
$$-9y + 12 + 7y = 6$$
$$-2y = -6$$
$$y = 3$$
$$x = -9 + 4 = -5$$
The solution is $(-5,3)$.

45. Adding the first and second equations yields:
$$2x - 2z = -2$$
$$x - z = -1$$
Adding the second and third equations yields $2x - 5z = -14$. So the system becomes:
$$x - z = -1$$
$$2x - 5z = -14$$
Multiply the first equation by -2:
$$-2x + 2z = 2$$
$$2x - 5z = -14$$
Adding yields:
$$-3z = -12$$
$$z = 4$$
Substitute to find x:
$$x - 4 = -1$$
$$x = 3$$
Substitute to find y:
$$3 + y + 4 = 6$$
$$y = -1$$
The solution is $(3,-1,4)$.

47. Multiply the second equation by 2 and add it to the third equation:
$$-6x + 8y - 2z = 4$$
$$6x - 8y + 2z = -4$$
Adding yields $0 = 0$. Since this is a true statement, there is no unique solution (dependent system).

49. Multiply the third equation by 2 and add it to the second equation:
$$3x - 2z = -2$$
$$10y + 2z = -2$$
Adding yields the equation $3x + 10y = -4$. So the system becomes:
$$2x - y = 5$$
$$3x + 10y = -4$$
Multiplying the first equation by 10:
$$20x - 10y = 50$$
$$3x + 10y = -4$$

Adding yields:
$$23x = 46$$
$$x = 2$$
Substituting to find y:
$$4 - y = 5$$
$$y = -1$$
Substituting to find z:
$$-5 + z = -1$$
$$z = 4$$
The solution is $(2, -1, 4)$.

Chapters 1-7 Cumulative Review

1. Simplifying: $15 - 12 \div 4 - 3 \cdot 2 = 15 - 3 - 6 = 6$

3. Simplifying: $\left(\frac{2}{5}\right)^{-2} = \left(\frac{5}{2}\right)^2 = \frac{25}{4}$

5. Simplifying: $5 - 3\left[2x - 4(x-2)\right] = 5 - 3(2x - 4x + 8) = 5 - 3(-2x + 8) = 5 + 6x - 24 = 6x - 19$

7. Simplifying: $(2x+3)\left(x^2 - 4x + 2\right) = 2x^3 - 8x^2 + 4x + 3x^2 - 12x + 6 = 2x^3 - 5x^2 - 8x + 6$

9. Solving the equation:
$$-6 + 2(2x+3) = 0$$
$$-6 + 4x + 6 = 0$$
$$4x = 0$$
$$x = 0$$

11. Solving the equation:
$$|2x - 3| + 7 = 1$$
$$|2x - 3| = -6$$
Since this statement is false, there is no solution, or $\varnothing$.

13. Multiply the first equation by 3 and the second equation by –2:
$$24x + 18y = 12$$
$$-24x - 18y = -16$$
Adding yields $0 = -4$. Since this statement is false, there is no solution (parallel lines).

15. Multiply the second equation by –2 and add it to the third equation:
$$-8y + 2z = 18$$
$$3x - 2z = -6$$
Adding yields the equation $3x - 8y = 12$. The system of equations becomes:
$$2x + y = 8$$
$$3x - 8y = 12$$
Multiply the first equation by 8:
$$16x + 8y = 64$$
$$3x - 8y = 12$$
Adding yields:
$$19x = 76$$
$$x = 4$$
Substituting to find y:
$$8 + y = 8$$
$$y = 0$$
Substituting to find z:
$$0 - z = -9$$
$$z = 9$$
The solution is $(4, 0, 9)$.

17.   Solving the inequality:
$$|2x - 7| \leq 3$$
$$-3 \leq 2x - 7 \leq 3$$
$$4 \leq 2x \leq 10$$
$$2 \leq x \leq 5$$
The solution set is $[2,5]$. Graphing the solution set:

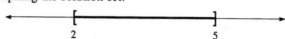

19.   Solving the inequality:
$$-3t \geq 12$$
$$t \leq -4$$
The solution set is $(-\infty, -4]$.

21.   Solving the compound inequality:
$$-2x > 6 \qquad \text{or} \qquad 3x - 7 > 2$$
$$x < -3 \qquad\qquad\qquad 3x > 9$$
$$x < -3 \qquad\qquad\qquad x > 3$$
The solution set is $(-\infty, -3) \cup (3, \infty)$.

23.   Finding the value: $-6 \cdot \frac{7}{54} - \frac{2}{9} = -\frac{7}{9} - \frac{2}{9} = -1$

25.   The rational numbers are: $-1, 0, 2.35, 4$

27.   Solving the equation:
$$3 + \frac{1}{x} = \frac{10}{x^2}$$
$$x^2\left(3 + \frac{1}{x}\right) = x^2\left(\frac{10}{x^2}\right)$$
$$3x^2 + x = 10$$
$$3x^2 + x - 10 = 0$$
$$(3x - 5)(x + 2) = 0$$
$$x = -2, \tfrac{5}{3}$$

29.   Simplifying: $\left(5x^{-3}y^2z^{-2}\right)\left(6x^{-5}y^4z^{-3}\right) = 30x^{-8}y^6z^{-5} = \dfrac{30y^6}{x^8z^5}$

31.   Simplifying: $\dfrac{\left(7 \times 10^{-5}\right)\left(21 \times 10^{-6}\right)}{3 \times 10^{-12}} = \dfrac{147 \times 10^{-11}}{3 \times 10^{-12}} = 49 \times 10 = 4.9 \times 10^2$

33.   Finding where the denominator is equal to 0:
$$x^2 - 25 = 0$$
$$(x + 5)(x - 5) = 0$$
$$x = -5, 5$$
The variable restrictions are $x \neq -5, 5$.

35.   Graphing the line:

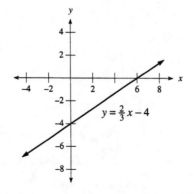

37. Factoring completely: $16y^2 + 2y + \frac{1}{16} = \left(4y + \frac{1}{4}\right)^2$

39. Factoring completely: $x^3 - 8 = (x-2)\left(x^2 + 2x + 4\right)$

41. Using long division:

$$
\begin{array}{r}
2x - 3 \\
x-2\overline{\smash{\big)}\,2x^2 - 7x + 9} \\
\underline{2x^2 - 4x} \\
-3x + 9 \\
\underline{-3x + 6} \\
3
\end{array}
$$

The quotient is $2x - 3 + \dfrac{3}{x-2}$.

43. Factor the denominators and add the fractions:

$$
\begin{aligned}
\frac{-2}{x^2 - 2x - 3} + \frac{3}{x^2 - 9} &= \frac{-2}{(x-3)(x+1)} \cdot \frac{x+3}{x+3} + \frac{3}{(x-3)(x+3)} \cdot \frac{x+1}{x+1} \\
&= \frac{-2x - 6}{(x-3)(x+3)(x+1)} + \frac{3x + 3}{(x-3)(x+3)(x+1)} \\
&= \frac{x - 3}{(x-3)(x+3)(x+1)} \\
&= \frac{1}{(x+3)(x+1)}
\end{aligned}
$$

45. Let b represent the rate of the boat and c the rate of the current. The system of equations is:

$$3(b + c) = 36$$
$$9(b - c) = 36$$

Simplifying the system:

$$b + c = 12$$
$$b - c = 4$$

Adding yields:

$$2b = 16$$
$$b = 8$$
$$c = 4$$

The boat's rate is 8 mph and the current's rate is 4 mph.

Chapter 7 Test

1. Solving the equation:
$$5 - \frac{4}{7}a = -11$$
$$7\left(5 - \frac{4}{7}a\right) = 7(-11)$$
$$35 - 4a = -77$$
$$-4a = -112$$
$$a = 28$$

2. Solving the equation:
$$3x^2 = 5x + 2$$
$$3x^2 - 5x - 2 = 0$$
$$(3x + 1)(x - 2) = 0$$
$$x = -\tfrac{1}{3}, 2$$

3. Solving the equation:
$$100x^3 = 500x^2$$
$$100x^3 - 500x^2 = 0$$
$$100x^2(x - 5) = 0$$
$$x = 0, 5$$

4. Solving the equation:
$$5(x - 1) - 2(2x + 3) = 5x - 4$$
$$5x - 5 - 4x - 6 = 5x - 4$$
$$x - 11 = 5x - 4$$
$$-4x = 7$$
$$x = -\tfrac{7}{4}$$

5.   Solving the equation:
$$(x+1)(x+2)=12$$
$$x^2+3x+2=12$$
$$x^2+3x-10=0$$
$$(x+5)(x-2)=0$$
$$x=-5,2$$

6.   Let w represent the width and $2w$ represent the length. Using the perimeter formula:
$$2(w)+2(2w)=36$$
$$2w+4w=36$$
$$6w=36$$
$$w=6$$
$$2w=12$$
The width is 6 inches and the length is 12 inches.

7.   Let $h=0$ in the equation:
$$32t-16t^2=0$$
$$16t(2-t)=0$$
$$t=0,2$$
The object is on the ground after 0 seconds and after 2 seconds.

8.   Solving the equation:
$$\left|\tfrac{1}{4}x-1\right|=\tfrac{1}{2}$$
$$\tfrac{1}{4}x-1=-\tfrac{1}{2},\tfrac{1}{2}$$
$$\tfrac{1}{4}x=\tfrac{1}{2},\tfrac{3}{2}$$
$$x=2,6$$

9.   Solving the equation:
$$|3-2x|+5=2$$
$$|3-2x|=-3$$
Since this statement is false, there is no solution, or $\varnothing$.

10.   Solving the inequality:
$$5-\tfrac{3}{2}x>-1 \qquad \text{or} \qquad 2x-5\geq 7$$
$$10-3x>-2 \qquad\qquad\qquad 2x\geq 12$$
$$-3x>-12 \qquad\qquad\qquad x\geq 6$$
$$x<4 \qquad\qquad\qquad\qquad x\geq 6$$
The solution set is $(-\infty,4)\cup[6,\infty)$. Graphing the solution set:

11.   Solving the inequality:
$$-3\leq 5x-1\leq 9$$
$$-2\leq 5x\leq 10$$
$$-\tfrac{2}{5}\leq x\leq 2$$

The solution set is $\left[-\tfrac{2}{5},2\right]$. Graphing the solution set:

12. Solving the inequality:

$$|6x - 1| > 7$$

$$6x - 1 < -7 \quad \text{or} \quad 6x - 1 > 7$$

$$6x < -6 \qquad\qquad\qquad 6x > 8$$

$$x < -1 \qquad\qquad\qquad x > \tfrac{4}{3}$$

Graphing the solution set:

$$-1 \qquad\qquad\qquad 4/3$$

13. Solving the inequality:

$$|3x - 5| - 4 \le 3$$

$$|3x - 5| \le 7$$

$$-7 \le 3x - 5 \le 7$$

$$-2 \le 3x \le 12$$

$$-\tfrac{2}{3} \le x \le 4$$

Graphing the solution set:

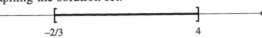

$$-2/3 \qquad\qquad\qquad 4$$

14. Factoring: $x^2 + x - 12 = (x + 4)(x - 3)$

15. Factoring: $16a^4 - 81y^4 = \left(4a^2 + 9y^2\right)\left(4a^2 - 9y^2\right) = \left(4a^2 + 9y^2\right)(2a + 3y)(2a - 3y)$

16. Factoring: $t^3 + \tfrac{1}{8} = t^3 + \left(\tfrac{1}{2}\right)^3 = \left(t + \tfrac{1}{2}\right)\left(t^2 - \tfrac{1}{2}t + \tfrac{1}{4}\right)$

17. Factoring: $4a^5b - 24a^4b^2 - 64a^3b^3 = 4a^3b\left(a^2 - 6ab - 16b^2\right) = 4a^3b(a - 8b)(a + 2b)$

18. Multiply the second equation by 5:

$$2x - 5y = -8$$

$$15x + 5y = 25$$

Adding yields:

$$17x = 17$$

$$x = 1$$

Substituting to find y:

$$3 + y = 5$$

$$y = 2$$

The solution is $(1,2)$.

19. To clear each equation of fractions, multiply the first equation by 6 and the second equation by 20:

$$2x - y = 18$$

$$-4x + 5y = 0$$

Multiply the first equation by 2:

$$4x - 2y = 36$$

$$-4x + 5y = 0$$

Adding yields:

$$3y = 36$$

$$y = 12$$

Substituting to find x:

$$2x - 12 = 18$$

$$2x = 30$$

$$x = 15$$

The solution is $(15,12)$.

20. Substituting into the first equation:

$$2x - 5(3x - 10) = 24$$

$$2x - 15x + 50 = 24$$

$$-13x = -26$$

$$x = 2$$

Substituting to find y: $y = 3(2) - 10 = -4$. The solution is $(2, -4)$.

21. Adding the first and third equations:

$$5x = 15$$

$$x = 3$$

Adding the first and second equations:

$$3x - 2z = 7$$

$$9 - 2z = 7$$

$$-2z = -2$$

$$z = 1$$

Substituting to find y:
$$3 + y - 3 = -2$$
$$y = -2$$
The solution is $(3, -2, 1)$.

22. Let x and $2x - 1$ represent the two numbers. The equation is:
$$x + 2x - 1 = 14$$
$$3x = 15$$
$$x = 5$$
$$2x - 1 = 9$$
The two numbers are 5 and 9.

23. Let x and $2x$ represent the two investments. The equation is:
$$0.05(x) + 0.06(2x) = 680$$
$$0.17x = 680$$
$$x = 4000$$
$$2x = 8000$$
John invested \$4000 at 5% and \$8000 at 6%.

24. Let n, d, and q represent the number of nickels, dimes, and quarters. The system of equations is:
$$n + d + q = 15$$
$$0.05n + 0.10d + 0.25q = 1.10$$
$$n = 4d - 1$$
Substituting into the first equation:
$$4d - 1 + d + q = 15$$
$$5d + q = 16$$
Substituting into the second equation:
$$0.05(4d - 1) + 0.10d + 0.25q = 1.10$$
$$0.3d + 0.25q = 1.15$$
The system of equations becomes:
$$5d + q = 16$$
$$0.3d + 0.25q = 1.15$$
Multiply the first equation by -0.25:
$$-1.25d - 0.25q = -4$$
$$0.3d + 0.25q = 1.15$$
Adding yields:
$$-0.95d = -2.85$$
$$d = 3$$
Substituting to find q:
$$15 + q = 16$$
$$q = 1$$
Substituting to find n:
$$n + 3 + 1 = 15$$
$$n = 11$$
The collection contains 11 nickels, 3 dimes, and 1 quarter.

Chapter 8
Equations and Inequalities in Two Variables

8.1 The Slope of a Line

1. The slope is $\frac{3}{2}$.

3. There is no slope (undefined).

5. The slope is $\frac{2}{3}$.

7. Finding the slope: $m = \dfrac{4-1}{4-2} = \dfrac{3}{2}$

Sketching the graph:

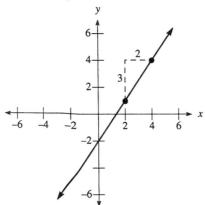

9. Finding the slope: $m = \dfrac{2-4}{5-1} = \dfrac{-2}{4} = -\dfrac{1}{2}$

Sketching the graph:

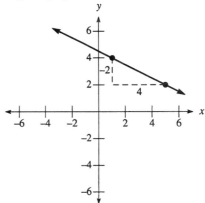

11. Finding the slope: $m = \dfrac{2-(-3)}{4-1} = \dfrac{2+3}{3} = \dfrac{5}{3}$

Sketching the graph:

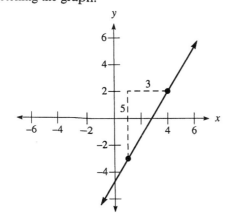

13. Finding the slope: $m = \dfrac{3-(-2)}{1-(-3)} = \dfrac{3+2}{1+3} = \dfrac{5}{4}$

Sketching the graph:

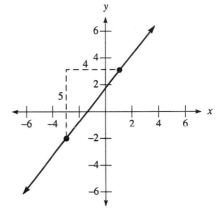

15. Finding the slope: $m = \dfrac{-2-2}{3-(-3)} = \dfrac{-4}{3+3} = -\dfrac{2}{3}$

Sketching the graph:

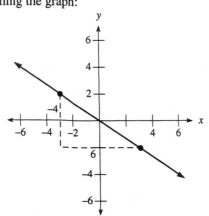

19. Completing the table:

x	y
0	2
3	0

Finding the slope: $m = \dfrac{2-0}{0-3} = -\dfrac{2}{3}$

23. Graphing the line:

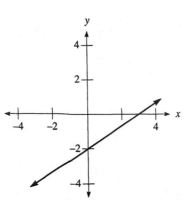

The slope is $m = \dfrac{2}{3}$.

17. Finding the slope: $m = \dfrac{-2-(-5)}{3-2} = \dfrac{-2+5}{1} = 3$

Sketching the graph:

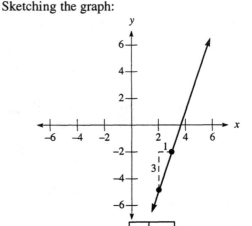

21. Completing the table:

x	y
0	−5
3	−3

Finding the slope: $m = \dfrac{-5-(-3)}{0-3} = \dfrac{-5+3}{-3} = \dfrac{2}{3}$

25. Graphing the line:

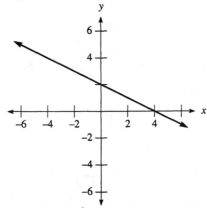

The slope is $m = -\dfrac{1}{2}$.

27. **a.**   Since the slopes between each successive pairs of points is 2, this could represent ordered pairs from a line.
b.   Since the slopes between each successive pairs of points is not the same, this could not represent ordered pairs from a line.

29. Finding the slope of this line: $m = \dfrac{1-3}{-8-2} = \dfrac{-2}{-10} = \dfrac{1}{5}$. Since the parallel slope is the same, its slope is $\dfrac{1}{5}$.

31. Finding the slope of this line: $m = \dfrac{2-(-6)}{5-5} = \dfrac{8}{0} =$ undefined

Since the perpendicular slope is a horizontal line, its slope is 0.

33. Using the slope formula:
$$\frac{\Delta y}{12} = \frac{2}{3}$$
$$3\Delta y = 24$$
$$\Delta y = 8$$

35. a. Let d represent the distance. Using the slope formula:

$$\frac{0-1106}{d-0}=-\frac{7}{100}$$
$$-7d=-110600$$
$$d=15800$$

The distance to point A is 15,800 feet.

b. The slope is $-\frac{7}{100}=-0.07$.

37. a. It takes 10 minutes for all the ice to melt.   **b.** It takes 20 minutes before the water boils.

c. The slope of A is 20°C per minute.   **d.** The slope of C is 10°C per minute.

e. It is changing faster during the first minute, since its slope is greater.

39. Substituting $x=4$:   **41.** Solving for y:

$$3(4)+2y=12$$
$$12+2y=12$$
$$2y=0$$
$$y=0$$

$$3x+2y=12$$
$$2y=-3x+12$$
$$y=-\frac{3}{2}x+6$$

43. Solving for t:

$$A=P+Prt$$
$$A-P=Prt$$
$$t=\frac{A-P}{Pr}$$

45. Graphing the curves:

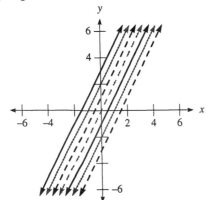

47. Graphing the curves:

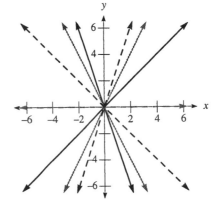

49. Graphing the curves:

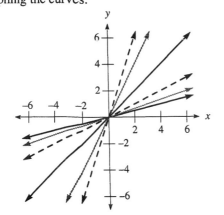

8.2 The Equation of a Line

1.   Using the slope-intercept formula: $y = 2x + 3$

3.   Using the slope-intercept formula: $y = x - 5$

5.   Using the slope-intercept formula: $y = \frac{1}{2}x + \frac{3}{2}$

7.   Using the slope-intercept formula: $y = 4$

9.   The slope is 3, the y-intercept is –2, and the perpendicular slope is $-\frac{1}{3}$.

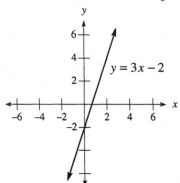

11.   The slope is $\frac{2}{3}$, the y-intercept is –4, and the perpendicular slope is $-\frac{3}{2}$.

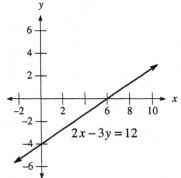

13.   The slope is $-\frac{4}{5}$, the y-intercept is 4, and the perpendicular slope is $\frac{5}{4}$.

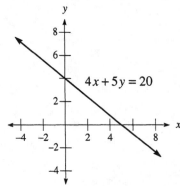

15.   The slope is $\frac{1}{2}$ and the y-intercept is –4. Using the slope-intercept form, the equation is $y = \frac{1}{2}x - 4$.

17.   The slope is $-\frac{2}{3}$ and the y-intercept = 3. Using the slope-intercept form, the equation is $y = -\frac{2}{3}x + 3$.

19. Using the point-slope formula:

$$y-(-5)=2(x-(-2))$$
$$y+5=2(x+2)$$
$$y+5=2x+4$$
$$y=2x-1$$

21. Using the point-slope formula:

$$y-1=-\tfrac{1}{2}(x-(-4))$$
$$y-1=-\tfrac{1}{2}(x+4)$$
$$y-1=-\tfrac{1}{2}x-2$$
$$y=-\tfrac{1}{2}x-1$$

23. Using the point-slope formula:

$$y-2=-3\left(x-\left(-\tfrac{1}{3}\right)\right)$$
$$y-2=-3\left(x+\tfrac{1}{3}\right)$$
$$y-2=-3x-1$$
$$y=-3x+1$$

25. First find the slope: $m=\dfrac{-1-(-4)}{1-(-2)}=\dfrac{-1+4}{1+2}=\dfrac{3}{3}=1$. Using the point-slope formula:

$$y-(-1)=1(x-1)$$
$$y+1=x-1$$
$$-x+y=-2$$
$$x-y=2$$

27. First find the slope: $m=\dfrac{6-(-2)}{3-(-3)}=\dfrac{6+2}{3+3}=\dfrac{8}{6}=\dfrac{4}{3}$. Using the point-slope formula:

$$y-6=\tfrac{4}{3}(x-3)$$
$$3y-18=4x-12$$
$$-4x+3y=6$$
$$4x-3y=-6$$

29. First find the slope: $m=\dfrac{-1-\left(-\tfrac{1}{5}\right)}{-\tfrac{1}{3}-\tfrac{1}{3}}=\dfrac{-1+\tfrac{1}{5}}{-\tfrac{2}{3}}=\dfrac{-\tfrac{4}{5}}{-\tfrac{2}{3}}=\tfrac{4}{5}\cdot\tfrac{3}{2}=\tfrac{6}{5}$. Using the point-slope formula:

$$y-(-1)=\tfrac{6}{5}\left(x-\left(-\tfrac{1}{3}\right)\right)$$
$$y+1=\tfrac{6}{5}\left(x+\tfrac{1}{3}\right)$$
$$5(y+1)=6\left(x+\tfrac{1}{3}\right)$$
$$5y+5=6x+2$$
$$6x-5y=3$$

31. Two points on the line are $(0,-4)$ and $(2,0)$. Finding the slope: $m=\dfrac{0-(-4)}{2-0}=\dfrac{4}{2}=2$

Using the slope-intercept form, the equation is $y=2x-4$.

33. Two points on the line are $(0,4)$ and $(-2,0)$. Finding the slope: $m=\dfrac{0-4}{-2-0}=\dfrac{-4}{-2}=2$

Using the slope-intercept form, the equation is $y=2x+4$.

35. The slope is 0 and the y-intercept is -2.

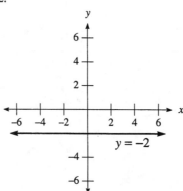

37. First find the slope:
$$3x - y = 5$$
$$-y = -3x + 5$$
$$y = 3x - 5$$
So the slope is 3. Using $(-1,4)$ in the point-slope formula:
$$y - 4 = 3(x - (-1))$$
$$y - 4 = 3(x + 1)$$
$$y - 4 = 3x + 3$$
$$y = 3x + 7$$

39. First find the slope:
$$2x - 5y = 10$$
$$-5y = -2x + 10$$
$$y = \tfrac{2}{5}x - 2$$
So the perpendicular slope is $-\tfrac{5}{2}$. Using $(-4,-3)$ in the point-slope formula:
$$y - (-3) = -\tfrac{5}{2}(x - (-4))$$
$$y + 3 = -\tfrac{5}{2}(x + 4)$$
$$y + 3 = -\tfrac{5}{2}x - 10$$
$$y = -\tfrac{5}{2}x - 13$$

41. The perpendicular slope is $\tfrac{1}{4}$. Using $(-1,0)$ in the point-slope formula:
$$y - 0 = \tfrac{1}{4}(x - (-1))$$
$$y = \tfrac{1}{4}(x + 1)$$
$$y = \tfrac{1}{4}x + \tfrac{1}{4}$$

43. Using the points $(3,0)$ and $(0,2)$, first find the slope: $m = \dfrac{2 - 0}{0 - 3} = -\tfrac{2}{3}$

Using the slope-intercept formula, the equation is: $y = -\tfrac{2}{3}x + 2$

45. **a.** Using the points $(0,32)$ and $(25,77)$, first find the slope: $m = \dfrac{77 - 32}{25 - 0} = \dfrac{45}{25} = \dfrac{9}{5}$

Using the slope-intercept formula, the equation is: $F = \tfrac{9}{5}C + 32$

b. Substituting $C = 30$: $F = \tfrac{9}{5}(30) + 32 = 54 + 32 = 86°$

47.   First find the slope: $m = \dfrac{1.5-1}{155-98} = \dfrac{\frac{1}{2}}{57} = \dfrac{1}{114}$. Using $(98,1)$ in the point-slope formula:

$$y - 1 = \tfrac{1}{114}(x - 98)$$
$$y - 1 = \tfrac{1}{114}x - \tfrac{49}{57}$$
$$y = \tfrac{1}{114}x + \tfrac{8}{57}$$

49.   a.   First finding the slope: $m = \dfrac{10 - 9.3}{5 - 3} = \dfrac{0.7}{2} = 0.35$. Using $(5,10)$ in the point-slope formula:

$$C - 10 = 0.35(x - 5)$$
$$C - 10 = 0.35x - 1.75$$
$$C = 0.35x + 8.25$$

   b.   The slope is 0.35; this means the variable cost increases by 0.35 cents/mile each year.

51.   Multiply the first equation by -1 and add it to the second equation:

$$-x - y - z = -6$$
$$2x - y + z = 3$$

Adding yields the equation $x - 2y = -3$. Multiply the first equation by 3 and add it to the third equation:

$$3x + 3y + 3z = 18$$
$$x + 2y - 3z = -4$$

Adding yields the equation $4x + 5y = 14$. So the system becomes:

$$x - 2y = -3$$
$$4x + 5y = 14$$

Multiply the first equation by -4 and add it to the second equation:

$$-4x + 8y = 12$$
$$4x + 5y = 14$$

Adding yields:

$$13y = 26$$
$$y = 2$$

Substituting to find x:

$$4x + 5(2) = 14$$
$$4x + 10 = 14$$
$$4x = 4$$
$$x = 1$$

Substituting into the original first equation:

$$1 + 2 + z = 6$$
$$z + 3 = 6$$
$$z = 3$$

The solution is $(1,2,3)$.

53.   Multiply the third equation by -1 and add it to the first equation:

$$3x + 4y = 15$$
$$-4y + 3z = -9$$

Adding yields the equation $3x + 3z = 6$, or $x + z = 2$. So the system becomes:

$$2x - 5z = -3$$
$$x + z = 2$$

Multiply the second equation by -2:

$$2x - 5z = -3$$
$$-2x - 2z = -4$$

Adding yields:

$$-7z = -7$$
$$z = 1$$

Substituting to find x:
$$2x - 5(1) = -3$$
$$2x - 5 = -3$$
$$2x = 2$$
$$x = 1$$
Substituting into the original first equation:
$$3(1) + 4y = 15$$
$$4y + 3 = 15$$
$$4y = 12$$
$$y = 3$$
The solution is $(1, 3, 1)$.

55.   Writing the equation in slope-intercept form:
$$\frac{x}{2} + \frac{y}{3} = 1$$
$$\frac{y}{3} = -\frac{x}{2} + 1$$
$$y = -\frac{3}{2}x + 3$$
The slope is $-\frac{3}{2}$, the x-intercept is 2, and the y-intercept is 3.

57.   Writing the equation in slope-intercept form:
$$\frac{x}{-2} + \frac{y}{3} = 1$$
$$\frac{y}{3} = \frac{x}{2} + 1$$
$$y = \frac{3}{2}x + 3$$
The slope is $\frac{3}{2}$, the x-intercept is -2, and the y-intercept is 3.

59.   The x-intercept is a, the y-intercept is b, and the slope is $-\frac{b}{a}$.

8.3   Linear Inequalities in Two Variables

1.   Graphing the solution set:

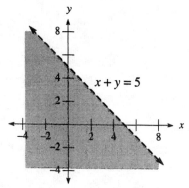

3.   Graphing the solution set:

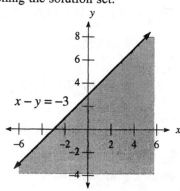

5. Graphing the solution set:

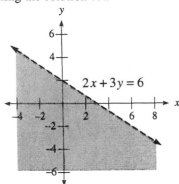

$2x + 3y = 6$

7. Graphing the solution set:

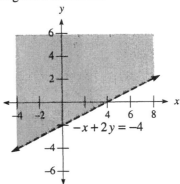

$-x + 2y = -4$

9. Graphing the solution set:

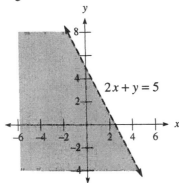

$2x + y = 5$

11. Graphing the solution set:

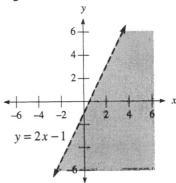

$y = 2x - 1$

13. The inequality is $x + y > 4$.

17. Graphing the inequality:

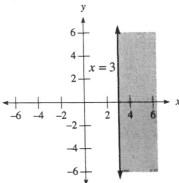

$x = 3$

15. The inequality is $-x + 2y \le 4$ or $y \le \frac{1}{2}x + 2$.

19. Graphing the inequality:

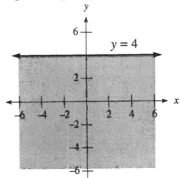

$y = 4$

21. Graphing the inequality:

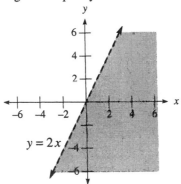

$y = 2x$

23. Graphing the inequality:

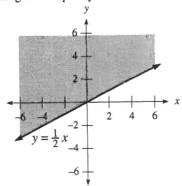

$y = \frac{1}{2}x$

25. Graphing the inequality:

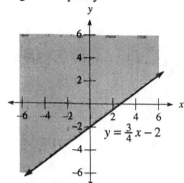

$y = \frac{3}{4}x - 2$

27. Graphing the inequality:

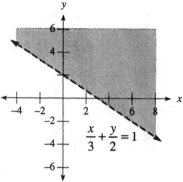

$\frac{x}{3} + \frac{y}{2} = 1$

29. Graphing the region:

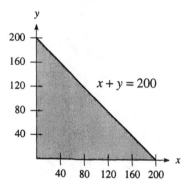

$x + y = 200$

31. Graphing the region:

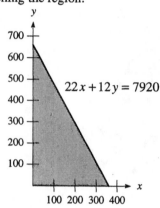

$22x + 12y = 7920$

33. Graphing the region:

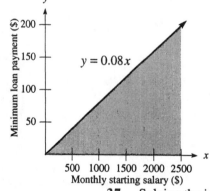

$y = 0.08x$

Monthly starting salary ($)

35. Solving the inequality:

$$\frac{1}{3} + \frac{y}{5} \le \frac{26}{15}$$

$$15\left(\frac{1}{3} + \frac{y}{5}\right) \le 15\left(\frac{26}{15}\right)$$

$$5 + 3y \le 26$$

$$3y \le 21$$

$$y \le 7$$

37. Solving the inequality:

$$5t - 4 > 3t - 8$$

$$2t - 4 > -8$$

$$2t > -4$$

$$t > -2$$

39. Solving the inequality:

$$-9 < -4 + 5t < 6$$

$$-5 < 5t < 10$$

$$-1 < t < 2$$

41. Graphing the inequality:

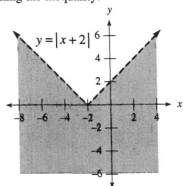

43. Graphing the inequality:

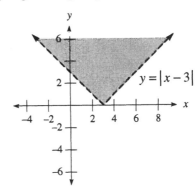

8.4 Introduction to Functions

1. The domain is {1,2,4} and the range is {1,3,5}. This is a function.
3. The domain is {−1,1,2} and the range is {−5,3}. This is a function.
5. The domain is {3,7} and the range is {−1,4}. This is not a function.
7. Yes, since it passes the vertical line test.
9. No, since it fails the vertical line test.
11. No, since it fails the vertical line test.
13. Yes, since it passes the vertical line test.
15. Yes, since it passes the vertical line test.
17. The domain is all real numbers and the range is $\{y \mid y \geq -1\}$. This is a function.

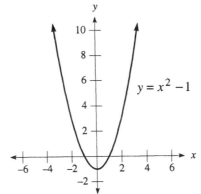

19. The domain is all real numbers and the range is $\{y \mid y \geq 4\}$. This is a function.

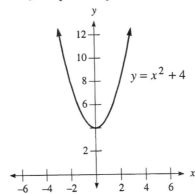

21.   The domain is $\{x \mid x \geq -1\}$ and the range is all real numbers. This is not a function.

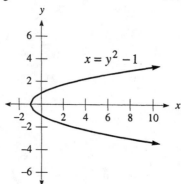

23.   The domain is $\{x \mid x \geq 4\}$ and the range is all real numbers. This is not a function.

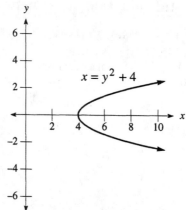

25.   The domain is all real numbers and the range is $\{y \mid y \geq 0\}$. This is a function.

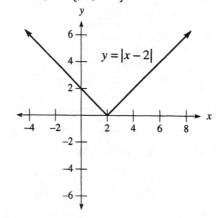

27. The domain is all real numbers and the range is $\{y \mid y \geq -2\}$. This is a function.

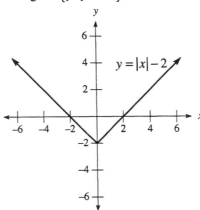

29. **a.** The equation is $y = 8.5x$ for $10 \leq x \leq 40$.

b. Completing the table:

Hours Worked	Function Rule	Gross Pay ($)
x	$y = 8.5x$	y
10	$y = 8.5(10) = 85$	85
20	$y = 8.5(20) = 170$	170
30	$y = 8.5(30) = 255$	255
40	$y = 8.5(40) = 340$	340

c. Constructing a line graph:

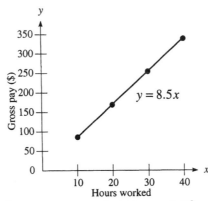

d. The domain is $\{x \mid 10 \leq x \leq 40\}$ and the range is $\{y \mid 85 \leq y \leq 340\}$.

e. The minimum is $85 and the maximum is $340.

Time (sec)	Function Rule	Distance (ft)
t	$h = 16t - 16t^2$	h
0	$h = 16(0) - 16(0)^2$	0
0.1	$h = 16(0.1) - 16(0.1)^2$	1.44
0.2	$h = 16(0.2) - 16(0.2)^2$	2.56
0.3	$h = 16(0.3) - 16(0.3)^2$	3.36
0.4	$h = 16(0.4) - 16(0.4)^2$	3.84
0.5	$h = 16(0.5) - 16(0.5)^2$	4
0.6	$h = 16(0.6) - 16(0.6)^2$	3.84
0.7	$h = 16(0.7) - 16(0.7)^2$	3.36
0.8	$h = 16(0.8) - 16(0.8)^2$	2.56
0.9	$h = 16(0.9) - 16(0.9)^2$	1.44
1	$h = 16(1) - 16(1)^2$	0

31. **a.** Completing the table:

b. The domain is $\{t \mid 0 \le t \le 1\}$ and the range is $\{h \mid 0 \le h \le 4\}$.

c. Graphing the function:

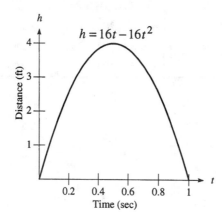

33. **a.** Graphing the function:

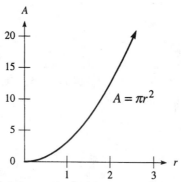

b. The domain is $\{r \mid 0 \le r \le 3\}$ and the range is $\{A \mid 0 \le A \le 9\pi\}$.

35. **a.** Yes, since it passes the vertical line test.

b. The domain is $\{t \mid 0 \le t \le 6\}$ and the range is $\{h \mid 0 \le h \le 60\}$.

c. At time $t = 3$ the ball reaches its maximum height.

d. The maximum height is $h = 60$.   **e.** At time $t = 6$ the ball hits the ground.

37. Substituting $x = 4$: $y = 3(4) - 2 = 12 - 2 = 10$ **39.** Substituting $x = -4$: $y = 3(-4) - 2 = -12 - 2 = -14$

41. Substituting $x = 2$: $y = (2)^2 - 3 = 4 - 3 = 1$ **43.** Substituting $x = 0$: $y = (0)^2 - 3 = 0 - 3 = -3$

45.   **a.**   Figure 11   **b.**   Figure 12
        c.   Figure 10   **d.**   Figure 9

8.5   Function   Notation

1.   Evaluating the function: $f(2) = 2(2) - 5 = 4 - 5 = -1$

3.   Evaluating the function: $f(-3) = 2(-3) - 5 = -6 - 5 = -11$

5.   Evaluating the function: $g(-1) = (-1)^2 + 3(-1) + 4 = 1 - 3 + 4 = 2$

7.   Evaluating the function: $g(-3) = (-3)^2 + 3(-3) + 4 = 9 - 9 + 4 = 4$

9.   First evaluate each function:

$$g(4) = (4)^2 + 3(4) + 4 = 16 + 12 + 4 = 32 \qquad f(4) = 2(4) - 5 = 8 - 5 = 3$$

Now evaluating: $g(4) + f(4) = 32 + 3 = 35$

11.   First evaluate each function:

$$f(3) = 2(3) - 5 = 6 - 5 = 1 \qquad g(2) = (2)^2 + 3(2) + 4 = 4 + 6 + 4 = 14$$

Now evaluating: $f(3) - g(2) = 1 - 14 = -13$

13.   Evaluating the function: $f(0) = 3(0)^2 - 4(0) + 1 = 0 - 0 + 1 = 1$

15.   Evaluating the function: $g(-4) = 2(-4) - 1 = -8 - 1 = -9$

17.   Evaluating the function: $f(-1) = 3(-1)^2 - 4(-1) + 1 = 3 + 4 + 1 = 8$

19.   Evaluating the function: $g(10) = 2(10) - 1 = 20 - 1 = 19$

21.   Evaluating the function: $f(3) = 3(3)^2 - 4(3) + 1 = 27 - 12 + 1 = 16$

23.   Evaluating the function: $g\left(\frac{1}{2}\right) = 2\left(\frac{1}{2}\right) - 1 = 1 - 1 = 0$ **25.**   Evaluating the function: $f(a) = 3a^2 - 4a + 1$

27. $f(1) = 4$ **29.** $g\left(\frac{1}{2}\right) = 0$

31. $g(-2) = 2$ **33.**   Evaluating the function: $f(0) = 2(0)^2 - 8 = 0 - 8 = -8$

35.   Evaluating the function: $g(-4) = \frac{1}{2}(-4) + 1 = -2 + 1 = -1$

37.   Evaluating the function: $f(a) = 2a^2 - 8$ **39.**   Evaluating the function: $f(b) = 2b^2 - 8$

41.   Evaluating the function: $f\big[g(2)\big] = f\left[\frac{1}{2}(2) + 1\right] = f(2) = 2(2)^2 - 8 = 8 - 8 = 0$

43.   Evaluating the function: $g\big[f(-1)\big] = g\left[2(-1)^2 - 8\right] = g(-6) = \frac{1}{2}(-6) + 1 = -3 + 1 = -2$

45.   Evaluating the function: $g\big[f(0)\big] = g\left[2(0)^2 - 8\right] = g(-8) = \frac{1}{2}(-8) + 1 = -4 + 1 = -3$

47.   Graphing the function:

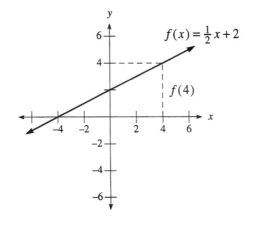

49.   Finding where $f(x) = x$:
$$\tfrac{1}{2}x + 2 = x$$
$$2 = \tfrac{1}{2}x$$
$$x = 4$$

51.   Graphing the function:

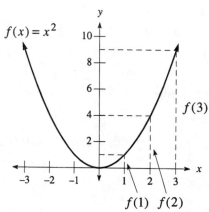

53.   Evaluating: $V(3) = 150 \cdot 2^{3/3} = 150 \cdot 2 = 300$; The painting is worth $300 in 3 years.
Evaluating: $V(6) = 150 \cdot 2^{6/3} = 150 \cdot 4 = 600$; The painting is worth $600 in 6 years.

55.   Let x represent the width and $2x + 3$ represent the length. Then the perimeter is given by:
$$P(x) = 2(x) + 2(2x + 3) = 2x + 4x + 6 = 6x + 6, \text{ where } x > 0$$

57.   Finding the values:
$$A(2) = 3.14(2)^2 = 3.14(4) = 12.56$$
$$A(5) = 3.14(5)^2 = 3.14(25) = 78.5$$
$$A(10) = 3.14(10)^2 = 3.14(100) = 314$$

59.   **a.**   Evaluating: $V(3.75) = -3300(3.75) + 18000 = \$5,625$
   b.   Evaluating: $V(5) = -3300(5) + 18000 = \$1,500$
   c.   The domain of this function is $\{t \mid 0 \le t \le 5\}$.
   d.   Sketching the graph:

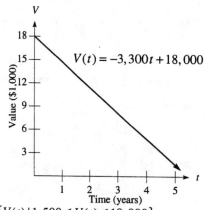

   e.   The range of this function is $\{V(t) \mid 1,500 \le V(t) \le 18,000\}$.
   f.   Solving $V(t) = 10000$:
$$-3300t + 18000 = 10000$$
$$-3300t = -8000$$
$$t \approx 2.42$$
The copier will be worth $10,000 after approximately 2.42 years.

61. Solving the equation:

$$|3x - 5| = 7$$
$$3x - 5 = -7, 7$$
$$3x = -2, 12$$
$$x = -\tfrac{2}{3}, 4$$

63. Solving the equation:

$$|4y + 2| - 8 = -2$$
$$|4y + 2| = 6$$
$$4y + 2 = -6, 6$$
$$4y = -8, 4$$
$$y = -2, 1$$

65. Solving the equation:

$$5 + |6t + 2| = 3$$
$$|6t + 2| = -2$$

Since this last equation is impossible, there is no solution, or $\varnothing$.

67. **a.** $f(2) = 2$ **b.** $f(-4) = 0$
c. $g(0) = 1$ **d.** $g(3) = 4$

69. **a.** Completing the table:

Weight (ounces)	0.6	1.0	1.1	2.5	3.0	4.8	5.0	5.3
Cost (cents)	32	32	55	78	78	124	124	147

b. The letter weighs over 2 ounces, but not over 3 ounces. As an inequality, this can be written as $2 < x \le 3$.
c. The domain is $\{x \mid 0 < x \le 6\}$. **d.** The range is $\{32, 55, 78, 101, 124, 147\}$.

8.6 Algebra and Composition with Functions

1. Writing the formula: $f + g = f(x) + g(x) = (4x - 3) + (2x + 5) = 6x + 2$
3. Writing the formula: $g - f = g(x) - f(x) = (2x + 5) - (4x - 3) = -2x + 8$
5. Writing the formula: $fg = f(x) \bullet g(x) = (4x - 3)(2x + 5) = 8x^2 + 14x - 15$
7. Writing the formula: $g / f = \dfrac{g(x)}{f(x)} = \dfrac{2x + 5}{4x - 3}$

9. Writing the formula: $f + g = f(x) + g(x) = (3x - 5) + (x - 2) = 4x - 7$
11. Writing the formula: $g + h = g(x) + h(x) = (x - 2) + (3x^2 - 11x + 10) = 3x^2 - 10x + 8$
13. Writing the formula: $g - f = g(x) - f(x) = (x - 2) - (3x - 5) = -2x + 3$
15. Writing the formula: $fg = f(x) \bullet g(x) = (3x - 5)(x - 2) = 3x^2 - 11x + 10$
17. Writing the formula:

$$fh = f(x) \bullet h(x) = (3x - 5)(3x^2 - 11x + 10) = 9x^3 - 33x^2 + 30x - 15x^2 + 55x - 50 = 9x^3 - 48x^2 + 85x - 50$$

19. Writing the formula: $h / f = \dfrac{h(x)}{f(x)} = \dfrac{3x^2 - 11x + 10}{3x - 5} = \dfrac{(3x - 5)(x - 2)}{3x - 5} = x - 2$

21. Writing the formula: $f / h = \dfrac{f(x)}{h(x)} = \dfrac{3x - 5}{3x^2 - 11x + 10} = \dfrac{3x - 5}{(3x - 5)(x - 2)} = \dfrac{1}{x - 2}$

23. Writing the formula: $f + g + h = f(x) + g(x) + h(x) = (3x - 5) + (x - 2) + (3x^2 - 11x + 10) = 3x^2 - 7x + 3$

25. Writing the formula:

$$h + fg = h(x) + f(x)g(x)$$
$$= (3x^2 - 11x + 10) + (3x - 5)(x - 2)$$
$$= 3x^2 - 11x + 10 + 3x^2 - 11x + 10$$
$$= 6x^2 - 22x + 20$$

27. Evaluating: $(f + g)(2) = f(2) + g(2) = (2 \bullet 2 + 1) + (4 \bullet 2 + 2) = 5 + 10 = 15$
29. Evaluating: $(fg)(3) = f(3) \bullet g(3) = (2 \bullet 3 + 1)(4 \bullet 3 + 2) = 7 \bullet 14 = 98$

31. Evaluating: $(h / g)(1) = \dfrac{h(1)}{g(1)} = \dfrac{4(1)^2 + 4(1) + 1}{4(1) + 2} = \dfrac{9}{6} = \dfrac{3}{2}$

33. Evaluating: $(fh)(0) = f(0) \bullet h(0) = (2(0)+1)(4(0)^2 + 4(0)+1) = (1)(1) = 1$

35. Evaluating: $(f+g+h)(2) = f(2) + g(2) + h(2) = (2(2)+1) + (4(2)+2) + (4(2)^2 + 4(2)+1) = 5+10+25 = 40$

37. Evaluating: $(h+fg)(3) = h(3) + f(3) \bullet g(3) = (4(3)^2 + 4(3)+1) + (2(3)+1) \bullet (4(3)+2) = 49+7 \bullet 14 = 49+98 = 147$

39. **a.**   Evaluating: $(f \circ g)(5) = f(g(5)) = f(5+4) = f(9) = 9^2 = 81$

     b.   Evaluating: $(g \circ f)(5) = g(f(5)) = g(5^2) = g(25) = 25+4 = 29$

     c.   Evaluating: $(f \circ g)(x) = f(g(x)) = f(x+4) = (x+4)^2$

     d.   Evaluating: $(g \circ f)(x) = g(f(x)) = g(x^2) = x^2 + 4$

41. **a.**   Evaluating: $(f \circ g)(0) = f(g(0)) = f(4 \bullet 0 - 1) = f(-1) = (-1)^2 + 3(-1) = 1-3 = -2$

     b.   Evaluating: $(g \circ f)(0) = g(f(0)) = g(0^2 + 3 \bullet 0) = g(0) = 4(0) - 1 = -1$

     c.   Evaluating: $(f \circ g)(x) = f(g(x)) = f(4x-1) = (4x-1)^2 + 3(4x-1) = 16x^2 - 8x + 1 + 12x - 3 = 16x^2 + 4x - 2$

     d.   Evaluating: $(g \circ f)(x) = g(f(x)) = g(x^2 + 3x) = 4(x^2 + 3x) - 1 = 4x^2 + 12x - 1$

43. **a.**   Finding the revenue: $R(x) = x(11.5 - 0.05x) = 11.5x - 0.05x^2$

     b.   Finding the cost: $C(x) = 2x + 200$

     c.   Finding the profit: $P(x) = R(x) - C(x) = (11.5x - 0.05x^2) - (2x + 200) = -0.05x^2 + 9.5x - 200$

     d.   Finding the average cost: $\overline{C}(x) = \dfrac{C(x)}{x} = \dfrac{2x+200}{x} = 2 + \dfrac{200}{x}$

45. **a.**   The function is $M(x) = 220 - x$.   **b.**   Evaluating: $M(24) = 220 - 24 = 196$ beats per minute

     c.   The training heart rate function is: $T(M) = 62 + 0.6(M - 62) = 0.6M + 24.8$

          Finding the composition: $T(M(x)) = T(220 - x) = 0.6(220 - x) + 24.8 = 156.8 - 0.6x$

          Evaluating: $T(M(24)) = 156.8 - 0.6(24) \approx 142$ beats per minute

     d.   Evaluating: $T(M(36)) = 156.8 - 0.6(36) \approx 135$ beats per minute

     e.   Evaluating: $T(M(48)) = 156.8 - 0.6(48) \approx 128$ beats per minute

47. Multiply the second equation by -2:

$$4x + 3y = 10$$
$$-4x - 2y = -8$$

Adding yields $y = 2$. Substituting into the first equation:

$$4x + 3(2) = 10$$
$$4x + 6 = 10$$
$$4x = 4$$
$$x = 1$$

The solution is $(1, 2)$.

49. Multiply the second equation by -5:

$$4x + 5y = 5$$
$$-6x - 5y = -10$$

Adding yields:

$$-2x = -5$$
$$x = \tfrac{5}{2}$$

Substituting into the first equation:

$$4\left(\tfrac{5}{2}\right) + 5y = 5$$
$$10 + 5y = 5$$
$$5y = -5$$
$$y = -1$$

The solution is $\left(\tfrac{5}{2}, -1\right)$.

51. Substituting into the first equation:

$$x + (x + 3) = 3$$
$$2x + 3 = 3$$
$$2x = 0$$
$$x = 0$$

The solution is $(0, 3)$.

53. Substituting into the first equation:

$$2x - 3(3x - 5) = -6$$
$$2x - 9x + 15 = -6$$
$$-7x + 15 = -6$$
$$-7x = -21$$
$$x = 3$$

The solution is $(3, 4)$.

8.7 Variation

1. The variation equation is $y = Kx$. Substituting $x = 2$ and $y = 10$:

$$10 = K \cdot 2$$
$$K = 5$$

So $y = 5x$. Substituting $x = 6$: $y = 5 \cdot 6 = 30$

3. The variation equation is $y = Kx$. Substituting $x = 4$ and $y = -32$:

$$-32 = K \cdot 4$$
$$K = -8$$

So $y = -8x$. Substituting $y = -40$:

$$-40 = -8x$$
$$x = 5$$

5. The variation equation is $r = \dfrac{K}{s}$. Substituting $s = 4$ and $r = -3$:

$$-3 = \frac{K}{4}$$
$$K = -12$$

So $r = \dfrac{-12}{s}$. Substituting $s = 2$: $r = \dfrac{-12}{2} = -6$

7. The variation equation is $r = \dfrac{K}{s}$. Substituting $s = 3$ and $r = 8$:

$$8 = \frac{K}{3}$$
$$K = 24$$

So $r = \dfrac{24}{s}$. Substituting $r = 48$:

$$48 = \frac{24}{s}$$
$$48s = 24$$
$$s = \tfrac{1}{2}$$

9. The variation equation is $d = Kr^2$. Substituting $r = 5$ and $d = 10$:

$$10 = K \cdot 5^2$$
$$10 = 25K$$
$$K = \tfrac{2}{5}$$

So $d = \tfrac{2}{5}r^2$. Substituting $r = 10$: $d = \tfrac{2}{5}(10)^2 = \tfrac{2}{5} \cdot 100 = 40$

11. The variation equation is $d = Kr^2$. Substituting $r = 2$ and $d = 100$:

$$100 = K \cdot 2^2$$
$$100 = 4K$$
$$K = 25$$

So $d = 25r^2$. Substituting $r = 3$: $d = 25(3)^2 = 25 \cdot 9 = 225$

13. The variation equation is $y = \dfrac{K}{x^2}$. Substituting $x = 3$ and $y = 45$:

$$45 = \frac{K}{3^2}$$
$$45 = \frac{K}{9}$$
$$K = 405$$

So $y = \dfrac{405}{x^2}$. Substituting $x = 5$: $y = \dfrac{405}{5^2} = \dfrac{405}{25} = \dfrac{81}{5}$

15. The variation equation is $y = \dfrac{K}{x^2}$. Substituting $x = 3$ and $y = 18$:

$$18 = \frac{K}{3^2}$$
$$18 = \frac{K}{9}$$
$$K = 162$$

So $y = \dfrac{162}{x^2}$. Substituting $x = 2$: $y = \dfrac{162}{2^2} = \dfrac{162}{4} = 40.5$

17. The variation equation is $z = Kxy^2$. Substituting $x = 3$, $y = 3$, and $z = 54$:

$$54 = K(3)(3)^2$$
$$54 = 27K$$
$$K = 2$$

So $z = 2xy^2$. Substituting $x = 2$ and $y = 4$: $z = 2(2)(4)^2 = 64$

19. The variation equation is $z = Kxy^2$. Substituting $x = 1$, $y = 4$, and $z = 64$:

$$64 = K(1)(4)^2$$
$$64 = 16K$$
$$K = 4$$

So $z = 4xy^2$. Substituting $z = 32$ and $y = 1$:

$$32 = 4x(1)^2$$
$$32 = 4x$$
$$x = 8$$

21. Let l represent the length and f represent the force. The variation equation is $l = Kf$. Substituting $f = 5$ and $l = 3$:

$$3 = K \cdot 5$$
$$K = \tfrac{3}{5}$$

So $l = \tfrac{3}{5}f$. Substituting $l = 10$:

$$10 = \tfrac{3}{5}f$$
$$50 = 3f$$
$$f = \tfrac{50}{3}$$

The force required is $\tfrac{50}{3}$ pounds.

23. **a.** The variation equation is $T = 4P$.

b. Graphing the equation:

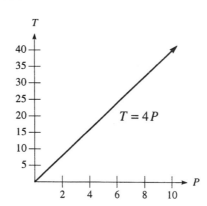

c. Substituting $T = 280$:
$$280 = 4P$$
$$P = 70$$
The pressure is 70 pounds per square inch.

26. Let f represent the frequency and w represent the wavelength. The variation equation is $f = \dfrac{K}{w}$.

Substituting $w = 200$ and $f = 800$:
$$800 = \dfrac{K}{200}$$
$$K = 160000$$

The equation is $f = \dfrac{160000}{w}$. Substituting $w = 500$: $f = \dfrac{160000}{500} = 320$. The frequency is 320 kilocycles per second.

27. **a.** The variation equation is $f = \dfrac{80}{d}$.

b. Graphing the equation:

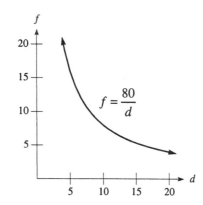

c. Substituting $d = 10$:
$$f = \dfrac{80}{10}$$
$$f = 8$$
The f-stop is 8.

29. Let A represent the surface area, h represent the height, and r represent the radius. The variation equation is $A = Khr$.
Substituting $A = 94$, $r = 3$, and $h = 5$:
$$94 = K(3)(5)$$
$$94 = 15K$$
$$K = \dfrac{94}{15}$$

The equation is $A = \dfrac{94}{15}hr$. Substituting $r = 2$ and $h = 8$: $A = \dfrac{94}{15}(8)(2) = \dfrac{1504}{15}$. The surface area is $\dfrac{1504}{15}$ square inches

31. Let R represent the resistance, l represent the length, and d represent the diameter. The variation equation is $R = \dfrac{Kl}{d^2}$.

Substituting $R = 10$, $l = 100$, and $d = 0.01$:

$$10 = \frac{K(100)}{(0.01)^2}$$
$$0.001 = 100K$$
$$K = 0.00001$$

The equation is $R = \dfrac{0.00001l}{d^2}$. Substituting $l = 60$ and $d = 0.02$: $R = \dfrac{0.00001(60)}{(0.02)^2} = 1.5$. The resistance is 1.5 ohms.

33. **a.** The variation equation is $P = 0.21\sqrt{L}$.

   b. Graphing the equation:

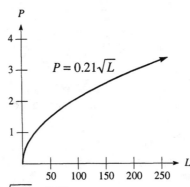

$$P = 0.21\sqrt{L}$$

   c. Substituting $L = 225$: $P = 0.21\sqrt{225} = 3.15$
   The period is 3.15 seconds.

35. Let p represent the pitch and w represent the wavelength. The variation equation is $p = \dfrac{K}{w}$.

Substituting $p = 420$ and $w = 2.2$:

$$420 = \frac{K}{2.2}$$
$$K = 924$$

The equation is $p = \dfrac{924}{w}$. Substituting $p = 720$:

$$720 = \frac{924}{w}$$
$$720w = 924$$
$$w \approx 1.28$$

The wavelength is approximately 1.28 meters.

37. The variation equation is $F = \dfrac{Gm_1 m_2}{d^2}$.

39. Solving the inequality:

$$\left|\frac{x}{5} + 1\right| \geq \frac{4}{5}$$

$$\frac{x}{5} + 1 \leq -\frac{4}{5} \qquad \text{or} \qquad \frac{x}{5} + 1 \geq \frac{4}{5}$$
$$x + 5 \leq -4 \qquad\qquad\qquad x + 5 \geq 4$$
$$x \leq -9 \qquad\qquad\qquad\quad x \geq -1$$

Graphing the solution set:

41. Since $|3-4t| > -5$ is always true, the solution set is all real numbers. Graphing the solution set:

43. Solving the inequality:
$$-8+|3y+5| < 5$$
$$|3y+5| < 13$$
$$-13 < 3y+5 < 13$$
$$-18 < 3y < 8$$
$$-6 < y < \frac{8}{3}$$

Graphing the solution set:

45. Substitute $W_1 = 85$, $d_1 = 4$, and $W_2 = 120$ into the law of levers:
$$W_1 \cdot d_1 = W_2 \cdot d_2$$
$$85 \cdot 4 = 120 \cdot d_2$$
$$340 = 120d_2$$
$$d_2 = \frac{17}{6} = 2\frac{5}{6}$$

Her brother should be $2\frac{5}{6}$ feet from the center of the balance.

Chapter 8 Review

1. Finding the slope: $m = \dfrac{2-2}{3-(-4)} = \dfrac{0}{7} = 0$

3. Graphing the line:

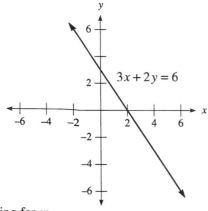

5. Graphing the line:

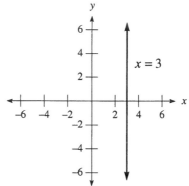

7. Solving for x:
$$\frac{x-7}{2+4} = -\frac{1}{3}$$
$$\frac{x-7}{6} = -\frac{1}{3}$$
$$x-7 = -2$$
$$x = 5$$

9. Solving for y:
$$\frac{y-3y}{2-5} = 4$$
$$\frac{-2y}{-3} = 4$$
$$-2y = -12$$
$$y = 6$$

11. Using the slope-intercept formula, the slope is $y = -2x$.

13.   Solving for y:
$$2x - 3y = 9$$
$$-3y = -2x + 9$$
$$y = \tfrac{2}{3}x - 3$$
The slope is $m = \tfrac{2}{3}$ and the y-intercept is $b = -3$.

15.   Using the point-slope formula:
$$y - 1 = -\tfrac{1}{3}(x + 3)$$
$$y - 1 = -\tfrac{1}{3}x - 1$$
$$y = -\tfrac{1}{3}x$$

17.   First find the slope: $m = \dfrac{7 - 7}{4 - (-3)} = \dfrac{0}{7} = 0$. Since the line is horizontal, its equation is $y = 7$.

19.   First find the slope by solving for y:
$$2x - y = 4$$
$$-y = -2x + 4$$
$$y = 2x - 4$$
The parallel slope is also $m = 2$. Now using the point-slope formula:
$$y + 3 = 2(x - 2)$$
$$y + 3 = 2x - 4$$
$$y = 2x - 7$$

21.   Graphing the inequality:

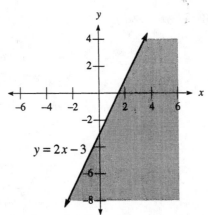

$$y = 2x - 3$$

23.   The domain is $\{2, 3, 4\}$ and the range is $\{2, 3, 4\}$. This is a function.

25. $f(-3) = 0$

27.   Evaluating the function: $f(0) = 2(0)^2 - 4(0) + 1 = 0 - 0 + 1 = 1$

29.   Evaluating the function: $f[g(0)] = f[3(0) + 2] = f(2) = 2(2)^2 - 4(2) + 1 = 8 - 8 + 1 = 1$

31.   Evaluating the function: $(f + g)(x) = f(x) + g(x) = (2x + 1) + (x^2 - 4) = x^2 + 2x - 3$

33.   Evaluating the function: $(fg)(1) = f(1) \cdot g(1) = (2 \cdot 1 + 1)(1^2 - 4) = (3)(-3) = -9$

35.   Evaluating the function: $(f \circ g)(-1) = f(g(-1)) = f((-1)^2 - 4) = f(-3) = 2(-3) + 1 = -5$

37.   The variation equation is $y = Kx$. Substituting $x = 2$ and $y = 6$:
$$6 = K \cdot 2$$
$$K = 3$$
The equation is $y = 3x$. Substituting $x = 8$: $y = 3 \cdot 8 = 24$

39. The variation equation is $y = \dfrac{K}{x^2}$. Substituting $x = 2$ and $y = 9$:

$$9 = \frac{K}{2^2}$$
$$9 = \frac{K}{4}$$
$$K = 36$$

The equation is $y = \dfrac{36}{x^2}$. Substituting $x = 3$: $y = \dfrac{36}{3^2} = \dfrac{36}{9} = 4$

Chapters 1-8 Cumulative Review

1. Simplifying: $11 - (-9) - 7 - (-5) = 11 + 9 - 7 + 5 = 18$ **3.**   Simplifying: $\dfrac{x^{-5}}{x^{-8}} = x^{-5+8} = x^3$

5. Simplifying: $-3(5x + 4) + 12x = -15x - 12 + 12x = -3x - 12$

7. Simplifying: $(x + 3)^2 - (x - 3)^2 = x^2 + 6x + 9 - x^2 + 6x - 9 = 12x$

9. Computing the value: $-3 \cdot \dfrac{5}{12} - \dfrac{3}{4} = -\dfrac{5}{4} - \dfrac{3}{4} = -\dfrac{8}{4} = -2$ **11.**   Factoring: $8x^3 - 27 = (2x - 3)\left(4x^2 + 6x + 9\right)$

13. These are the commutative and associative properties of addition.

15. Subtracting: $\dfrac{y}{x^2 - y^2} - \dfrac{x}{x^2 - y^2} = \dfrac{y - x}{x^2 - y^2} = \dfrac{-1(x - y)}{(x + y)(x - y)} = -\dfrac{1}{x + y}$

17. Multiplying: $\dfrac{x^4 - 16}{x^3 - 8} \cdot \dfrac{x^2 + 2x + 4}{x^2 + 4} = \dfrac{\left(x^2 + 4\right)(x + 2)(x - 2)}{(x - 2)\left(x^2 + 2x + 4\right)} \cdot \dfrac{x^2 + 2x + 4}{x^2 + 4} = x + 2$

19. Dividing using long division:

$$
\begin{array}{r}
a^3 - a^2 + 2a - 4 \\
a + 2 \overline{\smash{\big)}\, a^4 + a^3 + 0a^2 + 0a - 1} \\
\underline{a^4 + 2a^3} \\
-a^3 + 0a^2 \\
\underline{-a^3 - 2a^2} \\
2a^2 + 0a \\
\underline{2a^2 + 4a} \\
-4a - 1 \\
\underline{-4a - 8} \\
7
\end{array}
$$

The quotient is $a^3 - a^2 + 2a - 4 + \dfrac{7}{a + 2}$.

21. Solving the equation:
$$7y - 6 = 2y + 9$$
$$5y = 15$$
$$y = 3$$

23. Solving the equation:
$$|a| - 5 = 7$$
$$|a| = 12$$
$$a = -12, 12$$

25. Solving the equation:
$$\frac{3}{y - 2} = \frac{2}{y - 3}$$
$$3(y - 3) = 2(y - 2)$$
$$3y - 9 = 2y - 4$$
$$y = 5$$

27.   Substitute into the first equation:
$$5x - 2(3x + 2) = -1$$
$$5x - 6x - 4 = -1$$
$$-x = 3$$
$$x = -3$$
The solution is $(-3, -7)$.

29.   Multiply the second equation by 3:
$$-5x + 3y = 1$$
$$5x - 3y = 6$$
Adding yields $0 = 7$, which is false. There is no solution (lines are parallel).

31.   Multiply the first equation by 3 and the second equation by 7:
$$21x - 27y = 6$$
$$-21x + 77y = 7$$
Adding yields:
$$50y = 13$$
$$y = \frac{13}{50}$$
Substituting into the first equation:
$$7x - 9\left(\frac{13}{50}\right) = 2$$
$$7x - \frac{117}{50} = 2$$
$$7x = \frac{217}{50}$$
$$x = \frac{31}{50}$$
The solution is $\left(\frac{31}{50}, \frac{13}{50}\right)$.

33.   Solving the inequality:
$$-3(3x - 1) \le -2(3x - 3)$$
$$-9x + 3 \le -6x + 6$$
$$-3x \le 3$$
$$x \ge -1$$
Graphing the solution set:

35.   Solving for y:
$$x + y = -3$$
$$y = -x - 3$$
The slope is -1.

37.   Graphing the function:

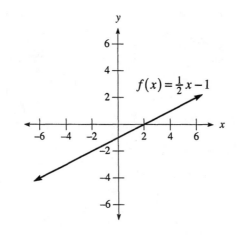

39. Factoring into primes: $168 = 8 \cdot 21 = 2^3 \cdot 3 \cdot 7$

41. Factoring: $x^2 + 10x + 25 - y^2 = (x+5)^2 - y^2 = (x+5-y)(x+5+y)$

43. Writing in scientific notation: $9,270,000.00 = 9.27 \times 10^6$

45. Evaluating the function: $f(-2) - g(3) = \left[(-2)^2 - 2(-2)\right] - [3+5] = 4+4-8 = 0$

47. Since the slope is undefined, the line is vertical and its equation is $x = -2$.

49. Let b represent the base and $2b - 5$ represent the height. Using the area formula:

$$\tfrac{1}{2}b(2b-5) = 75$$
$$2b^2 - 5b = 150$$
$$2b^2 - 5b - 150 = 0$$
$$(2b+15)(b-10) = 0$$
$$b = 10 \quad \left(b = -\tfrac{15}{2} \text{ is impossible}\right)$$

The base is 10 feet and the height is 15 feet.

Chapter 8 Test

1. The x-intercept is 3, the y-intercept is 6, and the slope is –2.

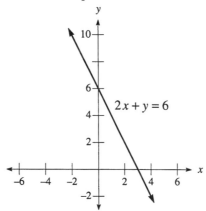

2. The x-intercept is $-\frac{3}{2}$, the y-intercept is –3, and the slope is –2.

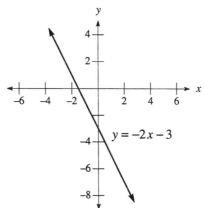

3.   The x-intercept is $-\frac{8}{3}$, the y-intercept is 4, and the slope is $\frac{3}{2}$.

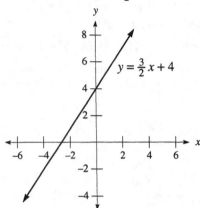

4.   The x-intercept is -2, there is no y-intercept, and there is no slope.

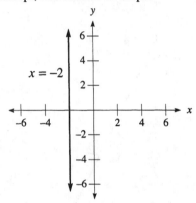

5.   Using the point-slope formula:
$$y - 3 = 2(x + 1)$$
$$y - 3 = 2x + 2$$
$$y = 2x + 5$$

6.   First find the slope: $m = \dfrac{-1-2}{4-(-3)} = \dfrac{-3}{4+3} = -\dfrac{3}{7}$. Using the point-slope formula:
$$y - 2 = -\frac{3}{7}(x + 3)$$
$$y - 2 = -\frac{3}{7}x - \frac{9}{7}$$
$$y = -\frac{3}{7}x + \frac{5}{7}$$

7.   First solve for y to find the slope:
$$2x - 5y = 10$$
$$-5y = -2x + 10$$
$$y = \frac{2}{5}x - 2$$

The parallel line will also have a slope of $\frac{2}{5}$. Now using the point-slope formula:
$$y - (-3) = \frac{2}{5}(x - 5)$$
$$y + 3 = \frac{2}{5}x - 2$$
$$y = \frac{2}{5}x - 5$$

8. The perpendicular slope is $-\frac{1}{3}$. Using the point-slope formula:
$$y - (-2) = -\frac{1}{3}(x - (-1))$$
$$y + 2 = -\frac{1}{3}x - \frac{1}{3}$$
$$y = -\frac{1}{3}x - \frac{7}{3}$$

9. Since the line is vertical, its equation is $x = 4$.

10. Graphing the inequality:

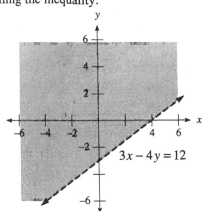

11. Graphing the inequality:

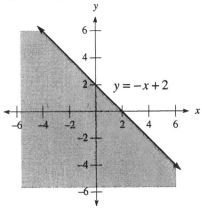

12. The domain is $\{-3, -2\}$ and the range is $\{0, 1\}$. This is not a function.

13. The domain is all real numbers and the range is $\{y \mid y \geq -9\}$. This is a function.

14. Evaluating the function: $f(3) + g(2) = [3 - 2] + [3 \cdot 2 + 4] = 1 + 10 = 11$

15. Evaluating the function: $h(0) + g(0) = [3 \cdot 0^2 - 2 \cdot 0 - 8] + [3 \cdot 0 + 4] = -8 + 4 = -4$

16. Evaluating the function: $(f \circ g)(2) = f[g(2)] = f(3 \cdot 2 + 4) = f(10) = 10 - 2 = 8$

17. Evaluating the function: $(g \circ f)(2) = g[f(2)] = g(2 - 2) = g(0) = 3 \cdot 0 + 4 = 4$

18. The restriction on the variable is $0 < x < 4$.

19. The height of the box is x, and the dimensions of the base are $8 - 2x$ by $8 - 2x$. Therefore the volume is given by $V(x) = x(8 - 2x)^2$.

20. $V(2) = 2(8 - 2 \cdot 2)^2 = 2(4)^2 = 32$ cubic inches
    This represents the volume of the box if a square with 2-inch sides is cut from each corner.

21. The variation equation is $y = Kx^2$. Substituting $x = 5$ and $y = 50$:
$$50 = K(5)^2$$
$$50 = 25K$$
$$K = 2$$
    The equation is $y = 2x^2$. Substituting $x = 3$: $y = 2(3)^2 = 2 \cdot 9 = 18$

22. The variation equation is $z = Kxy^3$. Substituting $x = 5$, $y = 2$, and $z = 15$:
$$15 = K(5)(2)^3$$
$$15 = 40K$$
$$K = \frac{3}{8}$$
    The equation is $z = \frac{3}{8}xy^3$. Substituting $x = 2$ and $y = 3$: $z = \frac{3}{8}(2)(3)^3 = \frac{3}{8} \cdot 54 = \frac{81}{4}$

23.   The variation equation is $L = \dfrac{Kwd^2}{l}$. Substituting $l = 10$, $w = 3$, $d = 4$, and $L = 800$:

$$800 = \frac{K(3)(4)^2}{10}$$
$$8000 = 48K$$
$$K = \frac{500}{3}$$

The equation is $L = \dfrac{500wd^2}{3l}$. Substituting $l = 12$, $w = 3$, and $d = 4$: $L = \dfrac{500(3)(4)^2}{3(12)} = \dfrac{2000}{3}$

The beam can safely hold $\dfrac{2000}{3}$ pounds.

Chapter 9
Rational Exponents and Roots

9.1 Rational Exponents

1. Finding the root: $\sqrt{144} = 12$

3. Finding the root: $\sqrt{-144}$ is not a real number

5. Finding the root: $-\sqrt{49} = -7$

7. Finding the root: $\sqrt[3]{-27} = -3$

9. Finding the root: $\sqrt[4]{16} = 2$

11. Finding the root: $\sqrt[4]{-16}$ is not a real number

13. Finding the root: $\sqrt{0.04} = 0.2$

15. Finding the root: $\sqrt[3]{0.008} = 0.2$

17. Simplifying: $\sqrt{36a^8} = 6a^4$

19. Simplifying: $\sqrt[3]{27a^{12}} = 3a^4$

21. Simplifying: $\sqrt[5]{32x^{10}y^5} = 2x^2y$

23. Simplifying: $\sqrt[4]{16a^{12}b^{20}} = 2a^3b^5$

25. Writing as a root and simplifying: $36^{1/2} = \sqrt{36} = 6$

27. Writing as a root and simplifying: $-9^{1/2} = -\sqrt{9} = -3$

29. Writing as a root and simplifying: $8^{1/3} = \sqrt[3]{8} = 2$

31. Writing as a root and simplifying: $(-8)^{1/3} = \sqrt[3]{-8} = -2$

33. Writing as a root and simplifying: $32^{1/5} = \sqrt[5]{32} = 2$

35. Writing as a root and simplifying: $\left(\frac{81}{25}\right)^{1/2} = \sqrt{\frac{81}{25}} = \frac{9}{5}$

37. Simplifying: $27^{2/3} = \left(27^{1/3}\right)^2 = 3^2 = 9$

39. Simplifying: $25^{3/2} = \left(25^{1/2}\right)^3 = 5^3 = 125$

41. Simplifying: $27^{-1/3} = \left(27^{1/3}\right)^{-1} = 3^{-1} = \frac{1}{3}$

43. Simplifying: $81^{-3/4} = \left(81^{1/4}\right)^{-3} = 3^{-3} = \frac{1}{3^3} = \frac{1}{27}$

45. Simplifying: $\left(\frac{25}{36}\right)^{-1/2} = \left(\frac{36}{25}\right)^{1/2} = \frac{6}{5}$

47. Simplifying: $\left(\frac{81}{16}\right)^{-3/4} = \left(\frac{16}{81}\right)^{3/4} = \left[\left(\frac{16}{81}\right)^{1/4}\right]^3 = \left(\frac{2}{3}\right)^3 = \frac{8}{27}$

49. Simplifying: $16^{1/2} + 27^{1/3} = 4 + 3 = 7$

51. Simplifying: $8^{-2/3} + 4^{-1/2} = \left(8^{1/3}\right)^{-2} + \left(4^{1/2}\right)^{-1} = 2^{-2} + 2^{-1} = \frac{1}{4} + \frac{1}{2} = \frac{3}{4}$

53. Using properties of exponents: $x^{3/5} \cdot x^{1/5} = x^{3/5+1/5} = x^{4/5}$

55. Using properties of exponents: $\left(a^{3/4}\right)^{4/3} = a^{3/4 \cdot 4/3} = a$

57. Using properties of exponents: $\dfrac{x^{1/5}}{x^{3/5}} = x^{1/5-3/5} = x^{-2/5} = \dfrac{1}{x^{2/5}}$

59.   Using properties of exponents: $\dfrac{x^{5/6}}{x^{2/3}} = x^{5/6-2/3} = x^{5/6-4/6} = x^{1/6}$

61.   Using properties of exponents: $\left(x^{3/5}y^{5/6}z^{1/3}\right)^{3/5} = x^{3/5\cdot3/5}y^{5/6\cdot3/5}z^{1/3\cdot3/5} = x^{9/25}y^{1/2}z^{1/5}$

63.   Using properties of exponents: $\dfrac{a^{3/4}b^2}{a^{7/8}b^{1/4}} = a^{3/4-7/8}b^{2-1/4} = a^{6/8-7/8}b^{8/4-1/4} = a^{-1/8}b^{7/4} = \dfrac{b^{7/4}}{a^{1/8}}$

65.   Using properties of exponents: $\dfrac{\left(y^{2/3}\right)^{3/4}}{\left(y^{1/3}\right)^{3/5}} = \dfrac{y^{1/2}}{y^{1/5}} = y^{1/2-1/5} = y^{5/10-2/10} = y^{3/10}$

67.   Using properties of exponents: $\left(\dfrac{a^{-1/4}}{b^{1/2}}\right)^8 = \dfrac{a^{-1/4\cdot8}}{b^{1/2\cdot8}} = \dfrac{a^{-2}}{b^4} = \dfrac{1}{a^2b^4}$

69.   Using properties of exponents: $\dfrac{\left(r^{-2}s^{1/3}\right)^6}{r^8s^{3/2}} = \dfrac{r^{-12}s^2}{r^8s^{3/2}} = r^{-12-8}s^{2-3/2} = r^{-20}s^{1/2} = \dfrac{s^{1/2}}{r^{20}}$

71.   Using properties of exponents: $\dfrac{\left(25a^6b^4\right)^{1/2}}{\left(8a^{-9}b^3\right)^{-1/3}} = \dfrac{25^{1/2}a^3b^2}{8^{-1/3}a^3b^{-1}} = \dfrac{5}{1/2}a^{3-3}b^{2+1} = 10b^3$

73.   Substituting $r = 250$: $v = \left(\dfrac{5\cdot250}{2}\right)^{1/2} = 625^{1/2} = 25$. The maximum speed is 25 mph.

75.   **a.**   The length of the side is $126 + 86 + 86 + 126 = 424$ pm

   b.   Let d represent the diagonal. Using the Pythagorean theorem:
$$d^2 = 424^2 + 424^2 = 359552$$
$$d = \sqrt{359552} \approx 600 \text{ pm}$$

   c.   Converting to meters: $600 \text{ pm} \cdot \dfrac{1 \text{ m}}{10^{12} \text{ pm}} = 6\times10^{-10} \text{ m}$

77.   **a.**   This graph is B.   **b.**   This graph is A.
   c.   This graph is C.   **d.**   The points of intersection are (0,0) and (1,1).

79.   Multiplying: $x^2\left(x^4 - x\right) = x^2 \cdot x^4 - x^2 \cdot x = x^6 - x^3$

81.   Multiplying: $(x-3)(x+5) = x^2 + 5x - 3x - 15 = x^2 + 2x - 15$

83.   Multiplying: $\left(x^2 - 5\right)^2 = \left(x^2\right)^2 - 2\left(x^2\right)(5) + 5^2 = x^4 - 10x^2 + 25$

85.   Multiplying: $(x-3)\left(x^2 + 3x + 9\right) = x^3 + 3x^2 + 9x - 3x^2 - 9x - 27 = x^3 - 27$

87.   Simplifying each expression:
$$\left(9^{1/2} + 4^{1/2}\right)^2 = (3+2)^2 = 5^2 = 25$$
$$9 + 4 = 13$$
Note that the values are not equal.

89.   Rewriting with exponents: $\sqrt{\sqrt{a}} = \sqrt{a^{1/2}} = \left(a^{1/2}\right)^{1/2} = a^{1/4} = \sqrt[4]{a}$

9.2   More Expressions Involving Rational Exponents

1.   Multiplying: $x^{2/3}\left(x^{1/3} + x^{4/3}\right) = x^{2/3} \bullet x^{1/3} + x^{2/3} \bullet x^{4/3} = x + x^2$

3.   Multiplying: $a^{1/2}\left(a^{3/2} - a^{1/2}\right) = a^{1/2} \bullet a^{3/2} - a^{1/2} \bullet a^{1/2} = a^2 - a$

5.   Multiplying: $2x^{1/3}\left(3x^{8/3} - 4x^{5/3} + 5x^{2/3}\right) = 2x^{1/3} \bullet 3x^{8/3} - 2x^{1/3} \bullet 4x^{5/3} + 2x^{1/3} \bullet 5x^{2/3} = 6x^3 - 8x^2 + 10x$

7.   Multiplying:
$$4x^{1/2}y^{3/5}\left(3x^{3/2}y^{-3/5} - 9x^{-1/2}y^{7/5}\right) = 4x^{1/2}y^{3/5} \bullet 3x^{3/2}y^{-3/5} - 4x^{1/2}y^{3/5} \bullet 9x^{-1/2}y^{7/5} = 12x^2 - 36y^2$$

9.   Multiplying: $\left(x^{2/3} - 4\right)\left(x^{2/3} + 2\right) = x^{2/3} \bullet x^{2/3} + 2x^{2/3} - 4x^{2/3} - 8 = x^{4/3} - 2x^{2/3} - 8$

11.   Multiplying: $\left(a^{1/2} - 3\right)\left(a^{1/2} - 7\right) = a^{1/2} \bullet a^{1/2} - 7a^{1/2} - 3a^{1/2} + 21 = a - 10a^{1/2} + 21$

13.   Multiplying: $\left(4y^{1/3} - 3\right)\left(5y^{1/3} + 2\right) = 20y^{2/3} + 8y^{1/3} - 15y^{1/3} - 6 = 20y^{2/3} - 7y^{1/3} - 6$

15.   Multiplying: $\left(5x^{2/3} + 3y^{1/2}\right)\left(2x^{2/3} + 3y^{1/2}\right) = 10x^{4/3} + 15x^{2/3}y^{1/2} + 6x^{2/3}y^{1/2} + 9y = 10x^{4/3} + 21x^{2/3}y^{1/2} + 9y$

17.   Multiplying: $\left(t^{1/2} + 5\right)^2 = \left(t^{1/2} + 5\right)\left(t^{1/2} + 5\right) = t + 5t^{1/2} + 5t^{1/2} + 25 = t + 10t^{1/2} + 25$

19.   Multiplying: $\left(x^{3/2} + 4\right)^2 = \left(x^{3/2} + 4\right)\left(x^{3/2} + 4\right) = x^3 + 4x^{3/2} + 4x^{3/2} + 16 = x^3 + 8x^{3/2} + 16$

21.   Multiplying: $\left(a^{1/2} - b^{1/2}\right)^2 = \left(a^{1/2} - b^{1/2}\right)\left(a^{1/2} - b^{1/2}\right) = a - a^{1/2}b^{1/2} - a^{1/2}b^{1/2} + b = a - 2a^{1/2}b^{1/2} + b$

23.   Multiplying:
$$\left(2x^{1/2} - 3y^{1/2}\right)^2 = \left(2x^{1/2} - 3y^{1/2}\right)\left(2x^{1/2} - 3y^{1/2}\right)$$
$$= 4x - 6x^{1/2}y^{1/2} - 6x^{1/2}y^{1/2} + 9y$$
$$= 4x - 12x^{1/2}y^{1/2} + 9y$$

25.   Multiplying: $\left(a^{1/2} - 3^{1/2}\right)\left(a^{1/2} + 3^{1/2}\right) = \left(a^{1/2}\right)^2 - \left(3^{1/2}\right)^2 = a - 3$

27.   Multiplying: $\left(x^{3/2} + y^{3/2}\right)\left(x^{3/2} - y^{3/2}\right) = \left(x^{3/2}\right)^2 - \left(y^{3/2}\right)^2 = x^3 - y^3$

29.   Multiplying: $\left(t^{1/2} - 2^{3/2}\right)\left(t^{1/2} + 2^{3/2}\right) = \left(t^{1/2}\right)^2 - \left(2^{3/2}\right)^2 = t - 2^3 = t - 8$

31.   Multiplying: $\left(2x^{3/2} + 3^{1/2}\right)\left(2x^{3/2} - 3^{1/2}\right) = \left(2x^{3/2}\right)^2 - \left(3^{1/2}\right)^2 = 4x^3 - 3$

33.   Multiplying: $\left(x^{1/3} + y^{1/3}\right)\left(x^{2/3} - x^{1/3}y^{1/3} + y^{2/3}\right) = \left(x^{1/3}\right)^3 + \left(y^{1/3}\right)^3 = x + y$

35.   Multiplying: $\left(a^{1/3} - 2\right)\left(a^{2/3} + 2a^{1/3} + 4\right) = \left(a^{1/3}\right)^3 - (2)^3 = a - 8$

37.   Multiplying: $\left(2x^{1/3} + 1\right)\left(4x^{2/3} - 2x^{1/3} + 1\right) = \left(2x^{1/3}\right)^3 + (1)^3 = 8x + 1$

39.   Multiplying: $\left(t^{1/4} - 1\right)\left(t^{1/4} + 1\right)\left(t^{1/2} + 1\right) = \left(t^{1/2} - 1\right)\left(t^{1/2} + 1\right) = t - 1$

41.   Dividing: $\dfrac{18x^{3/4} + 27x^{1/4}}{9x^{1/4}} = \dfrac{18x^{3/4}}{9x^{1/4}} + \dfrac{27x^{1/4}}{9x^{1/4}} = 2x^{1/2} + 3$

43.   Dividing: $\dfrac{12x^{2/3}y^{1/3} - 16x^{1/3}y^{2/3}}{4x^{1/3}y^{1/3}} = \dfrac{12x^{2/3}y^{1/3}}{4x^{1/3}y^{1/3}} - \dfrac{16x^{1/3}y^{2/3}}{4x^{1/3}y^{1/3}} = 3x^{1/3} - 4y^{1/3}$

45.   Dividing: $\dfrac{21a^{7/5}b^{3/5} - 14a^{2/5}b^{8/5}}{7a^{2/5}b^{3/5}} = \dfrac{21a^{7/5}b^{3/5}}{7a^{2/5}b^{3/5}} - \dfrac{14a^{2/5}b^{8/5}}{7a^{2/5}b^{3/5}} = 3a - 2b$

47.   Factoring: $12(x - 2)^{3/2} - 9(x - 2)^{1/2} = 3(x - 2)^{1/2}\left[4(x - 2) - 3\right] = 3(x - 2)^{1/2}(4x - 8 - 3) = 3(x - 2)^{1/2}(4x - 11)$

49.   Factoring: $5(x - 3)^{12/5} - 15(x - 3)^{7/5} = 5(x - 3)^{7/5}\left[(x - 3) - 3\right] = 5(x - 3)^{7/5}(x - 6)$

51.   Factoring: $9x(x + 1)^{3/2} + 6(x + 1)^{1/2} = 3(x + 1)^{1/2}\left[3x(x + 1) + 2\right] = 3(x + 1)^{1/2}\left(3x^2 + 3x + 2\right)$

53. Factoring: $x^{2/3} - 5x^{1/3} + 6 = \left(x^{1/3} - 2\right)\left(x^{1/3} - 3\right)$ **55.** Factoring: $a^{2/5} - 2a^{1/5} - 8 = \left(a^{1/5} - 4\right)\left(a^{1/5} + 2\right)$

57. Factoring: $2y^{2/3} - 5y^{1/3} - 3 = \left(2y^{1/3} + 1\right)\left(y^{1/3} - 3\right)$ **59.** Factoring: $9t^{2/5} - 25 = \left(3t^{1/5} + 5\right)\left(3t^{1/5} - 5\right)$

61. Factoring: $4x^{2/7} + 20x^{1/7} + 25 = \left(2x^{1/7} + 5\right)^2$

63. Writing as a single fraction: $\dfrac{3}{x^{1/2}} + x^{1/2} = \dfrac{3}{x^{1/2}} + x^{1/2} \cdot \dfrac{x^{1/2}}{x^{1/2}} = \dfrac{3+x}{x^{1/2}}$

65. Writing as a single fraction: $x^{2/3} + \dfrac{5}{x^{1/3}} = x^{2/3} \cdot \dfrac{x^{1/3}}{x^{1/3}} + \dfrac{5}{x^{1/3}} = \dfrac{x+5}{x^{1/3}}$

67. Writing as a single fraction:

$$\frac{3x^2}{\left(x^3 + 1\right)^{1/2}} + \left(x^3 + 1\right)^{1/2} = \frac{3x^2}{\left(x^3 + 1\right)^{1/2}} + \left(x^3 + 1\right)^{1/2} \cdot \frac{\left(x^3 + 1\right)^{1/2}}{\left(x^3 + 1\right)^{1/2}} = \frac{3x^2 + x^3 + 1}{\left(x^3 + 1\right)^{1/2}} = \frac{x^3 + 3x^2 + 1}{\left(x^3 + 1\right)^{1/2}}$$

69. Writing as a single fraction:

$$\frac{x^2}{\left(x^2 + 4\right)^{1/2}} - \left(x^2 + 4\right)^{1/2} = \frac{x^2}{\left(x^2 + 4\right)^{1/2}} - \left(x^2 + 4\right)^{1/2} \cdot \frac{\left(x^2 + 4\right)^{1/2}}{\left(x^2 + 4\right)^{1/2}}$$

$$= \frac{x^2 - \left(x^2 + 4\right)}{\left(x^2 + 4\right)^{1/2}}$$

$$= \frac{x^2 - x^2 - 4}{\left(x^2 + 4\right)^{1/2}}$$

$$= \frac{-4}{\left(x^2 + 4\right)^{1/2}}$$

71. Using the formula $r = \left(\dfrac{A}{P}\right)^{1/t} - 1$ to find the annual rate of return: $r = \left(\dfrac{900}{500}\right)^{1/4} - 1 \approx 0.158$

The annual rate of return is approximately 15.8%.

73. Substituting $k = 0.0408$ and $d = 240$: $v = \left(\dfrac{240}{0.0408}\right)^{1/2} \approx 76.7$ mph

75. Finding the slope: $m = \dfrac{5 - (-1)}{-2 - (-4)} = \dfrac{5 + 1}{-2 + 4} = \dfrac{6}{2} = 3$

77. Solving for y:

$$2x - 3y = 6$$
$$-3y = -2x + 6$$
$$y = \tfrac{2}{3}x - 2$$

The slope is $m = \tfrac{2}{3}$ and the y-intercept is $b = -2$.

79. Using the point-slope formula:

$$y - 2 = \tfrac{2}{3}(x + 6)$$
$$y - 2 = \tfrac{2}{3}x + 4$$
$$y = \tfrac{2}{3}x + 6$$

81. First find the slope: $m = \dfrac{-2 - 0}{0 - 3} = \tfrac{2}{3}$. The slope-intercept form is $y = \tfrac{2}{3}x - 2$.

9.3 Simplified Form for Radicals

1. Simplifying the radical: $\sqrt{8} = \sqrt{4 \cdot 2} = 2\sqrt{2}$ **3.** Simplifying the radical: $\sqrt{98} = \sqrt{49 \cdot 2} = 7\sqrt{2}$

5. Simplifying the radical: $\sqrt{288} = \sqrt{144 \cdot 2} = 12\sqrt{2}$ **7.** Simplifying the radical: $\sqrt{80} = \sqrt{16 \cdot 5} = 4\sqrt{5}$

9. Simplifying the radical: $\sqrt{48} = \sqrt{16 \cdot 3} = 4\sqrt{3}$ **11.** Simplifying the radical: $\sqrt{675} = \sqrt{225 \cdot 3} = 15\sqrt{3}$

13. Simplifying the radical: $\sqrt[3]{54} = \sqrt[3]{27 \cdot 2} = 3\sqrt[3]{2}$ **15.** Simplifying the radical: $\sqrt[3]{128} = \sqrt[3]{64 \cdot 2} = 4\sqrt[3]{2}$

17. Simplifying the radical: $\sqrt[3]{432} = \sqrt[3]{216 \cdot 2} = 6\sqrt[3]{2}$ **19.** Simplifying the radical: $\sqrt[5]{64} = \sqrt[5]{32 \cdot 2} = 2\sqrt[5]{2}$

21. Simplifying the radical: $\sqrt{18x^3} = \sqrt{9x^2 \cdot 2x} = 3x\sqrt{2x}$

23. Simplifying the radical: $\sqrt[4]{32y^7} = \sqrt[4]{16y^4 \cdot 2y^3} = 2y\sqrt[4]{2y^3}$

25. Simplifying the radical: $\sqrt[3]{40x^4y^7} = \sqrt[3]{8x^3y^6 \cdot 5xy} = 2xy^2\sqrt[3]{5xy}$

27. Simplifying the radical: $\sqrt{48a^2b^3c^4} = \sqrt{16a^2b^2c^4 \cdot 3b} = 4abc^2\sqrt{3b}$

29. Simplifying the radical: $\sqrt[3]{48a^2b^3c^4} = \sqrt[3]{8b^3c^3 \cdot 6a^2c} = 2bc\sqrt[3]{6a^2c}$

31. Simplifying the radical: $\sqrt[5]{64x^8y^{12}} = \sqrt[5]{32x^5y^{10} \cdot 2x^3y^2} = 2xy^2\sqrt[5]{2x^3y^2}$

33. Simplifying the radical: $\sqrt[5]{243x^7y^{10}z^5} = \sqrt[5]{243x^5y^{10}z^5 \cdot x^2} = 3xy^2z\sqrt[5]{x^2}$

35. Substituting into the expression: $\sqrt{b^2 - 4ac} = \sqrt{(-6)^2 - 4(2)(3)} = \sqrt{36 - 24} = \sqrt{12} = 2\sqrt{3}$

37. Substituting into the expression: $\sqrt{b^2 - 4ac} = \sqrt{(2)^2 - 4(1)(6)} = \sqrt{4 - 24} = \sqrt{-20}$, which is not a real number

39. Substituting into the expression: $\sqrt{b^2 - 4ac} = \sqrt{\left(-\frac{1}{2}\right)^2 - 4\left(\frac{1}{2}\right)\left(-\frac{5}{4}\right)} = \sqrt{\frac{1}{4} + \frac{5}{2}} = \sqrt{\frac{11}{4}} = \frac{\sqrt{11}}{2}$

41. Rationalizing the denominator: $\dfrac{2}{\sqrt{3}} = \dfrac{2}{\sqrt{3}} \cdot \dfrac{\sqrt{3}}{\sqrt{3}} = \dfrac{2\sqrt{3}}{3}$ **43.** Rationalizing the denominator: $\dfrac{5}{\sqrt{6}} = \dfrac{5}{\sqrt{6}} \cdot \dfrac{\sqrt{6}}{\sqrt{6}} = \dfrac{5\sqrt{6}}{6}$

45. Rationalizing the denominator: $\sqrt{\frac{1}{2}} = \dfrac{1}{\sqrt{2}} \cdot \dfrac{\sqrt{2}}{\sqrt{2}} = \dfrac{\sqrt{2}}{2}$ **47.** Rationalizing the denominator: $\sqrt{\frac{1}{5}} = \dfrac{1}{\sqrt{5}} \cdot \dfrac{\sqrt{5}}{\sqrt{5}} = \dfrac{\sqrt{5}}{5}$

49. Rationalizing the denominator: $\dfrac{4}{\sqrt[3]{2}} = \dfrac{4}{\sqrt[3]{2}} \cdot \dfrac{\sqrt[3]{4}}{\sqrt[3]{4}} = \dfrac{4\sqrt[3]{4}}{2} = 2\sqrt[3]{4}$

51. Rationalizing the denominator: $\dfrac{2}{\sqrt[3]{9}} = \dfrac{2}{\sqrt[3]{9}} \cdot \dfrac{\sqrt[3]{3}}{\sqrt[3]{3}} = \dfrac{2\sqrt[3]{3}}{3}$

53. Rationalizing the denominator: $\sqrt[4]{\dfrac{3}{2x^2}} = \dfrac{\sqrt[4]{3}}{\sqrt[4]{2x^2}} \cdot \dfrac{\sqrt[4]{8x^2}}{\sqrt[4]{8x^2}} = \dfrac{\sqrt[4]{24x^2}}{2x}$

55. Rationalizing the denominator: $\sqrt[4]{\dfrac{8}{y}} = \dfrac{\sqrt[4]{8}}{\sqrt[4]{y}} \cdot \dfrac{\sqrt[4]{y^3}}{\sqrt[4]{y^3}} = \dfrac{\sqrt[4]{8y^3}}{y}$

57. Rationalizing the denominator: $\sqrt[3]{\dfrac{4x}{3y}} = \dfrac{\sqrt[3]{4x}}{\sqrt[3]{3y}} \cdot \dfrac{\sqrt[3]{9y^2}}{\sqrt[3]{9y^2}} = \dfrac{\sqrt[3]{36xy^2}}{3y}$

59. Rationalizing the denominator: $\sqrt[3]{\dfrac{2x}{9y}} = \dfrac{\sqrt[3]{2x}}{\sqrt[3]{9y}} \cdot \dfrac{\sqrt[3]{3y^2}}{\sqrt[3]{3y^2}} = \dfrac{\sqrt[3]{6xy^2}}{3y}$

61. Rationalizing the denominator: $\sqrt[4]{\dfrac{1}{8x^3}} = \dfrac{1}{\sqrt[4]{8x^3}} \cdot \dfrac{\sqrt[4]{2x}}{\sqrt[4]{2x}} = \dfrac{\sqrt[4]{2x}}{2x}$

63. Simplifying: $\sqrt{\dfrac{27x^3}{5y}} = \dfrac{\sqrt{27x^3}}{\sqrt{5y}} \cdot \dfrac{\sqrt{5y}}{\sqrt{5y}} = \dfrac{\sqrt{135x^3y}}{5y} = \dfrac{3x\sqrt{15xy}}{5y}$

65. Simplifying: $\sqrt{\dfrac{75x^3y^2}{2z}} = \dfrac{\sqrt{75x^3y^2}}{\sqrt{2z}} \cdot \dfrac{\sqrt{2z}}{\sqrt{2z}} = \dfrac{\sqrt{150x^3y^2z}}{2z} = \dfrac{5xy\sqrt{6xz}}{2z}$

67. Simplifying: $\sqrt[3]{\dfrac{16a^4b^3}{9c}} = \dfrac{\sqrt[3]{16a^4b^3}}{\sqrt[3]{9c}} \cdot \dfrac{\sqrt[3]{3c^2}}{\sqrt[3]{3c^2}} = \dfrac{\sqrt[3]{48a^4b^3c^2}}{3c} = \dfrac{2ab\sqrt[3]{6ac^2}}{3c}$

69. Simplifying: $\sqrt[3]{\dfrac{8x^3y^6}{9z}} = \dfrac{\sqrt[3]{8x^3y^6}}{\sqrt[3]{9z}} \cdot \dfrac{\sqrt[3]{3z^2}}{\sqrt[3]{3z^2}} = \dfrac{\sqrt[3]{24x^3y^6z^2}}{3z} = \dfrac{2xy^2\sqrt[3]{3z^2}}{3z}$

71. Simplifying: $\sqrt{25x^2} = 5|x|$ **73.** Simplifying: $\sqrt{27x^3y^2} = \sqrt{9x^2y^2 \cdot 3x} = 3|xy|\sqrt{3x}$

75. Simplifying: $\sqrt{x^2-10x+25} = \sqrt{(x-5)^2} = |x-5|$ **77.** Simplifying: $\sqrt{4x^2+12x+9} = \sqrt{(2x+3)^2} = |2x+3|$

79. Simplifying: $\sqrt{4a^4+16a^3+16a^2} = \sqrt{4a^2(a^2+4a+4)} = \sqrt{4a^2(a+2)^2} = 2|a(a+2)|$

81. Simplifying: $\sqrt{4x^3-8x^2} = \sqrt{4x^2(x-2)} = 2|x|\sqrt{x-2}$

83. Substituting $a = 9$ and $b = 16$:
$$\sqrt{a+b} = \sqrt{9+16} = \sqrt{25} = 5$$
$$\sqrt{a}+\sqrt{b} = \sqrt{9}+\sqrt{16} = 3+4 = 7$$
Thus $\sqrt{a+b} \ne \sqrt{a}+\sqrt{b}$.

85. Substituting $w = 10$ and $l = 15$: $d = \sqrt{l^2+w^2} = \sqrt{15^2+10^2} = \sqrt{225+100} = \sqrt{325} = \sqrt{25 \cdot 13} = 5\sqrt{13}$ feet

87. **a.** Substituting $w = 3$, $l = 4$, and $h = 12$:
$$d = \sqrt{l^2+w^2+h^2} = \sqrt{4^2+3^2+12^2} = \sqrt{16+9+144} = \sqrt{169} = 13 \text{ feet}$$
b. Substituting $w = 2$, $l = 4$, and $h = 6$:
$$d = \sqrt{l^2+w^2+h^2} = \sqrt{4^2+2^2+6^2} = \sqrt{16+4+36} = \sqrt{56} = 2\sqrt{14} \approx 7.5 \text{ feet}$$

89. Answers will vary.

91. Finding the first six terms:
$$f(1) = \sqrt{1^2+1} = \sqrt{2}$$
$$f(f(1)) = \sqrt{\left(\sqrt{2}\right)^2+1} = \sqrt{2+1} = \sqrt{3}$$
$$f(f(f(1))) = \sqrt{\left(\sqrt{3}\right)^2+1} = \sqrt{3+1} = 2$$
$$f(f(f(f(1)))) = \sqrt{(2)^2+1} = \sqrt{4+1} = \sqrt{5}$$
$$f(f(f(f(f(1))))) = \sqrt{\left(\sqrt{5}\right)^2+1} = \sqrt{5+1} = \sqrt{6}$$
$$f\left(f\left(f\left(f\left(f(f(1))\right)\right)\right)\right) = \sqrt{\left(\sqrt{6}\right)^2+1} = \sqrt{6+1} = \sqrt{7}$$

Following this pattern, the 10th term will be $\sqrt{11}$ and the 100th term will be $\sqrt{101}$.

93. Evaluating the function: $f(0) = \frac{1}{2}(0)+3 = 3$ **95.** Evaluating the function: $g(2) = 2^2 - 4 = 0$

97. Evaluating the function: $f(-4) = \frac{1}{2}(-4)+3 = 1$

99. The prime factorization of 8,640 is: $8640 = 2^6 \cdot 3^3 \cdot 5$. Therefore: $\sqrt[3]{8640} = \sqrt[3]{2^6 \cdot 3^3 \cdot 5} = 2^2 \cdot 3\sqrt[3]{5} = 12\sqrt[3]{5}$

101. The prime factorization of 10,584 is: $10584 = 2^3 \cdot 3^3 \cdot 7^2$. Therefore: $\sqrt[3]{10584} = \sqrt[3]{2^3 \cdot 3^3 \cdot 7^2} = 2 \cdot 3\sqrt[3]{7^2} = 6\sqrt[3]{49}$

103. Rationalizing the denominator: $\dfrac{1}{\sqrt[10]{a^3}} = \dfrac{1}{\sqrt[10]{a^3}} \cdot \dfrac{\sqrt[10]{a^7}}{\sqrt[10]{a^7}} = \dfrac{\sqrt[10]{a^7}}{a}$

105. Rationalizing the denominator: $\dfrac{1}{\sqrt[20]{a^{11}}} = \dfrac{1}{\sqrt[20]{a^{11}}} \cdot \dfrac{\sqrt[20]{a^9}}{\sqrt[20]{a^9}} = \dfrac{\sqrt[20]{a^9}}{a}$

107. Graphing each function:

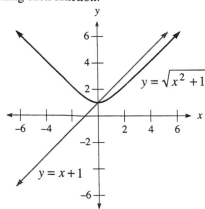

$y = \sqrt{x^2 + 1}$

$y = x + 1$

Note that the two graphs are not the same.

109. When $x = 2$, the distance apart is: $(2 + 1) - \sqrt{2^2 + 1} = 3 - \sqrt{5} \approx 0.76$

111. The two expressions are equal when $x = 0$.

9.4 Addition and Subtraction of Radical Expressions

1. Combining radicals: $3\sqrt{5} + 4\sqrt{5} = 7\sqrt{5}$

3. Combining radicals: $3x\sqrt{7} - 4x\sqrt{7} = -x\sqrt{7}$

5. Combining radicals: $5\sqrt[3]{10} - 4\sqrt[3]{10} = \sqrt[3]{10}$

7. Combining radicals: $8\sqrt[5]{6} - 2\sqrt[5]{6} + 3\sqrt[5]{6} = 9\sqrt[5]{6}$

9. Combining radicals: $3x\sqrt{2} - 4x\sqrt{2} + x\sqrt{2} = 0$

11. Combining radicals: $\sqrt{20} - \sqrt{80} + \sqrt{45} = 2\sqrt{5} - 4\sqrt{5} + 3\sqrt{5} = \sqrt{5}$

13. Combining radicals: $4\sqrt{8} - 2\sqrt{50} - 5\sqrt{72} = 8\sqrt{2} - 10\sqrt{2} - 30\sqrt{2} = -32\sqrt{2}$

15. Combining radicals: $5x\sqrt{8} + 3\sqrt{32x^2} - 5\sqrt{50x^2} = 10x\sqrt{2} + 12x\sqrt{2} - 25x\sqrt{2} = -3x\sqrt{2}$

17. Combining radicals: $5\sqrt[3]{16} - 4\sqrt[3]{54} = 10\sqrt[3]{2} - 12\sqrt[3]{2} = -2\sqrt[3]{2}$

19. Combining radicals: $\sqrt[3]{x^4 y^2} + 7x\sqrt[3]{xy^2} = x\sqrt[3]{xy^2} + 7x\sqrt[3]{xy^2} = 8x\sqrt[3]{xy^2}$

21. Combining radicals: $5a^2\sqrt{27ab^3} - 6b\sqrt{12a^5 b} = 15a^2 b\sqrt{3ab} - 12a^2 b\sqrt{3ab} = 3a^2 b\sqrt{3ab}$

23. Combining radicals: $b\sqrt[3]{24a^5 b} + 3a\sqrt[3]{81a^2 b^4} = 2ab\sqrt[3]{3a^2 b} + 9ab\sqrt[3]{3a^2 b} = 11ab\sqrt[3]{3a^2 b}$

25. Combining radicals: $5x\sqrt[4]{3y^5} + y\sqrt[4]{243x^4 y} + \sqrt[4]{48x^4 y^5} = 5xy\sqrt[4]{3y} + 3xy\sqrt[4]{3y} + 2xy\sqrt[4]{3y} = 10xy\sqrt[4]{3y}$

27. Combining radicals: $\dfrac{\sqrt{2}}{2} + \dfrac{1}{\sqrt{2}} = \dfrac{\sqrt{2}}{2} + \dfrac{1}{\sqrt{2}} \cdot \dfrac{\sqrt{2}}{\sqrt{2}} = \dfrac{\sqrt{2}}{2} + \dfrac{\sqrt{2}}{2} = \sqrt{2}$

29. Combining radicals: $\dfrac{\sqrt{5}}{3} + \dfrac{1}{\sqrt{5}} = \dfrac{\sqrt{5}}{3} + \dfrac{1}{\sqrt{5}} \cdot \dfrac{\sqrt{5}}{\sqrt{5}} = \dfrac{\sqrt{5}}{3} + \dfrac{\sqrt{5}}{5} = \dfrac{5\sqrt{5}}{15} + \dfrac{3\sqrt{5}}{15} = \dfrac{8\sqrt{5}}{15}$

31. Combining radicals: $\sqrt{x} - \dfrac{1}{\sqrt{x}} = \sqrt{x} - \dfrac{1}{\sqrt{x}} \cdot \dfrac{\sqrt{x}}{\sqrt{x}} = \sqrt{x} - \dfrac{\sqrt{x}}{x} = \dfrac{x\sqrt{x}}{x} - \dfrac{\sqrt{x}}{x} = \dfrac{(x-1)\sqrt{x}}{x}$

33. Combining radicals: $\dfrac{\sqrt{18}}{6} + \sqrt{\dfrac{1}{2}} + \dfrac{\sqrt{2}}{2} = \dfrac{3\sqrt{2}}{6} + \dfrac{1}{\sqrt{2}} \cdot \dfrac{\sqrt{2}}{\sqrt{2}} + \dfrac{\sqrt{2}}{2} = \dfrac{\sqrt{2}}{2} + \dfrac{\sqrt{2}}{2} + \dfrac{\sqrt{2}}{2} = \dfrac{3\sqrt{2}}{2}$

35. Combining radicals: $\sqrt{6} - \sqrt{\dfrac{2}{3}} + \sqrt{\dfrac{1}{6}} = \sqrt{6} - \dfrac{\sqrt{2}}{\sqrt{3}} \cdot \dfrac{\sqrt{3}}{\sqrt{3}} + \dfrac{1}{\sqrt{6}} \cdot \dfrac{\sqrt{6}}{\sqrt{6}} = \sqrt{6} - \dfrac{\sqrt{6}}{3} + \dfrac{\sqrt{6}}{6} = \dfrac{6\sqrt{6}}{6} - \dfrac{2\sqrt{6}}{6} + \dfrac{\sqrt{6}}{6} = \dfrac{5\sqrt{6}}{6}$

37. Combining radicals: $\sqrt[3]{25} + \dfrac{3}{\sqrt[3]{5}} = \sqrt[3]{25} + \dfrac{3}{\sqrt[3]{5}} \cdot \dfrac{\sqrt[3]{25}}{\sqrt[3]{25}} = \sqrt[3]{25} + \dfrac{3\sqrt[3]{25}}{5} = \dfrac{5\sqrt[3]{25}}{5} + \dfrac{3\sqrt[3]{25}}{5} = \dfrac{8\sqrt[3]{25}}{5}$

39. Using a calculator:
$$\sqrt{12} \approx 3.464 \qquad 2\sqrt{3} \approx 3.464$$

41. It is equal to the decimal approximation for $\sqrt{50}$:
$$\sqrt{8} + \sqrt{18} \approx 7.071 \approx \sqrt{50} \qquad \sqrt{26} \approx 5.099$$

43. The correct statement is: $3\sqrt{2x} + 5\sqrt{2x} = 8\sqrt{2x}$

45. The correct statement is: $\sqrt{9 + 16} = \sqrt{25} = 5$

47. Answers will vary.
49. Answers will vary.

51. First use the Pythagorean theorem to find the height h:

$$x^2 + h^2 = (2x)^2$$
$$x^2 + h^2 = 4x^2$$
$$h^2 = 3x^2$$
$$h = x\sqrt{3}$$

Therefore the ratio is: $\dfrac{x\sqrt{3}}{2x} = \dfrac{\sqrt{3}}{2}$

53. Graphing the inequality:

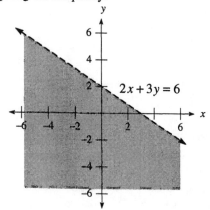

$2x + 3y = 6$

55. Graphing the inequality:

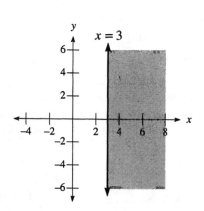

$x = 3$

9.5 Multiplication and Division of Radical Expressions

1. Multiplying: $\sqrt{6}\sqrt{3} = \sqrt{18} = 3\sqrt{2}$

3. Multiplying: $(2\sqrt{3})(5\sqrt{7}) = 10\sqrt{21}$

5. Multiplying: $(4\sqrt{6})(2\sqrt{15})(3\sqrt{10}) = 24\sqrt{900} = 24 \cdot 30 = 720$

7. Multiplying: $(3\sqrt[3]{3})(6\sqrt[3]{9}) = 18\sqrt[3]{27} = 18 \cdot 3 = 54$

9. Multiplying: $\sqrt{3}(\sqrt{2} - 3\sqrt{3}) = \sqrt{6} - 3\sqrt{9} = \sqrt{6} - 9$

11. Multiplying: $6\sqrt[3]{4}(2\sqrt[3]{2} + 1) = 12\sqrt[3]{8} + 6\sqrt[3]{4} = 24 + 6\sqrt[3]{4}$

13. Multiplying: $(\sqrt{3} + \sqrt{2})(3\sqrt{3} - \sqrt{2}) = 3\sqrt{9} - \sqrt{6} + 3\sqrt{6} - \sqrt{4} = 9 + 2\sqrt{6} - 2 = 7 + 2\sqrt{6}$

15. Multiplying: $(\sqrt{x} + 5)(\sqrt{x} - 3) = x - 3\sqrt{x} + 5\sqrt{x} - 15 = x + 2\sqrt{x} - 15$

17. Multiplying: $(3\sqrt{6} + 4\sqrt{2})(\sqrt{6} + 2\sqrt{2}) = 3\sqrt{36} + 4\sqrt{12} + 6\sqrt{12} + 8\sqrt{4} = 18 + 8\sqrt{3} + 12\sqrt{3} + 16 = 34 + 20\sqrt{3}$

19. Multiplying: $(\sqrt{3} + 4)^2 = (\sqrt{3} + 4)(\sqrt{3} + 4) = \sqrt{9} + 4\sqrt{3} + 4\sqrt{3} + 16 = 19 + 8\sqrt{3}$

21. Multiplying: $(\sqrt{x} - 3)^2 = (\sqrt{x} - 3)(\sqrt{x} - 3) = x - 3\sqrt{x} - 3\sqrt{x} + 9 = x - 6\sqrt{x} + 9$

23. Multiplying: $(2\sqrt{a} - 3\sqrt{b})^2 = (2\sqrt{a} - 3\sqrt{b})(2\sqrt{a} - 3\sqrt{b}) = 4a - 6\sqrt{ab} - 6\sqrt{ab} + 9b = 4a - 12\sqrt{ab} + 9b$

25. Multiplying: $(\sqrt{x-4} + 2)^2 = (\sqrt{x-4} + 2)(\sqrt{x-4} + 2) = x - 4 + 2\sqrt{x-4} + 2\sqrt{x-4} + 4 = x + 4\sqrt{x-4}$

27. Multiplying: $(\sqrt{x-5} - 3)^2 = (\sqrt{x-5} - 3)(\sqrt{x-5} - 3) = x - 5 - 3\sqrt{x-5} - 3\sqrt{x-5} + 9 = x + 4 - 6\sqrt{x-5}$

29. Multiplying: $(\sqrt{3} - \sqrt{2})(\sqrt{3} + \sqrt{2}) = (\sqrt{3})^2 - (\sqrt{2})^2 = 3 - 2 = 1$

31. Multiplying: $(\sqrt{a} + 7)(\sqrt{a} - 7) = (\sqrt{a})^2 - (7)^2 = a - 49$

33. Multiplying: $(5 - \sqrt{x})(5 + \sqrt{x}) = (5)^2 - (\sqrt{x})^2 = 25 - x$

35. Multiplying: $\left(\sqrt{x-4}+2\right)\left(\sqrt{x-4}-2\right)=\left(\sqrt{x-4}\right)^2-(2)^2=x-4-4=x-8$

37. Multiplying: $\left(\sqrt{3}+1\right)^3=\left(\sqrt{3}+1\right)\left(3+2\sqrt{3}+1\right)=\left(\sqrt{3}+1\right)\left(4+2\sqrt{3}\right)=4\sqrt{3}+4+6+2\sqrt{3}=10+6\sqrt{3}$

39. Rationalizing the denominator: $\dfrac{\sqrt{2}}{\sqrt{6}-\sqrt{2}}=\dfrac{\sqrt{2}}{\sqrt{6}-\sqrt{2}}\cdot\dfrac{\sqrt{6}+\sqrt{2}}{\sqrt{6}+\sqrt{2}}=\dfrac{\sqrt{12}+2}{6-2}=\dfrac{2\sqrt{3}+2}{4}=\dfrac{1+\sqrt{3}}{2}$

41. Rationalizing the denominator: $\dfrac{\sqrt{5}}{\sqrt{5}+1}=\dfrac{\sqrt{5}}{\sqrt{5}+1}\cdot\dfrac{\sqrt{5}-1}{\sqrt{5}-1}=\dfrac{5-\sqrt{5}}{5-1}=\dfrac{5-\sqrt{5}}{4}$

43. Rationalizing the denominator: $\dfrac{\sqrt{x}}{\sqrt{x}-3}=\dfrac{\sqrt{x}}{\sqrt{x}-3}\cdot\dfrac{\sqrt{x}+3}{\sqrt{x}+3}=\dfrac{x+3\sqrt{x}}{x-9}$

45. Rationalizing the denominator: $\dfrac{\sqrt{5}}{2\sqrt{5}-3}=\dfrac{\sqrt{5}}{2\sqrt{5}-3}\cdot\dfrac{2\sqrt{5}+3}{2\sqrt{5}+3}=\dfrac{2\sqrt{25}+3\sqrt{5}}{20-9}=\dfrac{10+3\sqrt{5}}{11}$

47. Rationalizing the denominator: $\dfrac{3}{\sqrt{x}-\sqrt{y}}=\dfrac{3}{\sqrt{x}-\sqrt{y}}\cdot\dfrac{\sqrt{x}+\sqrt{y}}{\sqrt{x}+\sqrt{y}}=\dfrac{3\sqrt{x}+3\sqrt{y}}{x-y}$

49. Rationalizing the denominator: $\dfrac{\sqrt{6}+\sqrt{2}}{\sqrt{6}-\sqrt{2}}=\dfrac{\sqrt{6}+\sqrt{2}}{\sqrt{6}-\sqrt{2}}\cdot\dfrac{\sqrt{6}+\sqrt{2}}{\sqrt{6}+\sqrt{2}}=\dfrac{6+2\sqrt{12}+2}{6-2}=\dfrac{8+4\sqrt{3}}{4}=2+\sqrt{3}$

51. Rationalizing the denominator: $\dfrac{\sqrt{7}-2}{\sqrt{7}+2}=\dfrac{\sqrt{7}-2}{\sqrt{7}+2}\cdot\dfrac{\sqrt{7}-2}{\sqrt{7}-2}=\dfrac{7-4\sqrt{7}+4}{7-4}=\dfrac{11-4\sqrt{7}}{3}$

53. Rationalizing the denominator: $\dfrac{\sqrt{a}+\sqrt{b}}{\sqrt{a}-\sqrt{b}}=\dfrac{\sqrt{a}+\sqrt{b}}{\sqrt{a}-\sqrt{b}}\cdot\dfrac{\sqrt{a}+\sqrt{b}}{\sqrt{a}+\sqrt{b}}=\dfrac{a+2\sqrt{ab}+b}{a-b}$

55. Rationalizing the denominator: $\dfrac{\sqrt{x}+2}{\sqrt{x}-2}=\dfrac{\sqrt{x}+2}{\sqrt{x}-2}\cdot\dfrac{\sqrt{x}+2}{\sqrt{x}+2}=\dfrac{x+4\sqrt{x}+4}{x-4}$

57. Rationalizing the denominator: $\dfrac{2\sqrt{3}-\sqrt{7}}{3\sqrt{3}+\sqrt{7}}=\dfrac{2\sqrt{3}-\sqrt{7}}{3\sqrt{3}+\sqrt{7}}\cdot\dfrac{3\sqrt{3}-\sqrt{7}}{3\sqrt{3}-\sqrt{7}}=\dfrac{18-3\sqrt{21}-2\sqrt{21}+7}{27-7}=\dfrac{25-5\sqrt{21}}{20}=\dfrac{5-\sqrt{21}}{4}$

59. Rationalizing the denominator: $\dfrac{3\sqrt{x}+2}{1+\sqrt{x}}=\dfrac{3\sqrt{x}+2}{1+\sqrt{x}}\cdot\dfrac{1-\sqrt{x}}{1-\sqrt{x}}=\dfrac{3\sqrt{x}+2-3x-2\sqrt{x}}{1-x}=\dfrac{\sqrt{x}-3x+2}{1-x}$

61. Simplifying the product: $\left(\sqrt[3]{2}+\sqrt[3]{3}\right)\left(\sqrt[3]{4}-\sqrt[3]{6}+\sqrt[3]{9}\right)=\sqrt[3]{8}-\sqrt[3]{12}+\sqrt[3]{18}+\sqrt[3]{12}-\sqrt[3]{18}+\sqrt[3]{27}=2+3=5$

63. The correct statement is: $5\left(2\sqrt{3}\right)=10\sqrt{3}$

65. The correct statement is: $\left(\sqrt{x}+3\right)^2=\left(\sqrt{x}+3\right)\left(\sqrt{x}+3\right)=x+6\sqrt{x}+9$

67. The correct statement is: $\left(5\sqrt{3}\right)^2=\left(5\sqrt{3}\right)\left(5\sqrt{3}\right)=25\cdot3=75$

69. Substituting $h=50$: $t=\dfrac{\sqrt{100-50}}{4}=\dfrac{\sqrt{50}}{4}=\dfrac{5\sqrt{2}}{4}$ second

Substituting $h=0$: $t=\dfrac{\sqrt{100-0}}{4}=\dfrac{\sqrt{100}}{4}=\dfrac{10}{4}=\dfrac{5}{2}$ second

71. Since the large rectangle is a golden rectangle and $AC=6$, then $CE=6\left(\dfrac{1+\sqrt{5}}{2}\right)=3+3\sqrt{5}$. Since $CD=6$, then

$DE=3+3\sqrt{5}-6=3\sqrt{5}-3$. Now computing the ratio:

$$\dfrac{EF}{DE}=\dfrac{6}{3\sqrt{5}-3}\cdot\dfrac{3\sqrt{5}+3}{3\sqrt{5}+3}=\dfrac{18\left(\sqrt{5}+1\right)}{45-9}=\dfrac{18\left(\sqrt{5}+1\right)}{36}=\dfrac{1+\sqrt{5}}{2}$$

Therefore the smaller rectangle $BDEF$ is also a golden rectangle.

73. Since the large rectangle is a golden rectangle and $AC = 2x$, then $CE = 2x\left(\dfrac{1+\sqrt{5}}{2}\right) = x\left(1+\sqrt{5}\right)$. Since $CD = 2x$, then

$DE = x\left(1+\sqrt{5}\right) - 2x = x\left(-1+\sqrt{5}\right)$. Now computing the ratio:

$$\frac{EF}{DE} = \frac{2x}{x\left(-1+\sqrt{5}\right)} = \frac{2}{-1+\sqrt{5}} \cdot \frac{-1-\sqrt{5}}{-1-\sqrt{5}} = \frac{-2\left(\sqrt{5}+1\right)}{1-5} = \frac{-2\left(\sqrt{5}+1\right)}{-4} = \frac{1+\sqrt{5}}{2}$$

Therefore the smaller rectangle $BDEF$ is also a golden rectangle.

75. The variation equation is $y = Kx^2$. Substituting $x = 5$ and $y = 75$:
$$75 = K \cdot 5^2$$
$$75 = 25K$$
$$K = 3$$
So $y = 3x^2$. Substituting $x = 7$: $y = 3 \cdot 7^2 = 3 \cdot 49 = 147$

77. The variation equation is $y = \dfrac{K}{x}$. Substituting $x = 25$ and $y = 10$:

$$10 = \frac{K}{25}$$
$$K = 250$$
So $y = \dfrac{250}{x}$. Substituting $y = 5$:
$$5 = \frac{250}{x}$$
$$5x = 250$$
$$x = 50$$

79. The variation equation is $z = Kxy^2$. Substituting $z = 40$, $x = 5$, and $y = 2$:
$$40 = K \cdot 5 \cdot 2^2$$
$$40 = 20K$$
$$K = 2$$
So $z = 2xy^2$. Substituting $x = 2$ and $y = 5$: $z = 2 \cdot 2 \cdot 5^2 = 100$

9.6   Equations with Radicals

1. Solving the equation:
$$\sqrt{2x+1} = 3$$
$$\left(\sqrt{2x+1}\right)^2 = 3^2$$
$$2x+1 = 9$$
$$2x = 8$$
$$x = 4$$

3. Solving the equation:
$$\sqrt{4x+1} = -5$$
$$\left(\sqrt{4x+1}\right)^2 = (-5)^2$$
$$4x+1 = 25$$
$$4x = 24$$
$$x = 6$$
Since this value does not check, the solution set is $\varnothing$.

5. Solving the equation:
$$\sqrt{2y-1} = 3$$
$$\left(\sqrt{2y-1}\right)^2 = 3^2$$
$$2y-1 = 9$$
$$2y = 10$$
$$y = 5$$

7. Solving the equation:
$$\sqrt{5x-7} = -1$$
$$\left(\sqrt{5x-7}\right)^2 = (-1)^2$$
$$5x-7 = 1$$
$$5x = 8$$
$$x = \tfrac{8}{5}$$
Since this value does not check, the solution set is $\varnothing$.

9. Solving the equation:
$$\sqrt{2x-3} - 2 = 4$$
$$\sqrt{2x-3} = 6$$
$$\left(\sqrt{2x-3}\right)^2 = 6^2$$
$$2x - 3 = 36$$
$$2x = 39$$
$$x = \frac{39}{2}$$

11. Solving the equation:
$$\sqrt{4a+1} + 3 = 2$$
$$\sqrt{4a+1} = -1$$
$$\left(\sqrt{4a+1}\right)^2 = (-1)^2$$
$$4a + 1 = 1$$
$$4a = 0$$
$$a = 0$$
Since this value does not check, the solution set is $\varnothing$.

13. Solving the equation:
$$\sqrt[4]{3x+1} = 2$$
$$\left(\sqrt[4]{3x+1}\right)^4 = 2^4$$
$$3x + 1 = 16$$
$$3x = 15$$
$$x = 5$$

15. Solving the equation:
$$\sqrt[3]{2x-5} = 1$$
$$\left(\sqrt[3]{2x-5}\right)^3 = 1^3$$
$$2x - 5 = 1$$
$$2x = 6$$
$$x = 3$$

17. Solving the equation:
$$\sqrt[3]{3a+5} = -3$$
$$\left(\sqrt[3]{3a+5}\right)^3 = (-3)^3$$
$$3a + 5 = -27$$
$$3a = -32$$
$$a = -\frac{32}{3}$$

19. Solving the equation:
$$\sqrt{y-3} = y - 3$$
$$\left(\sqrt{y-3}\right)^2 = (y-3)^2$$
$$y - 3 = y^2 - 6y + 9$$
$$0 = y^2 - 7y + 12$$
$$0 = (y-3)(y-4)$$
$$y = 3, 4$$

21. Solving the equation:
$$\sqrt{a+2} = a + 2$$
$$\left(\sqrt{a+2}\right)^2 = (a+2)^2$$
$$a + 2 = a^2 + 4a + 4$$
$$0 = a^2 + 3a + 2$$
$$0 = (a+2)(a+1)$$
$$a = -2, -1$$

23. Solving the equation:
$$\sqrt{2x+4} = \sqrt{1-x}$$
$$\left(\sqrt{2x+4}\right)^2 = \left(\sqrt{1-x}\right)^2$$
$$2x + 4 = 1 - x$$
$$3x = -3$$
$$x = -1$$

25. Solving the equation:
$$\sqrt{4a+7} = -\sqrt{a+2}$$
$$\left(\sqrt{4a+7}\right)^2 = \left(-\sqrt{a+2}\right)^2$$
$$4a + 7 = a + 2$$
$$3a = -5$$
$$a = -\frac{5}{3}$$
Since this value does not check, the solution set is $\varnothing$.

27. Solving the equation:
$$\sqrt[4]{5x-8} = \sqrt[4]{4x-1}$$
$$\left(\sqrt[4]{5x-8}\right)^4 = \left(\sqrt[4]{4x-1}\right)^4$$
$$5x - 8 = 4x - 1$$
$$x = 7$$

29. Solving the equation:
$$x + 1 = \sqrt{5x+1}$$
$$(x+1)^2 = \left(\sqrt{5x+1}\right)^2$$
$$x^2 + 2x + 1 = 5x + 1$$
$$x^2 - 3x = 0$$
$$x(x-3) = 0$$
$$x = 0, 3$$

31. Solving the equation:
$$t + 5 = \sqrt{2t+9}$$
$$(t+5)^2 = \left(\sqrt{2t+9}\right)^2$$
$$t^2 + 10t + 25 = 2t + 9$$
$$t^2 + 8t + 16 = 0$$
$$(t+4)^2 = 0$$
$$t = -4$$

33.   Solving the equation:
$$\sqrt{y-8} = \sqrt{8-y}$$
$$\left(\sqrt{y-8}\right)^2 = \left(\sqrt{8-y}\right)^2$$
$$y-8 = 8-y$$
$$2y = 16$$
$$y = 8$$

35.   Solving the equation:
$$\sqrt[3]{3x+5} = \sqrt[3]{5-2x}$$
$$\left(\sqrt[3]{3x+5}\right)^3 = \left(\sqrt[3]{5-2x}\right)^3$$
$$3x+5 = 5-2x$$
$$5x = 0$$
$$x = 0$$

37.   Solving the equation:
$$\sqrt{x-8} = \sqrt{x}-2$$
$$\left(\sqrt{x-8}\right)^2 = \left(\sqrt{x}-2\right)^2$$
$$x-8 = x-4\sqrt{x}+4$$
$$-12 = -4\sqrt{x}$$
$$\sqrt{x} = 3$$
$$x = 9$$

39.   Solving the equation:
$$\sqrt{x+1} = \sqrt{x}+1$$
$$\left(\sqrt{x+1}\right)^2 = \left(\sqrt{x}+1\right)^2$$
$$x+1 = x+2\sqrt{x}+1$$
$$0 = 2\sqrt{x}$$
$$\sqrt{x} = 0$$
$$x = 0$$

41.   Solving the equation:
$$\sqrt{x+8} = \sqrt{x-4}+2$$
$$\left(\sqrt{x+8}\right)^2 = \left(\sqrt{x-4}+2\right)^2$$
$$x+8 = x-4+4\sqrt{x-4}+4$$
$$8 = 4\sqrt{x-4}$$
$$\sqrt{x-4} = 2$$
$$x-4 = 4$$
$$x = 8$$

43.   Solving the equation:
$$\sqrt{x-5}-3 = \sqrt{x-8}$$
$$\left(\sqrt{x-5}-3\right)^2 = \left(\sqrt{x-8}\right)^2$$
$$x-5-6\sqrt{x-5}+9 = x-8$$
$$-6\sqrt{x-5} = -12$$
$$\sqrt{x-5} = 2$$
$$x-5 = 4$$
$$x = 9$$
Since this value does not check, the solution set is $\varnothing$.

45.   Solving the equation:
$$\sqrt{x+4} = 2-\sqrt{2x}$$
$$\left(\sqrt{x+4}\right)^2 = \left(2-\sqrt{2x}\right)^2$$
$$x+4 = 4-4\sqrt{2x}+2x$$
$$-x = -4\sqrt{2x}$$
$$(-x)^2 = \left(-4\sqrt{2x}\right)^2$$
$$x^2 = 32x$$
$$x^2 - 32x = 0$$
$$x(x-32) = 0$$
$$x = 0, 32$$
The solution is 0 (32 does not check).

47.   Solving the equation:
$$\sqrt{2x+4} = \sqrt{x+3}+1$$
$$\left(\sqrt{2x+4}\right)^2 = \left(\sqrt{x+3}+1\right)^2$$
$$2x+4 = x+3+2\sqrt{x+3}+1$$
$$x = 2\sqrt{x+3}$$
$$x^2 = \left(2\sqrt{x+3}\right)^2$$
$$x^2 = 4x+12$$
$$x^2 - 4x - 12 = 0$$
$$(x-6)(x+2) = 0$$
$$x = -2, 6$$
The solution is 6 (–2 does not check).

49. Graphing the equation:

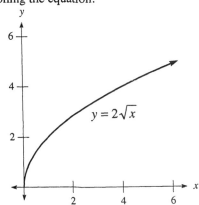

51. Graphing the equation:

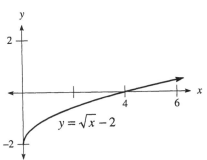

53. Graphing the equation:

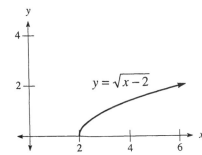

55. Graphing the equation:

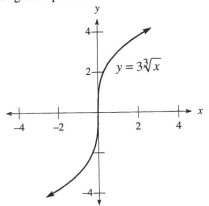

57. Graphing the equation:

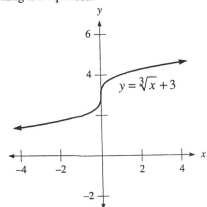

59. Graphing the equation:

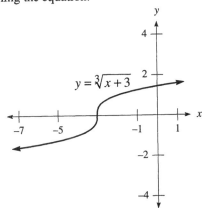

61. Solving for h:

$$t = \frac{\sqrt{100-h}}{4}$$
$$4t = \sqrt{100-h}$$
$$16t^2 = 100 - h$$
$$h = 100 - 16t^2$$

63. Solving for L:

$$2 = 2\left(\tfrac{22}{7}\right)\sqrt{\frac{L}{32}}$$
$$\tfrac{7}{22} = \sqrt{\frac{L}{32}}$$
$$\left(\tfrac{7}{22}\right)^2 = \frac{L}{32}$$
$$L = 32\left(\tfrac{7}{22}\right)^2 \approx 3.24 \text{ feet}$$

65.   The width is $\sqrt{25} = 5$ meters.

67.   Solving the equation:
$$\sqrt{x} = 50$$
$$x = 2500$$
The plume is 2,500 meters down river.

69.   Multiplying: $\sqrt{2}\left(\sqrt{3} - \sqrt{2}\right) = \sqrt{6} - \sqrt{4} = \sqrt{6} - 2$

71.   Multiplying: $\left(\sqrt{x} + 5\right)^2 = \left(\sqrt{x} + 5\right)\left(\sqrt{x} + 5\right) = x + 5\sqrt{x} + 5\sqrt{x} + 25 = x + 10\sqrt{x} + 25$

73.   Rationalizing the denominator: $\dfrac{\sqrt{x}}{\sqrt{x} + 3} = \dfrac{\sqrt{x}}{\sqrt{x} + 3} \cdot \dfrac{\sqrt{x} - 3}{\sqrt{x} - 3} = \dfrac{x - 3\sqrt{x}}{x - 9}$

75.   Solving the equation:
$$\frac{x}{3\sqrt{2x - 3}} - \frac{1}{\sqrt{2x - 3}} = \tfrac{1}{3}$$
$$3\sqrt{2x - 3}\left(\frac{x}{3\sqrt{2x - 3}} - \frac{1}{\sqrt{2x - 3}}\right) = 3\sqrt{2x - 3}\left(\tfrac{1}{3}\right)$$
$$x - 3 = \sqrt{2x - 3}$$
$$(x - 3)^2 = 2x - 3$$
$$x^2 - 6x + 9 = 2x - 3$$
$$x^2 - 8x + 12 = 0$$
$$(x - 6)(x - 2) = 0$$
$$x = 2, 6$$
The solution is 6 (2 does not check).

77.   Solving the equation:
$$x + 1 = \sqrt[3]{4x + 4}$$
$$(x + 1)^3 = \left(\sqrt[3]{4x + 4}\right)^3$$
$$x^3 + 3x^2 + 3x + 1 = 4x + 4$$
$$x^3 + 3x^2 - x - 3 = 0$$
$$x^2(x + 3) - 1(x + 3) = 0$$
$$(x + 3)\left(x^2 - 1\right) = 0$$
$$(x + 3)(x + 1)(x - 1) = 0$$
$$x = -3, -1, 1$$

79.   Solving for y:
$$y + 2 = \sqrt{x^2 + (y - 2)^2}$$
$$(y + 2)^2 = x^2 + (y - 2)^2$$
$$y^2 + 4y + 4 = x^2 + y^2 - 4y + 4$$
$$8y = x^2$$
$$y = \tfrac{1}{8}x^2$$

9.7 Complex Numbers

1.   Writing in terms of i: $\sqrt{-36} = 6i$

3.   Writing in terms of i: $-\sqrt{-25} = -5i$

5.   Writing in terms of i: $\sqrt{-72} = 6i\sqrt{2}$

7.   Writing in terms of i: $-\sqrt{-12} = -2i\sqrt{3}$

9.   Rewriting the expression: $i^{28} = \left(i^4\right)^7 = (1)^7 = 1$

11.   Rewriting the expression: $i^{26} = i^{24}i^2 = \left(i^4\right)^6 i^2 = (1)^6(-1) = -1$

13.   Rewriting the expression: $i^{75} = i^{72}i^3 = \left(i^4\right)^{18} i^2 i = (1)^{18}(-1)i = -i$

15.   Setting real and imaginary parts equal:
$$2x = 6 \qquad 3y = -3$$
$$x = 3 \qquad y = -1$$

17.   Setting real and imaginary parts equal:
$$-x = 2 \qquad 10y = -5$$
$$x = -2 \qquad y = -\tfrac{1}{2}$$

19. Setting real and imaginary parts equal:

$$2x = -16 \qquad -2y = 10$$
$$x = -8 \qquad\quad y = -5$$

21. Setting real and imaginary parts equal:

$$2x - 4 = 10 \qquad -6y = -3$$
$$2x - 14 \qquad\qquad y = \tfrac{1}{2}$$
$$x = 7$$

23. Setting real and imaginary parts equal:

$$7x - 1 = 2 \qquad 5y + 2 = 4$$
$$7x = 3 \qquad\quad 5y = 2$$
$$x = \tfrac{3}{7} \qquad\quad y = \tfrac{2}{5}$$

25. Combining the numbers: $(2 + 3i) + (3 + 6i) = 5 + 9i$ **27.** Combining the numbers: $(3 - 5i) + (2 + 4i) = 5 - i$

29. Combining the numbers: $(5 + 2i) - (3 + 6i) = 5 + 2i - 3 - 6i = 2 - 4i$

31. Combining the numbers: $(3 - 5i) - (2 + i) = 3 - 5i - 2 - i = 1 - 6i$

33. Combining the numbers: $[(3 + 2i) - (6 + i)] + (5 + i) = 3 + 2i - 6 - i + 5 + i = 2 + 2i$

35. Combining the numbers: $[(7 - i) - (2 + 4i)] - (6 + 2i) = 7 - i - 2 - 4i - 6 - 2i = -1 - 7i$

37. Combining the numbers:
$$(3 + 2i) - [(3 - 4i) - (6 + 2i)] = (3 + 2i) - (3 - 4i - 6 - 2i) = (3 + 2i) - (-3 - 6i) = 3 + 2i + 3 + 6i = 6 + 8i$$

39. Combining the numbers: $(4 - 9i) + [(2 - 7i) - (4 + 8i)] = (4 - 9i) + (2 - 7i - 4 - 8i) = (4 - 9i) + (-2 - 15i) = 2 - 24i$

41. Finding the product: $3i(4 + 5i) = 12i + 15i^2 = -15 + 12i$

43. Finding the product: $6i(4 - 3i) = 24i - 18i^2 = 18 + 24i$

45. Finding the product: $(3 + 2i)(4 + i) = 12 + 8i + 3i + 2i^2 = 12 + 11i - 2 = 10 + 11i$

47. Finding the product: $(4 + 9i)(3 - i) = 12 + 27i - 4i - 9i^2 = 12 + 23i + 9 = 21 + 23i$

49. Finding the product: $(1 + i)^3 = (1 + i)(1 + i)^2 = (1 + i)(1 + 2i - 1) = (1 + i)(2i) = -2 + 2i$

51. Finding the product: $(2 - i)^3 = (2 - i)(2 - i)^2 = (2 - i)(4 - 4i - 1) = (2 - i)(3 - 4i) = 6 - 11i - 4 = 2 - 11i$

53. Finding the product: $(2 + 5i)^2 = (2 + 5i)(2 + 5i) = 4 + 10i + 10i - 25 = -21 + 20i$

55. Finding the product: $(1 - i)^2 = (1 - i)(1 - i) = 1 - i - i - 1 = -2i$

57. Finding the product: $(3 - 4i)^2 = (3 - 4i)(3 - 4i) = 9 - 12i - 12i - 16 = -7 - 24i$

59. Finding the product: $(2 + i)(2 - i) = 4 - i^2 = 4 + 1 = 5$

61. Finding the product: $(6 - 2i)(6 + 2i) = 36 - 4i^2 = 36 + 4 = 40$

63. Finding the product: $(2 + 3i)(2 - 3i) = 4 - 9i^2 = 4 + 9 = 13$

65. Finding the product: $(10 + 8i)(10 - 8i) = 100 - 64i^2 = 100 + 64 = 164$

67. Finding the quotient: $\dfrac{2 - 3i}{i} = \dfrac{2 - 3i}{i} \cdot \dfrac{i}{i} = \dfrac{2i + 3}{-1} = -3 - 2i$

69. Finding the quotient: $\dfrac{5 + 2i}{-i} = \dfrac{5 + 2i}{-i} \cdot \dfrac{i}{i} = \dfrac{5i - 2}{1} = -2 + 5i$

71. Finding the quotient: $\dfrac{4}{2 - 3i} = \dfrac{4}{2 - 3i} \cdot \dfrac{2 + 3i}{2 + 3i} = \dfrac{8 + 12i}{4 + 9} = \dfrac{8 + 12i}{13} = \dfrac{8}{13} + \dfrac{12}{13}i$

73. Finding the quotient: $\dfrac{6}{-3 + 2i} = \dfrac{6}{-3 + 2i} \cdot \dfrac{-3 - 2i}{-3 - 2i} = \dfrac{-18 - 12i}{9 + 4} = \dfrac{-18 - 12i}{13} = -\dfrac{18}{13} - \dfrac{12}{13}i$

75. Finding the quotient: $\dfrac{2 + 3i}{2 - 3i} = \dfrac{2 + 3i}{2 - 3i} \cdot \dfrac{2 + 3i}{2 + 3i} = \dfrac{4 + 12i - 9}{4 + 9} = \dfrac{-5 + 12i}{13} = -\dfrac{5}{13} + \dfrac{12}{13}i$

77. Finding the quotient: $\dfrac{5 + 4i}{3 + 6i} = \dfrac{5 + 4i}{3 + 6i} \cdot \dfrac{3 - 6i}{3 - 6i} = \dfrac{15 - 18i + 24}{9 + 36} = \dfrac{39 - 18i}{45} = \dfrac{13}{15} - \dfrac{2}{5}i$

79. Dividing to find R: $R = \dfrac{80 + 20i}{-6 + 2i} = \dfrac{80 + 20i}{-6 + 2i} \cdot \dfrac{-6 - 2i}{-6 - 2i} = \dfrac{-480 - 280i + 40}{36 + 4} = \dfrac{-440 - 280i}{40} = (-11 - 7i)$ ohms

81. The domain is $\{1, 3, 4\}$ and the range is $\{2, 4\}$. This is a function.

83. The domain is $\{1, 2, 3\}$ and the range is $\{1, 2, 3\}$. This is a function.

85. Since this passes the vertical line test, it is a function.

87.   Since this fails the vertical line test, it is not a function.

89.   Simplifying: $\dfrac{1}{i} = \dfrac{1}{i} \cdot \dfrac{i}{i} = \dfrac{i}{i^2} = \dfrac{i}{-1} = -i$

91.   Substituting into the equation: $x^2 - 2x + 2 = (1+i)^2 - 2(1+i) + 2 = 1 + 2i - 1 - 2 - 2i + 2 = 0$
Thus $x = 1 + i$ is a solution to the equation.

93.   Substituting into the equation:
$$
\begin{aligned}
x^3 - 11x + 20 &= (2+i)^3 - 11(2+i) + 20 \\
&= (2+i)(4+4i-1) - 22 - 11i + 20 \\
&= (2+i)(3+4i) - 2 - 11i \\
&= 6 + 11i - 4 - 2 - 11i \\
&= 0
\end{aligned}
$$
Thus $x = 2 + i$ is a solution to the equation.

Chapter 9 Review

1.   Simplifying: $\sqrt{49} = 7$

3.   Simplifying: $16^{1/4} = 2$

5.   Simplifying: $\sqrt[5]{32x^{15}y^{10}} = 2x^3y^2$

7.   Simplifying: $x^{2/3} \cdot x^{4/3} = x^{2/3 + 4/3} = x^2$

9.   Simplifying: $\dfrac{a^{3/5}}{a^{1/4}} = a^{3/5 - 1/4} = a^{12/20 - 5/20} = a^{7/20}$

11.   Multiplying: $\left(3x^{1/2} + 5y^{1/2}\right)\left(4x^{1/2} - 3y^{1/2}\right) = 12x - 9x^{1/2}y^{1/2} + 20x^{1/2}y^{1/2} - 15y = 12x + 11x^{1/2}y^{1/2} - 15y$

13.   Dividing: $\dfrac{28x^{5/6} + 14x^{7/6}}{7x^{1/3}} = \dfrac{28x^{5/6}}{7x^{1/3}} + \dfrac{14x^{7/6}}{7x^{1/3}} = 4x^{5/6 - 1/3} + 2x^{7/6 - 1/3} = 4x^{1/2} + 2x^{5/6}$

15.   Simplifying: $x^{3/4} + \dfrac{5}{x^{1/4}} = x^{3/4} \cdot \dfrac{x^{1/4}}{x^{1/4}} + \dfrac{5}{x^{1/4}} = \dfrac{x+5}{x^{1/4}}$

17.   Simplifying: $\sqrt{50} = \sqrt{25 \cdot 2} = 5\sqrt{2}$

19.   Simplifying: $\sqrt{18x^2} = \sqrt{9x^2 \cdot 2} = 3x\sqrt{2}$

21.   Simplifying: $\sqrt[4]{32a^4b^5c^6} = \sqrt[4]{16a^4b^4c^4 \cdot 2bc^2} = 2abc\sqrt[4]{2bc^2}$

23.   Rationalizing the denominator: $\dfrac{6}{\sqrt[3]{2}} = \dfrac{6}{\sqrt[3]{2}} \cdot \dfrac{\sqrt[3]{4}}{\sqrt[3]{4}} = \dfrac{6\sqrt[3]{4}}{2} = 3\sqrt[3]{4}$

25.   Simplifying: $\sqrt[3]{\dfrac{40x^2y^3}{3z}} = \dfrac{2y\sqrt[3]{5x^2}}{\sqrt[3]{3z}} \cdot \dfrac{\sqrt[3]{9z^2}}{\sqrt[3]{9z^2}} = \dfrac{2y\sqrt[3]{45x^2z^2}}{3z}$

27.   Combining the expressions: $\sqrt{12} + \sqrt{3} = 2\sqrt{3} + \sqrt{3} = 3\sqrt{3}$

29.   Combining the expressions: $3\sqrt{8} - 4\sqrt{72} + 5\sqrt{50} = 6\sqrt{2} - 24\sqrt{2} + 25\sqrt{2} = 7\sqrt{2}$

31.   Combining the expressions: $2x\sqrt[3]{xy^3z^2} - 6y\sqrt[3]{x^4z^2} = 2xy\sqrt[3]{xz^2} - 6xy\sqrt[3]{xz^2} = -4xy\sqrt[3]{xz^2}$

33.   Multiplying: $\left(\sqrt{x} - 2\right)\left(\sqrt{x} - 3\right) = x - 2\sqrt{x} - 3\sqrt{x} + 6 = x - 5\sqrt{x} + 6$

35.   Rationalizing the denominator: $\dfrac{\sqrt{7}+\sqrt{5}}{\sqrt{7}-\sqrt{5}} = \dfrac{\sqrt{7}+\sqrt{5}}{\sqrt{7}-\sqrt{5}} \cdot \dfrac{\sqrt{7}+\sqrt{5}}{\sqrt{7}+\sqrt{5}} = \dfrac{7 + 2\sqrt{35} + 5}{7 - 5} = \dfrac{12 + 2\sqrt{35}}{2} = 6 + \sqrt{35}$

37.   Solving the equation:
$$
\begin{aligned}
\sqrt{4a+1} &= 1 \\
\left(\sqrt{4a+1}\right)^2 &= (1)^2 \\
4a + 1 &= 1 \\
4a &= 0 \\
a &= 0
\end{aligned}
$$

39.   Solving the equation:
$$
\begin{aligned}
\sqrt{3x+1} - 3 &= 1 \\
\sqrt{3x+1} &= 4 \\
\left(\sqrt{3x+1}\right)^2 &= (4)^2 \\
3x + 1 &= 16 \\
3x &= 15 \\
x &= 5
\end{aligned}
$$

41.   Graphing the equation:

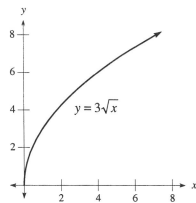

43.   Writing in terms of i: $i^{24} = \left(i^4\right)^6 = (1)^6 = 1$

45.   Setting real and imaginary parts equal:
$$-2x = 3 \qquad 8y = -4$$
$$x = -\tfrac{3}{2} \qquad y = -\tfrac{1}{2}$$

47.   Combining the numbers: $(3 + 5i) + (6 - 2i) = 9 + 3i$

49.   Multiplying: $3i(4 + 2i) = 12i + 6i^2 = -6 + 12i$

51.   Multiplying: $(4 + 2i)^2 = (4 + 2i)(4 + 2i) = 16 + 8i + 8i + 4i^2 = 16 + 16i - 4 = 12 + 16i$

53.   Dividing: $\dfrac{3 + i}{i} = \dfrac{3 + i}{i} \cdot \dfrac{i}{i} = \dfrac{3i - 1}{-1} = 1 - 3i$

55.   Let l represent the desired length. Using the Pythagorean theorem:
$$18^2 + 13.5^2 = l^2$$
$$l^2 = 506.25$$
$$l = \sqrt{506.25} = 22.5$$
The length of one side of the roof is 22.5 feet.

Chapters 1-9 Cumulative Review

1.   Simplifying: $33 - 22 - (-11) + 1 = 33 - 22 + 11 + 1 = 23$

3.   Simplifying: $-6 + 5\big[3 - 2(-4 - 1)\big] = -6 + 5\big[3 - 2(-5)\big] = -6 + 5(3 + 10) = -6 + 5(13) = -6 + 65 = 59$

5.   Simplifying: $\left(2y^{-3}\right)^{-1}\left(4y^{-3}\right)^2 = \tfrac{1}{2}y^3 \cdot 16y^{-6} = 8y^{-3} = \dfrac{8}{y^3}$

7.   Simplifying: $\sqrt{72y^5} = \sqrt{36y^4 \cdot 2y} = 6y^2\sqrt{2y}$

9.   The opposite is -12 and the reciprocal is $\tfrac{1}{12}$.

11.   Writing in scientific notation: $41,500 = 4.15 \times 10^4$

13.   Factoring completely: $625a^4 - 16b^4 = \left(25a^2 + 4b^2\right)\left(25a^2 - 4b^2\right) = (5a - 2b)(5a + 2b)\left(25a^2 + 4b^2\right)$

15.   Reducing to lowest terms: $\dfrac{246}{861} = \dfrac{2 \cdot 123}{7 \cdot 123} = \dfrac{2}{7}$

17.   Reducing to lowest terms: $\dfrac{x^2 - 9x + 20}{x^2 - 7x + 12} = \dfrac{(x - 4)(x - 5)}{(x - 4)(x - 3)} = \dfrac{x - 5}{x - 3}$

19.   Multiplying: $(2x - 5y)(3x - 2y) = 6x^2 - 4xy - 15xy + 10y^2 = 6x^2 - 19xy + 10y^2$

21.   Multiplying: $\left(\sqrt{x} - 2\right)^2 = \left(\sqrt{x} - 2\right)\left(\sqrt{x} - 2\right) = x - 2\sqrt{x} - 2\sqrt{x} + 4 = x - 4\sqrt{x} + 4$

23. Dividing: $\dfrac{27x^3y^2}{13x^2y^4} \div \dfrac{9xy}{26y} = \dfrac{27x^3y^2}{13x^2y^4} \cdot \dfrac{26y}{9xy} = \dfrac{27 \cdot 26 x^3 y^3}{9 \cdot 13 x^3 y^5} = \dfrac{6}{y^2}$

25. Subtracting:

$$\left(\tfrac{2}{3}x^3 + \tfrac{1}{6}x^2 + \tfrac{1}{2}\right) - \left(\tfrac{1}{4}x^2 - \tfrac{1}{3}x + \tfrac{1}{12}\right) = \tfrac{2}{3}x^3 + \tfrac{1}{6}x^2 + \tfrac{1}{2} - \tfrac{1}{4}x^2 + \tfrac{1}{3}x - \tfrac{1}{12}$$

$$= \tfrac{2}{3}x^3 + \left(\tfrac{1}{6} - \tfrac{1}{4}\right)x^2 + \tfrac{1}{3}x + \left(\tfrac{1}{2} - \tfrac{1}{12}\right)$$

$$= \tfrac{2}{3}x^3 - \tfrac{1}{12}x^2 + \tfrac{1}{3}x + \tfrac{5}{12}$$

27. Solving the equation:

$$3y - 8 = -4y + 6$$
$$7y = 14$$
$$y = 2$$

29. Solving the equation:

$$|3x - 1| - 2 = 6$$
$$|3x - 1| = 8$$
$$3x - 1 = -8, 8$$
$$3x = -7, 9$$
$$x = -\tfrac{7}{3}, 3$$

31. Solving the equation:

$$\frac{x+2}{x+1} - 2 = \frac{1}{x+1}$$
$$(x+1)\left(\frac{x+2}{x+1} - 2\right) = (x+1)\left(\frac{1}{x+1}\right)$$
$$x + 2 - 2(x+1) = 1$$
$$x + 2 - 2x - 2 = 1$$
$$-x = 1$$
$$x = -1$$

There is no solution (-1 does not check).

33. Solving the equation:

$$(x+3)^2 - 3(x+3) - 70 = 0$$
$$(x+3-10)(x+3+7) = 0$$
$$(x-7)(x+10) = 0$$
$$x = -10, 7$$

35. Solving for L:

$$\sqrt{y+7} - \sqrt{y+2} = 1$$
$$18 = 2L + 2(3)$$
$$18 = 2L + 6$$
$$2L = 12$$
$$L = 6$$

37. Solving the inequality:

$$|5x - 4| \geq 6$$

$$5x - 4 \leq -6 \qquad \text{or} \qquad 5x - 4 \geq 6$$
$$5x \leq -2 \qquad\qquad\qquad 5x \geq 10$$
$$x \leq -\tfrac{2}{5} \qquad\qquad\qquad x \geq 2$$

The solution set is $\left\{x \mid x \leq -\tfrac{2}{5} \text{ or } x \geq 2\right\}$. Graphing the solution set:

$-2/5$ 2

39. Substituting into the first equation:

$$4x + 9(2x - 12) = 2$$
$$4x + 18x - 108 = 2$$
$$22x = 110$$
$$x = 5$$

The solution is $(5, -2)$.

41. Multiply the first equation by -1 and add it to the first equation:
$$-x - y = 2$$
$$y + 10z = -1$$
Adding yields the equation $-x + 10z = 1$. So the system becomes:
$$-x + 10z = 1$$
$$2x - 13z = 5$$
Multiply the first equation by 2:
$$-2x + 20z = 2$$
$$2x - 13z = 5$$
Adding yields:
$$7z = 7$$
$$z = 1$$
Substituting to find x:
$$2x - 13 = 5$$
$$2x = 18$$
$$x = 9$$
Substituting to find y:
$$9 + y = -2$$
$$y = -11$$
The solution is $(9, -11, 1)$.

43. Graphing the equation:

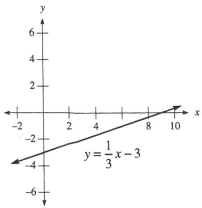

45. Graphing the equation:

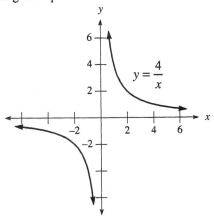

47. Solving for y:
$$4x - 5y = 15$$
$$-5y = -4x + 15$$
$$y = \frac{4}{5}x - 3$$

The slope is $\frac{4}{5}$ and the y-intercept is -3.

49. Using the point-slope formula:
$$y - 3 = \frac{2}{3}(x - 9)$$
$$y - 3 = \frac{2}{3}x - 6$$
$$y = \frac{2}{3}x - 3$$

Chapter 9 Test

1.   Simplifying: $27^{-2/3} = \left(27^{1/3}\right)^{-2} = 3^{-2} = \dfrac{1}{3^2} = \dfrac{1}{9}$ **2.**   Simplifying: $\left(\dfrac{25}{49}\right)^{-1/2} = \left(\dfrac{49}{25}\right)^{1/2} = \dfrac{7}{5}$

3.   Simplifying: $a^{3/4} \cdot a^{-1/3} = a^{3/4-1/3} = a^{9/12-4/12} = a^{5/12}$

4.   Simplifying: $\dfrac{\left(x^{2/3}y^{-3}\right)^{1/2}}{\left(x^{3/4}y^{1/2}\right)^{-1}} = \dfrac{x^{1/3}y^{-3/2}}{x^{-3/4}y^{-1/2}} = x^{1/3+3/4}\,y^{-3/2+1/2} = x^{13/12}y^{-1} = \dfrac{x^{13/12}}{y}$

5.   Simplifying: $\sqrt{49x^8y^{10}} = 7x^4y^5$ **6.**   Simplifying: $\sqrt[5]{32x^{10}y^{20}} = 2x^2y^4$

7.   Simplifying: $\dfrac{\left(36a^8b^4\right)^{1/2}}{\left(27a^9b^6\right)^{1/3}} = \dfrac{6a^4b^2}{3a^3b^2} = 2a$ **8.**   Simplifying: $\dfrac{\left(x^n y^{1/n}\right)^n}{\left(x^{1/n}y^n\right)^{n^2}} = \dfrac{x^{n^2}y}{x^n y^{n^3}} = x^{n^2-n}y^{1-n^3}$

9.   Multiplying: $2a^{1/2}\left(3a^{3/2} - 5a^{1/2}\right) = 6a^2 - 10a$

10.   Multiplying: $\left(4a^{3/2} - 5\right)^2 = \left(4a^{3/2} - 5\right)\left(4a^{3/2} - 5\right) = 16a^3 - 20a^{3/2} - 20a^{3/2} + 25 = 16a^3 - 40a^{3/2} + 25$

11.   Factoring: $3x^{2/3} + 5x^{1/3} - 2 = \left(3x^{1/3} - 1\right)\left(x^{1/3} + 2\right)$ **12.**   Factoring: $9x^{2/3} - 49 = \left(3x^{1/3} - 7\right)\left(3x^{1/3} + 7\right)$

13.   Combining: $\dfrac{4}{x^{1/2}} + x^{1/2} = \dfrac{4}{x^{1/2}} + x^{1/2} \cdot \dfrac{x^{1/2}}{x^{1/2}} = \dfrac{x+4}{x^{1/2}}$

14.   Combining: $\dfrac{x^2}{\left(x^2-3\right)^{1/2}} - \left(x^2-3\right)^{1/2} = \dfrac{x^2}{\left(x^2-3\right)^{1/2}} - \left(x^2-3\right)^{1/2} \cdot \dfrac{\left(x^2-3\right)^{1/2}}{\left(x^2-3\right)^{1/2}} = \dfrac{x^2 - x^2 + 3}{\left(x^2-3\right)^{1/2}} = \dfrac{3}{\left(x^2-3\right)^{1/2}}$

15.   Simplifying: $\sqrt{125x^3y^5} = \sqrt{25x^2y^4 \cdot 5xy} = 5xy^2\sqrt{5xy}$

16.   Simplifying: $\sqrt[3]{40x^7y^8} = \sqrt[3]{8x^6y^6 \cdot 5xy^2} = 2x^2y^2\sqrt[3]{5xy^2}$

17.   Simplifying: $\sqrt{\dfrac{2}{3}} = \dfrac{\sqrt{2}}{\sqrt{3}} \cdot \dfrac{\sqrt{3}}{\sqrt{3}} = \dfrac{\sqrt{6}}{3}$

18.   Simplifying: $\sqrt{\dfrac{12a^4b^3}{5c}} = \dfrac{2a^2b\sqrt{3b}}{\sqrt{5c}} \cdot \dfrac{\sqrt{5c}}{\sqrt{5c}} = \dfrac{2a^2b\sqrt{15bc}}{5c}$

19.   Combining: $3\sqrt{12} - 4\sqrt{27} = 6\sqrt{3} - 12\sqrt{3} = -6\sqrt{3}$

20.   Combining: $2\sqrt[3]{24a^3b^3} - 5a\sqrt[3]{3b^3} = 4ab\sqrt[3]{3} - 5ab\sqrt[3]{3} = -ab\sqrt[3]{3}$

21.   Multiplying: $\left(\sqrt{x} + 7\right)\left(\sqrt{x} - 4\right) = x - 4\sqrt{x} + 7\sqrt{x} - 28 = x + 3\sqrt{x} - 28$

22.   Multiplying: $\left(3\sqrt{2} - \sqrt{3}\right)^2 = \left(3\sqrt{2} - \sqrt{3}\right)\left(3\sqrt{2} - \sqrt{3}\right) = 18 - 3\sqrt{6} - 3\sqrt{6} + 3 = 21 - 6\sqrt{6}$

23.   Rationalizing the denominator: $\dfrac{5}{\sqrt{3}-1} = \dfrac{5}{\sqrt{3}-1} \cdot \dfrac{\sqrt{3}+1}{\sqrt{3}+1} = \dfrac{5\sqrt{3}+5}{3-1} = \dfrac{5+5\sqrt{3}}{2}$

24.   Rationalizing the denominator: $\dfrac{\sqrt{x}-\sqrt{2}}{\sqrt{x}+\sqrt{2}} = \dfrac{\sqrt{x}-\sqrt{2}}{\sqrt{x}+\sqrt{2}} \cdot \dfrac{\sqrt{x}-\sqrt{2}}{\sqrt{x}-\sqrt{2}} = \dfrac{x - \sqrt{2x} - \sqrt{2x} + 2}{x-2} = \dfrac{x - 2\sqrt{2x} + 2}{x-2}$

25. Solving the equation:
$$\sqrt{3x+1} = x-3$$
$$\left(\sqrt{3x+1}\right)^2 = (x-3)^2$$
$$3x+1 = x^2-6x+9$$
$$0 = x^2-9x+8$$
$$0 = (x-1)(x-8)$$
$$x = 1, 8$$
The solution is 8 (1 does not check).

26. Solving the equation:
$$\sqrt[3]{2x+7} = -1$$
$$\left(\sqrt[3]{2x+7}\right)^3 = (-1)^3$$
$$2x+7 = -1$$
$$2x = -8$$
$$x = -4$$

27. Solving the equation:
$$\sqrt{x+3} = \sqrt{x+4}-1$$
$$\left(\sqrt{x+3}\right)^2 = \left(\sqrt{x+4}-1\right)^2$$
$$x+3 = x+4-2\sqrt{x+4}+1$$
$$-2 = -2\sqrt{x+4}$$
$$\sqrt{x+4} = 1$$
$$x+4 = 1$$
$$x = -3$$

28. Graphing the equation:

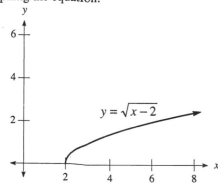

29. Graphing the equation:

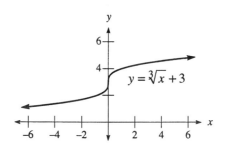

30. Setting the real and imaginary parts equal:
$$2x+5 = 6 \qquad -(y-3) = -4$$
$$2x = 1 \qquad y-3 = 4$$
$$x = \tfrac{1}{2} \qquad y = 7$$

31. Performing the operations:
$$(3+2i)-\left[(7-i)-(4+3i)\right] = (3+2i)-(7-i-4-3i) = (3+2i)-(3-4i) = 3+2i-3+4i = 6i$$

32. Performing the operations: $(2-3i)(4+3i) = 8+6i-12i+9 = 17-6i$

33. Performing the operations: $(5-4i)^2 = (5-4i)(5-4i) = 25-20i-20i-16 = 9-40i$

34. Performing the operations: $\dfrac{2-3i}{2+3i} = \dfrac{2-3i}{2+3i} \cdot \dfrac{2-3i}{2-3i} = \dfrac{4-12i-9}{4+9} = \dfrac{-5-12i}{13} = -\dfrac{5}{13}-\dfrac{12}{13}i$

35. Rewriting the exponent: $i^{38} = \left(i^2\right)^{19} = (-1)^{19} = -1$

Chapter 10
Quadratic Functions

10.1 Completing the Square

1. Solving the equation:

$$x^2 = 25$$
$$x = \pm\sqrt{25} = \pm 5$$

3. Solving the equation:

$$y^2 = \frac{3}{4}$$
$$y = \pm\sqrt{\frac{3}{4}} = \pm\frac{\sqrt{3}}{2}$$

5. Solving the equation:

$$x^2 + 12 = 0$$
$$x^2 = -12$$
$$x = \pm\sqrt{-12} = \pm 2i\sqrt{3}$$

7. Solving the equation:

$$4a^2 - 45 = 0$$
$$4a^2 = 45$$
$$a^2 = \frac{45}{4}$$
$$a = \pm\sqrt{\frac{45}{4}} = \pm\frac{3\sqrt{5}}{2}$$

9. Solving the equation:

$$(2y-1)^2 = 25$$
$$2y - 1 = \pm\sqrt{25} = \pm 5$$
$$2y - 1 = -5, 5$$
$$2y = -4, 6$$
$$y = -2, 3$$

11. Solving the equation:

$$(2a+3)^2 = -9$$
$$2a + 3 = \pm\sqrt{-9} = \pm 3i$$
$$2a = -3 \pm 3i$$
$$a = \frac{-3 \pm 3i}{2}$$

13. Solving the equation:

$$x^2 + 8x + 16 = -27$$
$$(x+4)^2 = -27$$
$$x + 4 = \pm\sqrt{-27} = \pm 3i\sqrt{3}$$
$$x = -4 \pm 3i\sqrt{3}$$

15. Solving the equation:

$$4a^2 - 12a + 9 = -4$$
$$(2a-3)^2 = -4$$
$$2a - 3 = \pm\sqrt{-4} = \pm 2i$$
$$2a = 3 \pm 2i$$
$$a = \frac{3 \pm 2i}{2}$$

17.   Solving the equation:
$$(x+5)^2 + (x-5)^2 = 52$$
$$x^2 + 10x + 25 + x^2 - 10x + 25 = 52$$
$$2x^2 + 50 = 52$$
$$2x^2 = 2$$
$$x^2 = 1$$
$$x = \pm\sqrt{1} = \pm 1$$

19.   Completing the square: $x^2 + 12x + 36 = (x+6)^2$

21.   Completing the square: $x^2 - 4x + 4 = (x-2)^2$

23.   Completing the square: $a^2 - 10a + 25 = (a-5)^2$

25.   Completing the square: $x^2 + 5x + \frac{25}{4} = \left(x + \frac{5}{2}\right)^2$

27.   Completing the square: $y^2 - 7y + \frac{49}{4} = \left(y - \frac{7}{2}\right)^2$

29.   Solving the equation:
$$x^2 + 4x = 12$$
$$x^2 + 4x + 4 = 12 + 4$$
$$(x+2)^2 = 16$$
$$x + 2 = \pm\sqrt{16} = \pm 4$$
$$x + 2 = -4, 4$$
$$x = -6, 2$$

31.   Solving the equation:
$$x^2 + 12x = -27$$
$$x^2 + 12x + 36 = -27 + 36$$
$$(x+6)^2 = 9$$
$$x + 6 = \pm\sqrt{9} = \pm 3$$
$$x + 6 = -3, 3$$
$$x = -9, -3$$

33.   Solving the equation:
$$a^2 - 2a + 5 = 0$$
$$a^2 - 2a + 1 = -5 + 1$$
$$(a-1)^2 = -4$$
$$a - 1 = \pm\sqrt{-4} = \pm 2i$$
$$a = 1 \pm 2i$$

35.   Solving the equation:
$$y^2 - 8y + 1 = 0$$
$$y^2 - 8y + 16 = -1 + 16$$
$$(y-4)^2 = 15$$
$$y - 4 = \pm\sqrt{15}$$
$$y = 4 \pm \sqrt{15}$$

37.   Solving the equation:
$$x^2 - 5x - 3 = 0$$
$$x^2 - 5x + \frac{25}{4} = 3 + \frac{25}{4}$$
$$\left(x - \frac{5}{2}\right)^2 = \frac{37}{4}$$
$$x - \frac{5}{2} = \pm\frac{\sqrt{37}}{2}$$
$$x = \frac{5 \pm \sqrt{37}}{2}$$

39.   Solving the equation:
$$2x^2 - 4x - 8 = 0$$
$$x^2 - 2x - 4 = 0$$
$$x^2 - 2x + 1 = 4 + 1$$
$$(x-1)^2 = 5$$
$$x - 1 = \pm\sqrt{5}$$
$$x = 1 \pm \sqrt{5}$$

41.   Solving the equation:
$$3t^2 - 8t + 1 = 0$$
$$t^2 - \frac{8}{3}t + \frac{1}{3} = 0$$
$$t^2 - \frac{8}{3}t + \frac{16}{9} = -\frac{1}{3} + \frac{16}{9}$$
$$\left(t - \frac{4}{3}\right)^2 = \frac{13}{9}$$
$$t - \frac{4}{3} = \pm\sqrt{\frac{13}{9}} = \pm\frac{\sqrt{13}}{3}$$
$$t = \frac{4 \pm \sqrt{13}}{3}$$

43.   Solving the equation:
$$4x^2 - 3x + 5 = 0$$
$$x^2 - \frac{3}{4}x + \frac{5}{4} = 0$$
$$x^2 - \frac{3}{4}x + \frac{9}{64} = -\frac{5}{4} + \frac{9}{64}$$
$$\left(x - \frac{3}{8}\right)^2 = -\frac{71}{64}$$
$$x - \frac{3}{8} = \pm\sqrt{-\frac{71}{64}} = \pm\frac{i\sqrt{71}}{8}$$
$$x = \frac{3 \pm i\sqrt{71}}{8}$$

45. The other two sides are $\dfrac{\sqrt{3}}{2}$ inch, 1 inch.

47. The hypotenuse is $\sqrt{2}$ inches.

49. The hypotenuse is $x\sqrt{2}$.

51. Let x represent the horizontal distance. Using the Pythagorean theorem:
$$x^2 + 120^2 = 790^2$$
$$x^2 + 14400 = 624100$$
$$x^2 = 609700$$
$$x = \sqrt{609700} \approx 781 \text{ feet}$$

53. Solving for r:
$$3456 = 3000(1 + r)^2$$
$$(1 + r)^2 = 1.152$$
$$1 + r = \sqrt{1.152}$$
$$r = \sqrt{1.152} - 1 \approx 0.073$$
The annual interest rate is 7.3%.

55. Its length is $20\sqrt{2} \approx 28$ feet.

57. Simplifying: $\sqrt{45} = \sqrt{9 \cdot 5} = 3\sqrt{5}$

59. Simplifying: $\sqrt{27y^5} = \sqrt{9y^4 \cdot 3y} = 3y^2\sqrt{3y}$

61. Simplifying: $\sqrt[3]{54x^6y^5} = \sqrt[3]{27x^6y^3 \cdot 2y^2} = 3x^2y\sqrt[3]{2y^2}$

63. Substituting values: $\sqrt{b^2 - 4ac} = \sqrt{7^2 - 4(6)(-5)} = \sqrt{49 + 120} = \sqrt{169} = 13$

65. Rationalizing the denominator: $\dfrac{3}{\sqrt{2}} = \dfrac{3}{\sqrt{2}} \cdot \dfrac{\sqrt{2}}{\sqrt{2}} = \dfrac{3\sqrt{2}}{2}$

67. Rationalizing the denominator: $\dfrac{2}{\sqrt[3]{4}} = \dfrac{2}{\sqrt[3]{4}} \cdot \dfrac{\sqrt[3]{2}}{\sqrt[3]{2}} = \dfrac{2\sqrt[3]{2}}{2} = \sqrt[3]{2}$

69. **a.** Using the formula:
$$z^n = x^n + y^n$$
$$z^2 = 6^2 + 8^2$$
$$z^2 = 36 + 64 = 100$$
$$z = \pm\sqrt{100} = \pm 10$$

b. Using the formula:
$$x^n + y^n = z^n$$
$$5^2 + y^2 = 13^2$$
$$25 + y^2 = 169$$
$$y^2 = 144$$
$$y = \pm\sqrt{144} = \pm 12$$

10.2 The Quadratic Formula

1. Solving the equation:
$$x^2 + 5x + 6 = 0$$
$$(x + 3)(x + 2) = 0$$
$$x = -3, -2$$

3. Solving the equation:
$$a^2 - 4a + 1 = 0$$
$$a = \frac{4 \pm \sqrt{16 - 4}}{2} = \frac{4 \pm \sqrt{12}}{2} = \frac{4 \pm 2\sqrt{3}}{2} = 2 \pm \sqrt{3}$$

5. Solving the equation:
$$\tfrac{1}{6}x^2 - \tfrac{1}{2}x + \tfrac{1}{3} = 0$$
$$x^2 - 3x + 2 = 0$$
$$(x - 1)(x - 2) = 0$$
$$x = 1, 2$$

7. Solving the equation:
$$\frac{x^2}{2}+1=\frac{2x}{3}$$
$$3x^2+6=4x$$
$$3x^2-4x+6=0$$
$$x=\frac{4\pm\sqrt{16-72}}{6}=\frac{4\pm\sqrt{-56}}{6}=\frac{4\pm2i\sqrt{14}}{6}=\frac{2\pm i\sqrt{14}}{3}$$

9. Solving the equation:
$$y^2-5y=0$$
$$y(y-5)=0$$
$$y=0,5$$

11. Solving the equation:
$$30x^2+40x=0$$
$$10x(3x+4)=0$$
$$x=-\tfrac{4}{3},0$$

13. Solving the equation:
$$\frac{2t^2}{3}-t=-\tfrac{1}{6}$$
$$4t^2-6t=-1$$
$$4t^2-6t+1=0$$
$$t=\frac{6\pm\sqrt{36-16}}{8}=\frac{6\pm\sqrt{20}}{8}=\frac{6\pm2\sqrt{5}}{8}=\frac{3\pm\sqrt{5}}{4}$$

15. Solving the equation:
$$0.01x^2+0.06x-0.08=0$$
$$x^2+6x-8=0$$
$$x=\frac{-6\pm\sqrt{36+32}}{2}=\frac{-6\pm\sqrt{68}}{2}=\frac{-6\pm2\sqrt{17}}{2}=-3\pm\sqrt{17}$$

17. Solving the equation:
$$2x+3=-2x^2$$
$$2x^2+2x+3=0$$
$$x=\frac{-2\pm\sqrt{4-24}}{4}=\frac{-2\pm\sqrt{-20}}{4}=\frac{-2\pm2i\sqrt{5}}{4}=\frac{-1\pm i\sqrt{5}}{2}$$

19. Solving the equation:
$$100x^2-200x+100=0$$
$$100\left(x^2-2x+1\right)=0$$
$$100(x-1)^2=0$$
$$x=1$$

21. Solving the equation:
$$\tfrac{1}{2}r^2=\tfrac{1}{6}r-\tfrac{2}{3}$$
$$3r^2=r-4$$
$$3r^2-r+4=0$$
$$r=\frac{1\pm\sqrt{1-48}}{6}=\frac{1\pm\sqrt{-47}}{6}=\frac{1\pm i\sqrt{47}}{6}$$

23. Solving the equation:
$$(x-3)(x-5)=1$$
$$x^2-8x+15=1$$
$$x^2-8x+14=0$$
$$x=\frac{8\pm\sqrt{64-56}}{2}=\frac{8\pm\sqrt{8}}{2}=\frac{8\pm2\sqrt{2}}{2}=4\pm\sqrt{2}$$

25. Solving the equation:

$$(x+3)^2 + (x-8)(x-1) = 16$$
$$x^2 + 6x + 9 + x^2 - 9x + 8 = 16$$
$$2x^2 - 3x + 1 = 0$$
$$(2x-1)(x-1) = 0$$
$$x = \tfrac{1}{2}, 1$$

27. Solving the equation:

$$\frac{x^2}{3} - \frac{5x}{6} = \frac{1}{2}$$
$$2x^2 - 5x = 3$$
$$2x^2 - 5x - 3 = 0$$
$$(2x+1)(x-3) = 0$$
$$x = -\tfrac{1}{2}, 3$$

29. Solving the equation:

$$\frac{1}{x+1} - \frac{1}{x} = \frac{1}{2}$$
$$2x(x+1)\left(\frac{1}{x+1} - \frac{1}{x}\right) = 2x(x+1) \cdot \frac{1}{2}$$
$$2x - (2x+2) = x^2 + x$$
$$2x - 2x - 2 = x^2 + x$$
$$x^2 + x + 2 = 0$$
$$x = \frac{-1 \pm \sqrt{1-8}}{2} = \frac{-1 \pm \sqrt{-7}}{2} = \frac{-1 \pm i\sqrt{7}}{2}$$

31. Solving the equation:

$$\frac{1}{y-1} + \frac{1}{y+1} = 1$$
$$(y+1)(y-1)\left(\frac{1}{y-1} + \frac{1}{y+1}\right) = (y+1)(y-1) \cdot 1$$
$$y+1 + y-1 = y^2 - 1$$
$$2y = y^2 - 1$$
$$y^2 - 2y - 1 = 0$$
$$y = \frac{2 \pm \sqrt{4+4}}{2} = \frac{2 \pm \sqrt{8}}{2} = \frac{2 \pm 2\sqrt{2}}{2} = 1 \pm \sqrt{2}$$

33. Solving the equation:

$$\frac{1}{x+2} + \frac{1}{x+3} = 1$$
$$(x+2)(x+3)\left(\frac{1}{x+2} + \frac{1}{x+3}\right) = (x+2)(x+3) \cdot 1$$
$$x+3 + x+2 = x^2 + 5x + 6$$
$$2x + 5 = x^2 + 5x + 6$$
$$x^2 + 3x + 1 = 0$$
$$x = \frac{-3 \pm \sqrt{9-4}}{2} = \frac{-3 \pm \sqrt{5}}{2}$$

35.   Solving the equation:

$$\frac{6}{r^2-1}-\frac{1}{2}=\frac{1}{r+1}$$

$$2(r+1)(r-1)\left(\frac{6}{(r+1)(r-1)}-\frac{1}{2}\right)=2(r+1)(r-1)\cdot\frac{1}{r+1}$$

$$12-\left(r^2-1\right)=2r-2$$

$$12-r^2+1=2r-2$$

$$r^2+2r-15=0$$

$$(r+5)(r-3)=0$$

$$r=-5,3$$

37.   Solving the equation:

$$x^3-8=0$$

$$(x-2)\left(x^2+2x+4\right)=0$$

$$x=2 \quad \text{or} \quad x=\frac{-2\pm\sqrt{4-16}}{2}=\frac{-2\pm\sqrt{-12}}{2}=\frac{-2\pm2i\sqrt{3}}{2}=-1\pm i\sqrt{3}$$

$$x=2,-1\pm i\sqrt{3}$$

39.   Solving the equation:

$$8a^3+27=0$$

$$(2a+3)\left(4a^2-6a+9\right)=0$$

$$a=-\frac{3}{2} \quad \text{or} \quad a=\frac{6\pm\sqrt{36-144}}{8}=\frac{6\pm\sqrt{-108}}{8}=\frac{6\pm6i\sqrt{3}}{8}=\frac{3\pm3i\sqrt{3}}{4}$$

$$a=-\frac{3}{2},\frac{3\pm3i\sqrt{3}}{4}$$

41.   Solving the equation:

$$125t^3-1=0$$

$$(5t-1)\left(25t^2+5t+1\right)=0$$

$$t=\frac{1}{5} \quad \text{or} \quad t=\frac{-5\pm\sqrt{25-100}}{50}=\frac{-5\pm\sqrt{-75}}{50}=\frac{-5\pm5i\sqrt{3}}{50}=\frac{-1\pm i\sqrt{3}}{10}$$

$$t=\frac{1}{5},\frac{-1\pm i\sqrt{3}}{10}$$

43.   Solving the equation:

$$2x^3+2x^2+3x=0$$

$$x\left(2x^2+2x+3\right)=0$$

$$x=0 \quad \text{or} \quad x=\frac{-2\pm\sqrt{4-24}}{4}=\frac{-2\pm\sqrt{-20}}{4}=\frac{-2\pm2i\sqrt{5}}{4}=\frac{-1\pm i\sqrt{5}}{2}$$

$$x=0,\frac{-1\pm i\sqrt{5}}{2}$$

45.   Solving the equation:

$$3y^4=6y^3-6y^2$$

$$3y^4-6y^3+6y^2=0$$

$$3y^2\left(y^2-2y+2\right)=0$$

$$y=0 \quad \text{or} \quad y=\frac{2\pm\sqrt{4-8}}{2}=\frac{2\pm\sqrt{-4}}{2}=\frac{2\pm2i}{2}=1\pm i$$

$$y=0,1\pm i$$

47. Solving the equation:
$$6t^5 + 4t^4 = -2t^3$$
$$6t^5 + 4t^4 + 2t^3 = 0$$
$$2t^3\left(3t^2 + 2t + 1\right) = 0$$

$t = 0$ or $t = \dfrac{-2 \pm \sqrt{4-12}}{6} = \dfrac{-2 \pm \sqrt{-8}}{6} = \dfrac{-2 \pm 2i\sqrt{2}}{6} = \dfrac{-1 \pm i\sqrt{2}}{3}$

$$t = 0, \dfrac{-1 \pm i\sqrt{2}}{3}$$

49. Substituting $s = 74$:
$$5t + 16t^2 = 74$$
$$16t^2 + 5t - 74 = 0$$
$$(16t + 37)(t - 2) = 0$$
$$t = 2 \quad \left(t = -\tfrac{37}{16} \text{ is impossible}\right)$$

It will take 2 seconds for the object to fall 74 feet.

51. Since profit is revenue minus the cost, the equation is:
$$100x - 0.5x^2 - (60x + 300) = 300$$
$$100x - 0.5x^2 - 60x - 300 = 300$$
$$-0.5x^2 + 40x - 600 = 0$$
$$x^2 - 80x + 1200 = 0$$
$$(x - 20)(x - 60) = 0$$
$$x = 20, 60$$

The weekly profit is $300 if 20 items or 60 items are sold.

53. Let x represent the width of strip being cut off. After removing the strip, the overall area is 80% of its original area. The equation is:
$$(10.5 - 2x)(8.2 - 2x) = 0.80(10.5 \times 8.2)$$
$$86.1 - 37.4x + 4x^2 = 68.88$$
$$4x^2 - 37.4x + 17.22 = 0$$

$$x = \frac{37.4 \pm \sqrt{(-37.4)^2 - 4(4)(17.22)}}{8} = \frac{37.4 \pm \sqrt{1123.24}}{8} \approx \frac{37.4 \pm 33.5}{8} \approx 0.49, 8.86$$

The width of strip is 0.49 centimeter (8.86 cm is impossible).

55.   a.   The two equations are: $l + w = 10, lw = 15$

     b.   Since $l = 10 - w$, the equation is:
$$w(10 - w) = 15$$
$$10w - w^2 = 15$$
$$w^2 - 10w + 15 = 0$$

$$w = \frac{10 \pm \sqrt{100 - 60}}{2} = \frac{10 \pm \sqrt{40}}{2} = \frac{10 \pm 2\sqrt{10}}{2} = 5 \pm \sqrt{10} \approx 1.84, 8.16$$

The length and width are 8.16 yards and 1.84 yards.

     c.   Two answers are possible because either dimension (long or short) may be considered the length.

57. Dividing using long division:

$$
\begin{array}{r}
4y+1 \\
2y-7{\overline{\smash{\big)}\,8y^2-26y-9}} \\
\underline{8y^2-28y} \\
2y-9 \\
\underline{2y-7} \\
-2
\end{array}
$$

The quotient is $4y+1-\dfrac{2}{2y-7}$.

59. Dividing using long division:

$$
\begin{array}{r}
x^2+7x+12 \\
x+2{\overline{\smash{\big)}\,x^3+9x^2+26x+24}} \\
\underline{x^3+2x^2} \\
7x^2+26x \\
\underline{7x^2+14x} \\
12x+24 \\
\underline{12x+24} \\
0
\end{array}
$$

The quotient is $x^2+7x+12$.

61. Simplifying: $25^{1/2}=\sqrt{25}=5$

63. Simplifying: $\left(\frac{9}{25}\right)^{3/2}=\left(\frac{3}{5}\right)^3=\frac{27}{125}$

65. Simplifying: $8^{-2/3}=\left(8^{-1/3}\right)^2=\left(\frac{1}{2}\right)^2=\frac{1}{4}$

67. Simplifying: $\dfrac{\left(49x^8y^{-4}\right)^{1/2}}{\left(27x^{-3}y^9\right)^{-1/3}}=\dfrac{7x^4y^{-2}}{\frac{1}{3}xy^{-3}}=21x^{4-1}y^{-2+3}=21x^3y$

69. Solving the equation:
$$
\begin{aligned}
x^2+\sqrt{3}x-6&=0 \\
\left(x+2\sqrt{3}\right)\left(x-\sqrt{3}\right)&=0 \\
x&=-2\sqrt{3},\sqrt{3}
\end{aligned}
$$

71. Solving the equation:
$$
\begin{aligned}
\sqrt{2}x^2+2x-\sqrt{2}&=0 \\
x=\frac{-2\pm\sqrt{4+4(2)}}{2\sqrt{2}}=\frac{-2\pm\sqrt{12}}{2\sqrt{2}}&=\frac{-2\pm2\sqrt{3}}{2\sqrt{2}}\cdot\frac{\sqrt{2}}{\sqrt{2}}=\frac{-2\sqrt{2}\pm2\sqrt{6}}{4}=\frac{-\sqrt{2}\pm\sqrt{6}}{2}
\end{aligned}
$$

73. Solving the equation:
$$
\begin{aligned}
x^2+ix+2&=0 \\
x=\frac{-i\pm\sqrt{-1-8}}{2}&=\frac{-i\pm3i}{2}=-2i,i
\end{aligned}
$$

10.3   Additional Items Involving Solutions to Equations

1. Computing the discriminant: $D=(-6)^2-4(1)(5)=36-20=16$. The equation will have two rational solutions.

3. First write the equation as $4x^2-4x+1=0$. Computing the discriminant: $D=(-4)^2-4(4)(1)=16-16=0$
The equation will have one rational solution.

5. Computing the discriminant: $D=1^2-4(1)(-1)=1+4=5$. The equation will have two irrational solutions.

7. First write the equation as $2y^2-3y-1=0$. Computing the discriminant: $D=(-3)^2-4(2)(-1)=9+8=17$
The equation will have two irrational solutions.

9. Computing the discriminant: $D=0^2-4(1)(-9)=36$. The equation will have two rational solutions.

11. First write the equation as $5a^2-4a-5=0$. Computing the discriminant: $D=(-4)^2-4(5)(-5)=16+100=116$
The equation will have two irrational solutions.

13. Setting the discriminant equal to 0:
$$
\begin{aligned}
(-k)^2-4(1)(25)&=0 \\
k^2-100&=0 \\
k^2&=100 \\
k&=\pm10
\end{aligned}
$$

15. First write the equation as $x^2 - kx + 36 = 0$. Setting the discriminant equal to 0:
$$(-k)^2 - 4(1)(36) = 0$$
$$k^2 - 144 = 0$$
$$k^2 = 144$$
$$k = \pm 12$$

17. Setting the discriminant equal to 0:
$$(-12)^2 - 4(4)(k) = 0$$
$$144 - 16k = 0$$
$$16k = 144$$
$$k = 9$$

19. First write the equation as $kx^2 - 40x - 25 = 0$. Setting the discriminant equal to 0:
$$(-40)^2 - 4(k)(-25) = 0$$
$$1600 + 100k = 0$$
$$100k = -1600$$
$$k = -16$$

21. Setting the discriminant equal to 0:
$$(-k)^2 - 4(3)(2) = 0$$
$$k^2 - 24 = 0$$
$$k^2 = 24$$
$$k = \pm\sqrt{24} = \pm 2\sqrt{6}$$

23. Writing the equation:
$$(x - 5)(x - 2) = 0$$
$$x^2 - 7x + 10 = 0$$

25. Writing the equation:
$$(t + 3)(t - 6) = 0$$
$$t^2 - 3t - 18 = 0$$

27. Writing the equation:
$$(y - 2)(y + 2)(y - 4) = 0$$
$$(y^2 - 4)(y - 4) = 0$$
$$y^3 - 4y^2 - 4y + 16 = 0$$

29. Writing the equation:
$$(2x - 1)(x - 3) = 0$$
$$2x^2 - 7x + 3 = 0$$

31. Writing the equation:
$$(4t + 3)(t - 3) = 0$$
$$4t^2 - 9t - 9 = 0$$

33. Writing the equation:
$$(x - 3)(x + 3)(6x - 5) = 0$$
$$(x^2 - 9)(6x - 5) = 0$$
$$6x^3 - 5x^2 - 54x + 45 = 0$$

35. Writing the equation:
$$(2a + 1)(5a - 3) = 0$$
$$10a^2 - a - 3 = 0$$

37. Writing the equation:
$$(3x + 2)(3x - 2)(x - 1) = 0$$
$$(9x^2 - 4)(x - 1) = 0$$
$$9x^3 - 9x^2 - 4x + 4 = 0$$

39. Writing the equation:
$$(x - 2)(x + 2)(x - 3)(x + 3) = 0$$
$$(x^2 - 4)(x^2 - 9) = 0$$
$$x^4 - 13x^2 + 36 = 0$$

41. Multiplying: $a^4\left(a^{3/2} - a^{1/2}\right) = a^{11/2} - a^{9/2}$

43. Multiplying: $\left(x^{3/2} - 3\right)^2 = \left(x^{3/2} - 3\right)\left(x^{3/2} - 3\right) = x^3 - 6x^{3/2} + 9$

45. Dividing: $\dfrac{30x^{3/4} - 25x^{5/4}}{5x^{1/4}} = \dfrac{30x^{3/4}}{5x^{1/4}} - \dfrac{25x^{5/4}}{5x^{1/4}} = 6x^{1/2} - 5x$

47. Factoring: $10(x - 3)^{3/2} - 15(x - 3)^{1/2} = 5(x - 3)^{1/2}[2(x - 3) - 3] = 5(x - 3)^{1/2}(2x - 9)$

49. Factoring: $2x^{2/3} - 11x^{1/3} + 12 = \left(2x^{1/3} - 3\right)\left(x^{1/3} - 4\right)$

51. Solving the equation:
$$x^2 = 4x + 5$$
$$x^2 - 4x - 5 = 0$$
$$(x-5)(x+1) = 0$$
$$x = -1, 5$$

53. Solving the equation:
$$x^2 - 1 = 2x$$
$$x^2 - 2x - 1 = 0$$
$$x = \frac{2 \pm \sqrt{4+4}}{2} = \frac{2 \pm \sqrt{8}}{2} = \frac{2 \pm 2\sqrt{2}}{2} = 1 \pm \sqrt{2} \approx -0.41, 2.41$$

55. Solving the equation:
$$2x^3 - x^2 - 2x + 1 = 0$$
$$x^2(2x-1) - 1(2x-1) = 0$$
$$(2x-1)\left(x^2 - 1\right) = 0$$
$$(2x-1)(x+1)(x-1) = 0$$
$$x = -1, \tfrac{1}{2}, 1$$

57. Solving the equation:
$$2x^3 + 2 = x^2 + 4x$$
$$2x^3 - x^2 - 4x + 2 = 0$$
$$x^2(2x-1) - 2(2x-1) = 0$$
$$(2x-1)\left(x^2 - 2\right) = 0$$
$$x = \tfrac{1}{2}, \pm\sqrt{2}$$
$$x \approx -1.41, \tfrac{1}{2}, 1.41$$

10.4   Equations Quadratic in Form

1. Solving the equation:
$$(x-3)^2 + 3(x-3) + 2 = 0$$
$$(x-3+2)(x-3+1) = 0$$
$$(x-1)(x-2) = 0$$
$$x = 1, 2$$

3. Solving the equation:
$$2(x+4)^2 + 5(x+4) - 12 = 0$$
$$[2(x+4) - 3][(x+4) + 4] = 0$$
$$(2x+5)(x+8) = 0$$
$$x = -8, -\tfrac{5}{2}$$

5. Solving the equation:
$$x^4 - 6x^2 - 27 = 0$$
$$\left(x^2 - 9\right)\left(x^2 + 3\right) = 0$$
$$x^2 = 9, -3$$
$$x = \pm 3, \pm i\sqrt{3}$$

7. Solving the equation:
$$x^4 + 9x^2 = -20$$
$$x^4 + 9x^2 + 20 = 0$$
$$\left(x^2 + 4\right)\left(x^2 + 5\right) = 0$$
$$x^2 = -4, -5$$
$$x = \pm 2i, \pm i\sqrt{5}$$

9. Solving the equation:
$$(2a-3)^2 - 9(2a-3) = -20$$
$$(2a-3)^2 - 9(2a-3) + 20 = 0$$
$$(2a-3-4)(2a-3-5) = 0$$
$$(2a-7)(2a-8) = 0$$
$$a = \tfrac{7}{2}, 4$$

11. Solving the equation:
$$2(4a+2)^2 = 3(4a+2) + 20$$
$$2(4a+2)^2 - 3(4a+2) - 20 = 0$$
$$[2(4a+2) + 5][(4a+2) - 4] = 0$$
$$(8a+9)(4a-2) = 0$$
$$a = -\tfrac{9}{8}, \tfrac{1}{2}$$

13. Solving the equation:
$$6t^4 = -t^2 + 5$$
$$6t^4 + t^2 - 5 = 0$$
$$\left(6t^2 - 5\right)\left(t^2 + 1\right) = 0$$
$$t^2 = \tfrac{5}{6}, -1$$
$$t = \pm\sqrt{\tfrac{5}{6}} = \pm\frac{\sqrt{30}}{6}, \pm i$$

15. Solving the equation:
$$9x^4 - 49 = 0$$
$$\left(3x^2 - 7\right)\left(3x^2 + 7\right) = 0$$
$$x^2 = \tfrac{7}{3}, -\tfrac{7}{3}$$
$$x = \pm\sqrt{\tfrac{7}{3}}, \pm\sqrt{-\tfrac{7}{3}}$$
$$t = \pm\frac{\sqrt{21}}{3}, \pm\frac{i\sqrt{21}}{3}$$

17. Solving the equation:
$$x - 7\sqrt{x} + 10 = 0$$
$$\left(\sqrt{x} - 5\right)\left(\sqrt{x} - 2\right) = 0$$
$$\sqrt{x} = 2, 5$$
$$x = 4, 25$$
Both values check in the original equation.

19. Solving the equation:
$$t - 2\sqrt{t} - 15 = 0$$
$$\left(\sqrt{t} - 5\right)\left(\sqrt{t} + 3\right) = 0$$
$$\sqrt{t} = -3, 5$$
$$t = 9, 25$$
Only $t = 25$ checks in the original equation.

21. Solving the equation:
$$6x + 11\sqrt{x} = 35$$
$$6x + 11\sqrt{x} - 35 = 0$$
$$\left(3\sqrt{x} - 5\right)\left(2\sqrt{x} + 7\right) = 0$$
$$\sqrt{x} = \tfrac{5}{3}, -\tfrac{7}{2}$$
$$x = \tfrac{25}{9}, \tfrac{49}{4}$$
Only $x = \tfrac{25}{9}$ checks in the original equation.

23. Solving the equation:
$$(a - 2) - 11\sqrt{a - 2} + 30 = 0$$
$$\left(\sqrt{a - 2} - 6\right)\left(\sqrt{a - 2} - 5\right) = 0$$
$$\sqrt{a - 2} = 5, 6$$
$$a - 2 = 25, 36$$
$$a = 27, 38$$

25. Solving the equation:
$$(2x + 1) - 8\sqrt{2x + 1} + 15 = 0$$
$$\left(\sqrt{2x + 1} - 3\right)\left(\sqrt{2x + 1} - 5\right) = 0$$
$$\sqrt{2x + 1} = 3, 5$$
$$2x + 1 = 9, 25$$
$$2x = 8, 24$$
$$x = 4, 12$$

27. Solving for t:
$$16t^2 - vt - h = 0$$
$$t = \frac{v \pm \sqrt{v^2 - 4(16)(-h)}}{32} = \frac{v \pm \sqrt{v^2 + 64h}}{32}$$

29. Solving for x:
$$kx^2 + 8x + 4 = 0$$
$$x = \frac{-8 \pm \sqrt{64 - 16k}}{2k} = \frac{-8 \pm 4\sqrt{4 - k}}{2k} = \frac{-4 \pm 2\sqrt{4 - k}}{k}$$

31. Solving for x:
$$x^2 + 2xy + y^2 = 0$$
$$x = \frac{-2y \pm \sqrt{4y^2 - 4y^2}}{2} = \frac{-2y}{2} = -y$$

33. Solving for t (note that $t > 0$):
$$16t^2 - 8t - h = 0$$
$$t = \frac{8 + \sqrt{64 + 64h}}{32} = \frac{8 + 8\sqrt{1 + h}}{32} = \frac{1 + \sqrt{1 + h}}{4}$$

35.   **a.**   Sketching the graph:

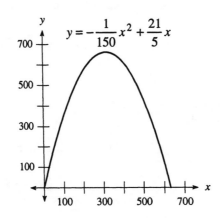

$$y = -\frac{1}{150}x^2 + \frac{21}{5}x$$

b.   Finding the x-intercepts:

$$-\frac{1}{150}x^2 + \frac{21}{5}x = 0$$

$$-\frac{1}{150}x(x - 630) = 0$$

$$x = 0, 630$$

The width is 630 feet.

37.   Let $x = BC$. Solving the proportion:

$$\frac{AB}{BC} = \frac{BC}{AC}$$

$$\frac{4}{x} = \frac{x}{4 + x}$$

$$16 + 4x = x^2$$

$$0 = x^2 - 4x - 16$$

$$x = \frac{4 \pm \sqrt{16 + 64}}{2} = \frac{4 \pm \sqrt{80}}{2} = \frac{4 \pm 4\sqrt{5}}{2} = 2 \pm 2\sqrt{5}$$

Thus $BC = 2 + 2\sqrt{5} = 4\left(\dfrac{1 + \sqrt{5}}{2}\right)$.

39.   Combining: $5\sqrt{7} - 2\sqrt{7} = 3\sqrt{7}$

41.   Combining: $\sqrt{18} - \sqrt{8} + \sqrt{32} = 3\sqrt{2} - 2\sqrt{2} + 4\sqrt{2} = 5\sqrt{2}$

43.   Combining: $9x\sqrt{20x^3y^2} + 7y\sqrt{45x^5} = 9x \cdot 2xy\sqrt{5x} + 7y \cdot 3x^2\sqrt{5x} = 18x^2y\sqrt{5x} + 21x^2y\sqrt{5x} = 39x^2y\sqrt{5x}$

45.   Multiplying: $\left(\sqrt{5} - 2\right)\left(\sqrt{5} + 8\right) = 5 - 2\sqrt{5} + 8\sqrt{5} - 16 = -11 + 6\sqrt{5}$

47.   Multiplying: $\left(\sqrt{x} + 2\right)^2 = \left(\sqrt{x}\right)^2 + 4\sqrt{x} + 4 = x + 4\sqrt{x} + 4$

49.   Rationalizing the denominator: $\dfrac{\sqrt{7}}{\sqrt{7} - 2} \cdot \dfrac{\sqrt{7} + 2}{\sqrt{7} + 2} = \dfrac{7 + 2\sqrt{7}}{7 - 4} = \dfrac{7 + 2\sqrt{7}}{3}$

51.   To find the x-intercepts, set $y = 0$:

$$x^3 - 4x = 0$$

$$x\left(x^2 - 4\right) = 0$$

$$x(x + 2)(x - 2) = 0$$

$$x = -2, 0, 2$$

To find the y-intercept, set $x = 0$: $y = 0^3 - 4(0) = 0$. The x-intercepts are $-2, 0, 2$ and the y-intercept is 0.

53. To find the x-intercepts, set $y = 0$:
$$3x^3 + x^2 - 27x - 9 = 0$$
$$x^2(3x+1) - 9(3x+1) = 0$$
$$(3x+1)(x^2 - 9) = 0$$
$$(3x+1)(x+3)(x-3) = 0$$
$$x = -3, -\tfrac{1}{3}, 3$$

To find the y-intercept, set $x = 0$: $y = 3(0)^3 + (0)^2 - 27(0) - 9 = -9$

The x-intercepts are $-3, -\tfrac{1}{3}, 3$ and the y-intercept is -9.

55. Using long division:

$$\begin{array}{r} 2x^2 + x - 1 \\ x-4 \overline{\smash{\big)}\ 2x^3 - 7x^2 - 5x + 4} \\ \underline{2x^3 - 8x^2} \\ x^2 - 5x \\ \underline{x^2 - 4x} \\ -x + 4 \\ \underline{-x + 4} \\ 0 \end{array}$$

Now solve the equation:
$$2x^2 + x - 1 = 0$$
$$(2x-1)(x+1) = 0$$
$$x = -1, \tfrac{1}{2}$$

It also crosses the x-axis at $\tfrac{1}{2}, -1$.

10.5 Graphing Parabolas

1. First complete the square: $y = x^2 + 2x - 3 = (x^2 + 2x + 1) - 1 - 3 = (x+1)^2 - 4$

The x-intercepts are $-3, 1$ and the vertex is $(-1, -4)$. Graphing the parabola:

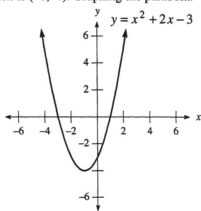

3. First complete the square: $y = -x^2 - 4x + 5 = -\left(x^2 + 4x + 4\right) + 4 + 5 = -(x+2)^2 + 9$

The x-intercepts are $-5, 1$ and the vertex is $(-2, 9)$. Graphing the parabola:

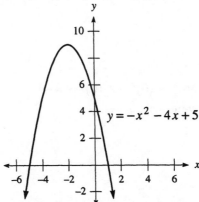

5. The x-intercepts are $-1, 1$ and the vertex is $(0, -1)$. Graphing the parabola:

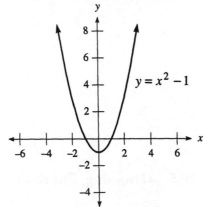

7. The x-intercepts are $-3, 3$ and the vertex is $(0, 9)$. Graphing the parabola:

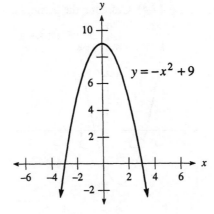

9. First complete the square: $f(x) = 2x^2 - 4x - 6 = 2(x^2 - 2x + 1) - 2 - 6 = 2(x-1)^2 - 8$

The x-intercepts are $-1, 3$ and the vertex is $(1, -8)$. Graphing the parabola:

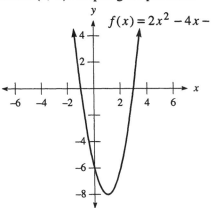

$f(x) = 2x^2 - 4x - 6$

11. First complete the square: $f(x) = x^2 - 2x - 4 = (x^2 - 2x + 1) - 1 - 4 = (x-1)^2 - 5$

The x-intercepts are $1 \pm \sqrt{5}$ and the vertex is $(1, -5)$. Graphing the parabola:

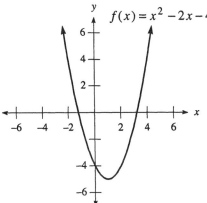

$f(x) = x^2 - 2x - 4$

13. First complete the square: $y = x^2 - 4x - 4 = (x^2 - 4x + 4) - 4 - 4 = (x-2)^2 - 8$

The vertex is $(2, -8)$. Graphing the parabola:

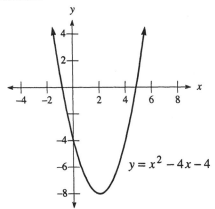

$y = x^2 - 4x - 4$

15.   First complete the square: $y = -x^2 + 2x - 5 = -\left(x^2 - 2x + 1\right) + 1 - 5 = -(x-1)^2 - 4$

The vertex is (1,–4). Graphing the parabola:

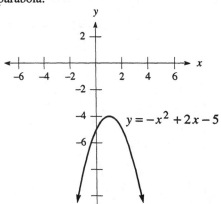

17.   The vertex is (0,1). Graphing the parabola:

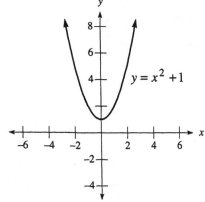

19.   The vertex is (0,–3). Graphing the parabola:

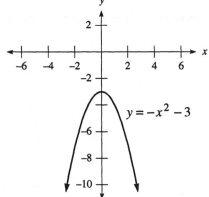

21.   Completing the square: $y = x^2 - 6x + 5 = \left(x^2 - 6x + 9\right) - 9 + 5 = (x-3)^2 - 4$

The vertex is (3,–4), which is the lowest point on the graph.

23.   Completing the square: $y = -x^2 + 2x + 8 = -\left(x^2 - 2x + 1\right) + 1 + 8 = -(x-1)^2 + 9$

The vertex is (1,9), which is the highest point on the graph.

25.   Completing the square: $y = -x^2 + 4x + 12 = -\left(x^2 - 4x + 4\right) + 4 + 12 = -(x-2)^2 + 16$

The vertex is (2,16), which is the highest point on the graph.

27.   Completing the square: $y = -x^2 - 8x = -\left(x^2 + 8x + 16\right) + 16 = -(x+4)^2 + 16$

The vertex is (–4,16), which is the highest point on the graph.

29.   First complete the square:

$$P(x) = -0.002x^2 + 3.5x - 800 = -0.002\left(x^2 - 1750x + 765625\right) + 1531.25 - 800 = -0.002(x - 875)^2 + 731.25$$

It must sell 875 patterns to obtain a maximum profit of $731.25.

31.   The ball is in her hand at times 0 sec and 2 sec.

Completing the square: $h(t) = -16t^2 + 32t = -16\left(t^2 - 2t + 1\right) + 16 = -16(t-1)^2 + 16$

The maximum height of the ball is 16 feet.

33.   Let w represent the width, and $80 - 2w$ represent the length. Finding the area:

$$A(w) = w(80 - 2w) = -2w^2 + 80w = -2\left(w^2 - 40w + 400\right) + 800 = -2(w - 20)^2 + 800$$

The dimensions are 40 feet by 20 feet.

35. Completing the square: $R = xp = 1200p - 100p^2 = -100\left(p^2 - 12p + 36\right) + 3600 = -100(p-6)^2 + 3600$

The price is $6.00 and the maximum revenue is $3,600. Sketching the graph:

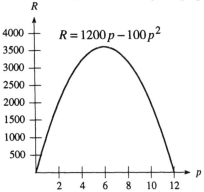

37. Completing the square: $R = xp = 1700p - 100p^2 = -100\left(p^2 - 17p + 72.25\right) + 7225 = -100(p-8.5)^2 + 7225$

The price is $8.50 and the maximum revenue is $7,225. Sketching the graph:

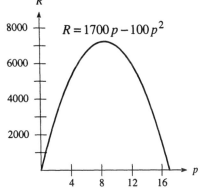

39. Let x represent the number of $10 increases in dues. Then the revenue is given by:

$$y = (10000 - 200x)(100 + 10x)$$
$$= -2000x^2 + 80000x + 1000000$$
$$= -2000\left(x^2 - 40x + 400\right) + 1000000 + 800000$$
$$= -2000(x-20)^2 + 1,800,000$$

The dues should be increased by $200, so the dues should be $300 to result in a maximum income of $1,800,000.

41. Performing the operations: $(3 - 5i) - (2 - 4i) = 3 - 5i - 2 + 4i = 1 - i$

43. Performing the operations: $(3 + 2i)(7 - 3i) = 21 + 5i - 6i^2 = 27 + 5i$

45. Performing the operations: $\dfrac{i}{3+i} = \dfrac{i}{3+i} \cdot \dfrac{3-i}{3-i} = \dfrac{3i - i^2}{9 - i^2} = \dfrac{3i + 1}{9 + 1} = \dfrac{1}{10} + \dfrac{3}{10}i$

47. The equation is $y = (x-2)^2 - 4$.

49.   The equation is given on the graph:

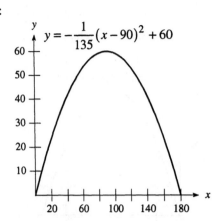

$$y = -\frac{1}{135}(x-90)^2 + 60$$

10.6   Quadratic Inequalities

1.   Factoring the inequality:
$$x^2 + x - 6 > 0$$
$$(x+3)(x-2) > 0$$
Forming the sign chart:

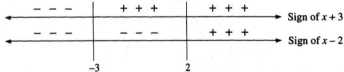

The solution set is $x < -3$ or $x > 2$. Graphing the solution set:

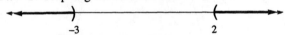

3.   Factoring the inequality:
$$x^2 - x - 12 \leq 0$$
$$(x+3)(x-4) \leq 0$$
Forming the sign chart:

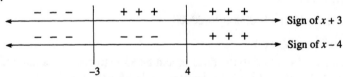

The solution set is $-3 \leq x \leq 4$. Graphing the solution set:

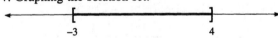

5.   Factoring the inequality:
$$x^2 + 5x \geq -6$$
$$x^2 + 5x + 6 \geq 0$$
$$(x+2)(x+3) \geq 0$$
Forming the sign chart:

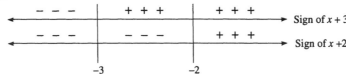

The solution set is $x \leq -3$ or $x \geq -2$. Graphing the solution set:

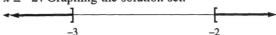

7.   Factoring the inequality:
$$6x^2 < 5x - 1$$
$$6x^2 - 5x + 1 < 0$$
$$(3x-1)(2x-1) < 0$$
Forming the sign chart:

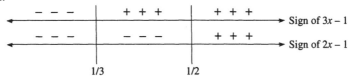

The solution set is $\frac{1}{3} < x < \frac{1}{2}$. Graphing the solution set:

9.   Factoring the inequality:
$$x^2 - 9 < 0$$
$$(x+3)(x-3) < 0$$
Forming the sign chart:

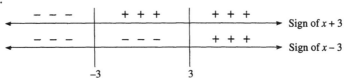

The solution set is $-3 < x < 3$. Graphing the solution set:

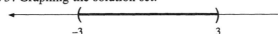

11.   Factoring the inequality:
$$4x^2 - 9 \geq 0$$
$$(2x+3)(2x-3) \geq 0$$
Forming the sign chart:

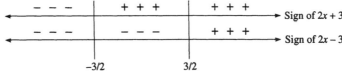

The solution set is $x \leq -\frac{3}{2}$ or $x \geq \frac{3}{2}$. Graphing the solution set:

13. Factoring the inequality:
$$2x^2 - x - 3 < 0$$
$$(2x-3)(x+1) < 0$$
Forming the sign chart:

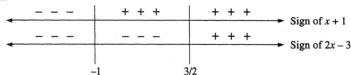

The solution set is $-1 < x < \frac{3}{2}$. Graphing the solution set:

15. Factoring the inequality:
$$x^2 - 4x + 4 \geq 0$$
$$(x-2)^2 \geq 0$$
Since this inequality is always true, the solution set is all real numbers. Graphing the solution set:

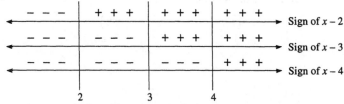

17. Factoring the inequality:
$$x^2 - 10x + 25 < 0$$
$$(x-5)^2 < 0$$
Since this inequality is never true, there is no solution.

19. Forming the sign chart:

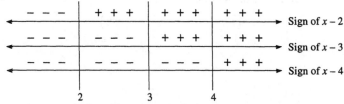

The solution set is $2 < x < 3$ or $x > 4$. Graphing the solution set:

21. Forming the sign chart:

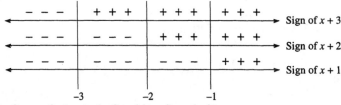

The solution set is $x \leq -3$ or $-2 \leq x \leq -1$. Graphing the solution set:

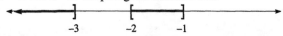

23. a. The solution set is $-2 < x < 2$.
    b. The solution set is $x < -2$ or $x > 2$.
    c. The solution set is $x = -2, 2$.

25. a. The solution set is $-2 < x < 5$.
    b. The solution set is $x < -2$ or $x > 5$.
    c. The solution set is $x = -2, 5$.

27. a. The solution set is $x < -1$ or $1 < x < 3$.
    b. The solution set is $-1 < x < 1$ or $x > 3$.
    c. The solution set is $x = -1, 1, 3$.

29. Let w represent the width and $2w + 3$ represent the length. Using the area formula:

$$w(2w + 3) \geq 44$$
$$2w^2 + 3w \geq 44$$
$$2w^2 + 3w - 44 \geq 0$$
$$(2w + 11)(w - 4) \geq 0$$

Forming the sign chart:

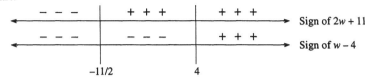

The width is at least 4 inches.

31. Solving the inequality:

$$1300p - 100p^2 \geq 4000$$
$$-100p^2 + 1300p - 4000 \geq 0$$
$$p^2 - 13p + 40 \leq 0$$
$$(p - 8)(p - 5) \leq 0$$

Forming the sign chart:

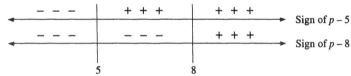

Charge at least $5 but no more than $8 per radio.

33. Solving the equation:

$$\sqrt{3t - 1} = 2$$
$$\left(\sqrt{3t - 1}\right)^2 = (2)^2$$
$$3t - 1 = 4$$
$$3t = 5$$
$$t = \frac{5}{3}$$

The solution is $\frac{5}{3}$.

35. Solving the equation:

$$\sqrt{x + 3} = x - 3$$
$$\left(\sqrt{x + 3}\right)^2 = (x - 3)^2$$
$$x + 3 = x^2 - 6x + 9$$
$$0 = x^2 - 7x + 6$$
$$0 = (x - 6)(x - 1)$$
$$x = 1, 6 \qquad (x = 1 \text{ does not check})$$

The solution is 6.

37. Graphing the equation:

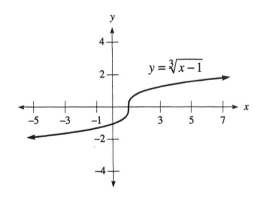

39. Using the quadratic formula: $x = \dfrac{2 \pm \sqrt{4 - 4(-1)}}{2(1)} = \dfrac{2 \pm \sqrt{8}}{2} = \dfrac{2 \pm 2\sqrt{2}}{2} = 1 \pm \sqrt{2}$

The inequality is satisfied when $1 - \sqrt{2} < x < 1 + \sqrt{2}$.

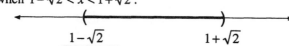

$$1 - \sqrt{2} \qquad\qquad 1 + \sqrt{2}$$

41. Using the quadratic formula: $x = \dfrac{8 \pm \sqrt{64 - 4(13)}}{2(1)} = \dfrac{8 \pm \sqrt{12}}{2} = \dfrac{8 \pm 2\sqrt{3}}{2} = 4 \pm \sqrt{3}$

The inequality is satisfied when $x < 4 - \sqrt{3}$ or $x > 4 + \sqrt{3}$.

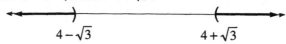

$$4 - \sqrt{3} \qquad\qquad 4 + \sqrt{3}$$

Chapter 10 Review

1. Solving the equation:
$$(2t - 5)^2 = 25$$
$$2t - 5 = \pm 5$$
$$2t - 5 = -5, 5$$
$$2t = 0, 10$$
$$t = 0, 5$$

3. Solving the equation:
$$(3y - 4)^2 = -49$$
$$3y - 4 = \pm\sqrt{-49}$$
$$3y - 4 = \pm 7i$$
$$3y = 4 \pm 7i$$
$$y = \dfrac{4 \pm 7i}{3}$$

5. Solving by completing the square:
$$2x^2 + 6x - 20 = 0$$
$$x^2 + 3x = 10$$
$$x^2 + 3x + \tfrac{9}{4} = 10 + \tfrac{9}{4}$$
$$\left(x + \tfrac{3}{2}\right)^2 = \tfrac{49}{4}$$
$$x + \tfrac{3}{2} = -\tfrac{7}{2}, \tfrac{7}{2}$$
$$x = -5, 2$$

7. Solving by completing the square:
$$a^2 + 9 = 6a$$
$$a^2 - 6a = -9$$
$$a^2 - 6a + 9 = -9 + 9$$
$$(a - 3)^2 = 0$$
$$a - 3 = 0$$
$$a = 3$$

9. Solving by completing the square:
$$2y^2 + 6y = -3$$
$$y^2 + 3y = -\tfrac{3}{2}$$
$$y^2 + 3y + \tfrac{9}{4} = -\tfrac{3}{2} + \tfrac{9}{4}$$
$$\left(y + \tfrac{3}{2}\right)^2 = \tfrac{3}{4}$$
$$y + \tfrac{3}{2} = \pm\dfrac{\sqrt{3}}{2}$$
$$y = \dfrac{-3 \pm \sqrt{3}}{2}$$

11. Solving by completing the square:
$$\tfrac{1}{6}x^2 + \tfrac{1}{2}x - \tfrac{5}{3} = 0$$
$$x^2 + 3x - 10 = 0$$
$$(x + 5)(x - 2) = 0$$
$$x = -5, 2$$

13. Solving the equation:
$$4t^2 - 8t + 19 = 0$$
$$t = \dfrac{8 \pm \sqrt{64 - 304}}{8} = \dfrac{8 \pm \sqrt{-240}}{8} = \dfrac{8 \pm 4i\sqrt{15}}{8} = \dfrac{2 \pm i\sqrt{15}}{2}$$

15. Solving the equation:
$$0.06a^2 + 0.05a = 0.04$$
$$0.06a^2 + 0.05a - 0.04 = 0$$
$$6a^2 + 5a - 4 = 0$$
$$(2a - 1)(3a + 4) = 0$$
$$a = -\frac{4}{3}, \frac{1}{2}$$

17. Solving the equation:
$$(2x + 1)(x - 5) - (x + 3)(x - 2) = -17$$
$$2x^2 - 9x - 5 - x^2 - x + 6 = -17$$
$$x^2 - 10x + 1 = -17$$
$$x^2 - 10x + 18 = 0$$
$$x = \frac{10 \pm \sqrt{100 - 72}}{2} = \frac{10 \pm \sqrt{28}}{2} = \frac{10 \pm 2\sqrt{7}}{2} = 5 \pm \sqrt{7}$$

19. Solving the equation:
$$5x^2 = -2x + 3$$
$$5x^2 + 2x - 3 = 0$$
$$(x + 1)(5x - 3) = 0$$
$$x = -1, \frac{3}{5}$$

21. Solving the equation:
$$3 - \frac{2}{x} + \frac{1}{x^2} = 0$$
$$3x^2 - 2x + 1 = 0$$
$$x = \frac{2 \pm \sqrt{4 - 12}}{6} = \frac{2 \pm \sqrt{-8}}{6} = \frac{2 \pm 2i\sqrt{2}}{6} = \frac{1 \pm i\sqrt{2}}{3}$$

23. The profit equation is given by:
$$34x - 0.1x^2 - 7x - 400 = 1300$$
$$-0.1x^2 + 27x - 1700 = 0$$
$$x^2 - 270x + 17000 = 0$$
$$(x - 100)(x - 170) = 0$$
$$x = 100, 170$$
The company must sell either 100 or 170 items for its weekly profit to be \$1,300.

25. First write the equation as $2x^2 - 8x + 8 = 0$. Computing the discriminant: $D = (-8)^2 - 4(2)(8) = 64 - 64 = 0$
The equation will have one rational solution.

27. Computing the discriminant: $D = (1)^2 - 4(2)(-3) = 1 + 24 = 25$. The equation will have two rational solutions.

29. First write the equation as $x^2 - x - 1 = 0$. Computing the discriminant: $D = (-1)^2 - 4(1)(-1) = 1 + 4 = 5$
The equation will have two irrational solutions.

31. First write the equation as $3x^2 + 5x + 4 = 0$. Computing the discriminant: $D = (5)^2 - 4(3)(4) = 25 - 48 = -23$
The equation will have two complex solutions.

33. Setting the discriminant equal to 0:
$$(-k)^2 - 4(25)(4) = 0$$
$$k^2 - 400 = 0$$
$$k^2 = 400$$
$$k = \pm 20$$

35. Setting the discriminant equal to 0:
$$(12)^2 - 4(k)(9) = 0$$
$$144 - 36k = 0$$
$$36k = 144$$
$$k = 4$$

37. Setting the discriminant equal to 0:
$$30^2 - 4(9)(k) = 0$$
$$900 - 36k = 0$$
$$36k = 900$$
$$k = 25$$

39. Writing the equation:
$$(x-3)(x-5) = 0$$
$$x^2 - 8x + 15 = 0$$

41. Writing the equation:
$$(2y-1)(y+4) = 0$$
$$2y^2 + 7y - 4 = 0$$

43. Solving the equation:
$$(x-2)^2 - 4(x-2) - 60 = 0$$
$$(x-2-10)(x-2+6) = 0$$
$$(x-12)(x+4) = 0$$
$$x = -4, 12$$

45. Solving the equation:
$$x^4 - x^2 = 12$$
$$x^4 - x^2 - 12 = 0$$
$$(x^2 - 4)(x^2 + 3) = 0$$
$$x^2 = 4, -3$$
$$x = \pm 2, \pm i\sqrt{3}$$

47. Solving the equation:
$$2x - 11\sqrt{x} = -12$$
$$2x - 11\sqrt{x} + 12 = 0$$
$$(2\sqrt{x} - 3)(\sqrt{x} - 4) = 0$$
$$\sqrt{x} = \tfrac{3}{2}, 4$$
$$x = \tfrac{9}{4}, 16$$

49. Solving the equation:
$$\sqrt{y+21} + \sqrt{y} = 7$$
$$\left(\sqrt{y+21}\right)^2 = \left(7 - \sqrt{y}\right)^2$$
$$y + 21 = 49 - 14\sqrt{y} + y$$
$$14\sqrt{y} = 28$$
$$\sqrt{y} = 2$$
$$y = 4$$

51. Solving for t:
$$16t^2 - 10t - h = 0$$
$$t = \frac{10 \pm \sqrt{100 + 64h}}{32} = \frac{10 \pm 2\sqrt{25 + 16h}}{32} = \frac{5 \pm \sqrt{25 + 16h}}{16}$$

53. Factoring the inequality:
$$x^2 - x - 2 < 0$$
$$(x-2)(x+1) < 0$$
Forming a sign chart:

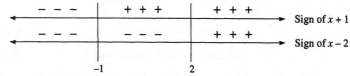

The solution set is $-1 < x < 2$. Graphing the solution set:

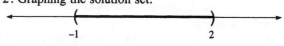

55. Factoring the inequality:
$$2x^2 + 5x - 12 \geq 0$$
$$(2x - 3)(x + 4) \geq 0$$
Forming a sign chart:

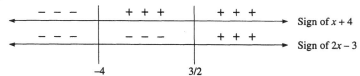

The solution set is $x \leq -4$ or $x \geq \frac{3}{2}$. Graphing the solution set:

57. First complete the square: $y = x^2 - 6x + 8 = \left(x^2 - 6x + 9\right) + 8 - 9 = (x - 3)^2 - 1$

The x-intercepts are 2,4, and the vertex is $(3,-1)$. Graphing the parabola:

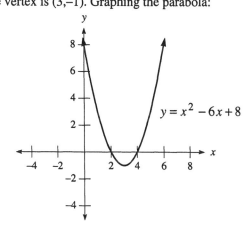

Chapters 1-10 Cumulative Review

1. Simplifying: $11 + 20 \div 5 - 3 \bullet 5 = 11 + 4 - 15 = 0$

3. Simplifying: $4(15 - 19)^2 - 3(17 - 19)^3 = 4(-4)^2 - 3(-2)^3 = 4(16) - 3(-8) = 64 + 24 = 88$

5. Simplifying: $3 - 5[2x - 4(x - 2)] = 3 - 5[2x - 4x + 8] = 3 - 5(-2x + 8) = 3 + 10x - 40 = 10x - 37$

7. Simplifying: $\sqrt[3]{32} = \sqrt[3]{8 \bullet 4} = 2\sqrt[3]{4}$

9. Simplifying: $\dfrac{1 - \frac{3}{4}}{1 + \frac{3}{4}} = \dfrac{1 - \frac{3}{4}}{1 + \frac{3}{4}} \bullet \dfrac{4}{4} = \dfrac{4 - 3}{4 + 3} = \frac{1}{7}$

11. Reducing the fraction: $\dfrac{5x^2 - 26xy - 24y^2}{5x + 4y} = \dfrac{(5x + 4y)(x - 6y)}{5x + 4y} = x - 6y$

13. Multiplying: $(3x - 2)\left(x^2 - 3x - 2\right) = 3x^3 - 9x^2 - 6x - 2x^2 + 6x + 4 = 3x^3 - 11x^2 + 4$

15. Dividing: $\dfrac{7 - i}{3 - 2i} = \dfrac{7 - i}{3 - 2i} \bullet \dfrac{3 + 2i}{3 + 2i} = \dfrac{21 + 11i - 2i^2}{9 + 4} = \dfrac{23 + 11i}{13} = \frac{23}{13} + \frac{11}{13}i$

17. Solving the equation:
$$\frac{7}{5}a - 6 = 15$$
$$\frac{7}{5}a = 21$$
$$7a = 105$$
$$a = 15$$

19.   Solving the equation:

$$\frac{a}{2}+\frac{3}{a-3}=\frac{a}{a-3}$$

$$2(a-3)\left(\frac{a}{2}+\frac{3}{a-3}\right)=2(a-3)\left(\frac{a}{a-3}\right)$$

$$a(a-3)+6=2a$$

$$a^2-3a+6=2a$$

$$a^2-5a+6=0$$

$$(a-2)(a-3)=0$$

$$a=2,3 \qquad (a=3 \text{ does not check})$$

21.   Solving the equation:

$$(3x-4)^2=18$$

$$3x-4=\pm\sqrt{18}$$

$$3x-4=\pm3\sqrt{2}$$

$$3x=4\pm3\sqrt{2}$$

$$x=\frac{4\pm3\sqrt{2}}{3}$$

23.   Solving the equation:

$$3y^3-y=5y^2$$

$$3y^3-5y^2-y=0$$

$$y\left(3y^2-5y-1\right)=0$$

$$y=0,\frac{5\pm\sqrt{25+12}}{6}=\frac{5\pm\sqrt{37}}{6}$$

25.   Solving the equation:

$$\sqrt{x-2}=2-\sqrt{x}$$

$$\left(\sqrt{x-2}\right)^2=\left(2-\sqrt{x}\right)^2$$

$$x-2=4-4\sqrt{x}+x$$

$$-4\sqrt{x}=-6$$

$$\sqrt{x}=\frac{3}{2}$$

$$x=\frac{9}{4}$$

27.   Solving the inequality:

$$5\le\frac{1}{4}x+3\le8$$

$$2\le\frac{1}{4}x\le5$$

$$8\le x\le20$$

Graphing the solution set:

29.   Multiply the first equation by 2:

$$6x-2y=4$$

$$-6x+2y=-4$$

Adding yields $0=0$, which is true. The two lines coincide.

31.   Substituting into the first equation:

$$3x-2(3x-7)=5$$

$$3x-6x+14=5$$

$$-3x+14=5$$

$$-3x=-9$$

$$x=3$$

$$y=3\bullet3-7=2$$

The solution is $(3,2)$.

33. Graphing the line:

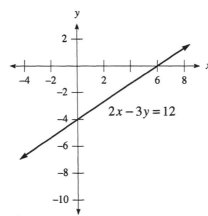

$2x - 3y = 12$

35. First find the slope: $m = \dfrac{-\frac{1}{3} - \frac{4}{3}}{\frac{1}{4} - \frac{3}{2}} = \dfrac{-\frac{5}{3}}{-\frac{5}{4}} = \dfrac{4}{3}$. Using the point-slope formula:

$$y - \tfrac{4}{3} = \tfrac{4}{3}\left(x - \tfrac{3}{2}\right)$$
$$y - \tfrac{4}{3} = \tfrac{4}{3}x - 2$$
$$y = \tfrac{4}{3}x - \tfrac{2}{3}$$

37. Rationalizing the denominator: $\dfrac{7}{\sqrt[3]{9}} = \dfrac{7}{\sqrt[3]{9}} \cdot \dfrac{\sqrt[3]{3}}{\sqrt[3]{3}} = \dfrac{7\sqrt[3]{3}}{3}$

39. Let x represent the largest angle, $\frac{1}{4}x$ represent the smallest angle, and $\frac{1}{4}x + 30$ represent the remaining angle. The equation is:

$$x + \tfrac{1}{4}x + \tfrac{1}{4}x + 30 = 180$$
$$\tfrac{3}{2}x + 30 = 180$$
$$\tfrac{3}{2}x = 150$$
$$3x = 300$$
$$x = 100$$

The angles are $25°, 55°$, and $100°$.

Chapter 10 Test

1. Solving the equation:

$$(2x + 4)^2 = 25$$
$$2x + 4 = \pm 5$$
$$2x + 4 = -5, 5$$
$$2x = -9, 1$$
$$x = -\tfrac{9}{2}, \tfrac{1}{2}$$

2. Solving the equation:

$$(2x - 6)^2 = -8$$
$$2x - 6 = \pm\sqrt{-8}$$
$$2x - 6 = \pm 2i\sqrt{2}$$
$$2x = 6 \pm 2i\sqrt{2}$$
$$x = 3 \pm i\sqrt{2}$$

3. Solving the equation:

$$y^2 - 10y + 25 = -4$$
$$(y - 5)^2 = -4$$
$$y - 5 = \pm 2i$$
$$y = 5 \pm 2i$$

4.   Solving the equation:
$$(y+1)(y-3) = -6$$
$$y^2 - 2y - 3 = -6$$
$$y^2 - 2y + 3 = 0$$
$$y = \frac{2 \pm \sqrt{4-12}}{2} = \frac{2 \pm \sqrt{-8}}{2} = \frac{2 \pm 2i\sqrt{2}}{2} = 1 \pm i\sqrt{2}$$

5.   Solving the equation:
$$8t^3 - 125 = 0$$
$$(2t-5)(4t^2 + 10t + 25) = 0$$
$$t = \tfrac{5}{2}, \frac{-10 \pm \sqrt{100-400}}{8} = \frac{-10 \pm \sqrt{-300}}{8} = \frac{-10 \pm 10i\sqrt{3}}{8} = \frac{-5 \pm 5i\sqrt{3}}{4}$$

6.   Solving the equation:
$$\frac{1}{a+2} - \tfrac{1}{3} = \frac{1}{a}$$
$$3a(a+2)\left(\frac{1}{a+2} - \tfrac{1}{3}\right) = 3a(a+2)\left(\frac{1}{a}\right)$$
$$3a - a(a+2) = 3(a+2)$$
$$3a - a^2 - 2a = 3a + 6$$
$$a^2 + 2a + 6 = 0$$
$$a = \frac{-2 \pm \sqrt{4-24}}{2} = \frac{-2 \pm \sqrt{-20}}{2} = \frac{-2 \pm 2i\sqrt{5}}{2} = -1 \pm i\sqrt{5}$$

7.   Solving for r:
$$64(1+r)^2 = A$$
$$(1+r)^2 = \frac{A}{64}$$
$$1 + r = \pm\frac{\sqrt{A}}{8}$$
$$r = \pm\frac{\sqrt{A}}{8} - 1$$

8.   Solving by completing the square:
$$x^2 - 4x = -2$$
$$x^2 - 4x + 4 = -2 + 4$$
$$(x-2)^2 = 2$$
$$x - 2 = \pm\sqrt{2}$$
$$x = 2 \pm \sqrt{2}$$

9.   Solving the equation:
$$32t - 16t^2 = 12$$
$$-16t^2 + 32t - 12 = 0$$
$$4t^2 - 8t + 3 = 0$$
$$(2t-1)(2t-3) = 0$$
$$t = \tfrac{1}{2}, \tfrac{3}{2}$$
The object will be 12 feet above the ground after $\tfrac{1}{2}$ or $\tfrac{3}{2}$ sec.

10.   Setting the profit equal to $200:
$$25x - 0.2x^2 - 2x - 100 = 200$$
$$-0.2x^2 + 23x - 300 = 0$$
$$x^2 - 115x + 1500 = 0$$
$$(x-15)(x-100) = 0$$
$$x = 15, 100$$
The company must sell 15 or 100 cups to make a weekly profit of $200.

11. First write the equation as $kx^2 - 12x + 4 = 0$. Setting the discriminant equal to 0:
$$(-12)^2 - 4(k)(4) = 0$$
$$144 - 16k = 0$$
$$16k = 144$$
$$k = 9$$

12. First write the equation as $2x^2 - 5x - 7 = 0$. Finding the discriminant: $D = (-5)^2 - 4(2)(-7) = 25 + 56 = 81$
The equation has two rational solutions.

13. Finding the equation:
$$(x - 5)(3x + 2) = 0$$
$$3x^2 - 13x - 10 = 0$$

14. Finding the equation:
$$(x - 2)(x + 2)(x - 7) = 0$$
$$(x^2 - 4)(x - 7) = 0$$
$$x^3 - 7x^2 - 4x + 28 = 0$$

15. Solving the equation:
$$4x^4 - 7x^2 - 2 = 0$$
$$(x^2 - 2)(4x^2 + 1) = 0$$
$$x^2 = 2, -\tfrac{1}{4}$$
$$x = \pm\sqrt{2}, \pm\tfrac{1}{2}i$$

16. Solving the equation:
$$(2t + 1)^2 - 5(2t + 1) + 6 = 0$$
$$(2t + 1 - 3)(2t + 1 - 2) = 0$$
$$(2t - 2)(2t - 1) = 0$$
$$t = \tfrac{1}{2}, 1$$

17. Solving the equation:
$$2t - 7\sqrt{t} + 3 = 0$$
$$(2\sqrt{t} - 1)(\sqrt{t} - 3) = 0$$
$$\sqrt{t} = \tfrac{1}{2}, 3$$
$$t = \tfrac{1}{4}, 9$$

18. Solving for t:
$$16t^2 - 14t - h = 0$$
$$t = \frac{14 + \sqrt{196 - 4(16)(-h)}}{32} = \frac{14 + \sqrt{196 + 64h}}{32} = \frac{14 + 4\sqrt{49 + 16h}}{32} = \frac{7 + \sqrt{49 + 16h}}{16}$$

19. Completing the square: $y = x^2 - 2x - 3 = (x^2 - 2x + 1) - 1 - 3 = (x - 1)^2 - 4$

The vertex is $(1, -4)$. Graphing the parabola:

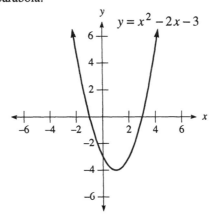

20.   Completing the square: $y = -x^2 + 2x + 8 = -\left(x^2 - 2x + 1\right) + 1 + 8 = -(x-1)^2 + 9$

The vertex is (1,9). Graphing the parabola:

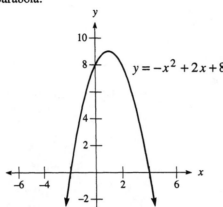

21.   Factoring the inequality:

$$x^2 - x - 6 \le 0$$
$$(x-3)(x+2) \le 0$$

Forming a sign chart:

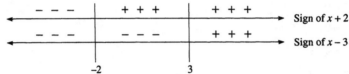

The solution set is $-2 \le x \le 3$. Graphing the solution set:

22.   Factoring the inequality:

$$2x^2 + 5x > 3$$
$$2x^2 + 5x - 3 > 0$$
$$(x+3)(2x-1) > 0$$

Forming a sign chart:

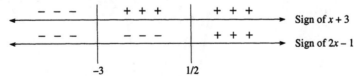

The solution set is $x < -3$ or $x > \frac{1}{2}$. Graphing the solution set:

23.   Finding the profit and completing the square:

$$P = 25x - 0.1x^2 - 5x - 100$$
$$= -0.1x^2 + 20x - 100$$
$$= -0.1\left(x^2 - 200x + 10000\right) + 1000 - 100$$
$$= -0.1(x - 100)^2 + 900$$

The maximum weekly profit is $900, obtained by selling 100 items per week.

Chapter 11
Exponential and Logarithmic Functions

11.1 Exponential Functions

1. Evaluating: $g(0) = \left(\frac{1}{2}\right)^0 = 1$

3. Evaluating: $g(-1) = \left(\frac{1}{2}\right)^{-1} = 2$

5. Evaluating: $f(-3) = 3^{-3} = \frac{1}{27}$

7. Evaluating: $f(2) + g(-2) = 3^2 + \left(\frac{1}{2}\right)^{-2} = 9 + 4 = 13$

9. Graphing the function:

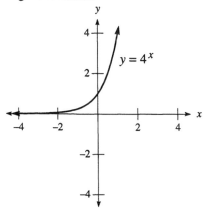

11. Graphing the function:

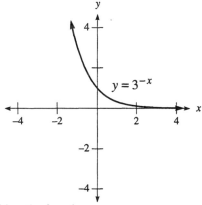

13. Graphing the function:

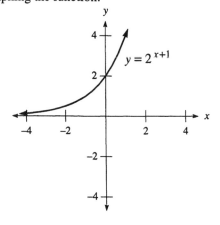

15. Graphing the function:

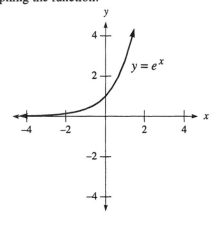

17. Graphing the function:

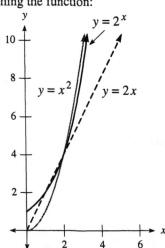

19. Graphing the function:

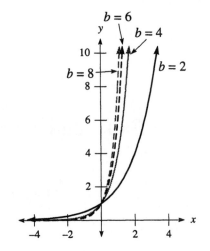

21. The equation is: $h(n) = 6\left(\frac{2}{3}\right)^n$

Substituting $n = 5$: $h(5) = 6\left(\frac{2}{3}\right)^5 \approx 0.79$ feet

23. After 8 days, there will be: $1400 \cdot 2^{-8/8} = 1400 \cdot \frac{1}{2} = 700$ micrograms

After 11 days, there will be: $1400 \cdot 2^{-11/8} \approx 539.8$ micrograms

25. **a.** The equation is $A(t) = 1200\left(1 + \dfrac{0.06}{4}\right)^{4t}$.   **b.** Substitute $t = 8$: $A(8) = 1200\left(1 + \dfrac{0.06}{4}\right)^{32} \approx \$1,932.39$

c. Using a graphing calculator, the time is approximately 11.6 years.

d. Substitute $t = 8$ into the compound interest formula: $A(8) = 1200e^{0.06 \times 8} \approx \$1,939.29$

27. **a.** Substitute $t = 20$: $E(20) = 78.16(1.11)^{20} \approx \$630$ billion. The estimate is $69 billion too low.

b. For 2005, substitute $t = 35$: $E(35) = 78.16(1.11)^{35} \approx \$3,015$ billion

For 2006, substitute $t = 36$: $E(36) = 78.16(1.11)^{36} \approx \$3,347$ billion

For 2007, substitute $t = 37$: $E(37) = 78.16(1.11)^{37} \approx \$3,715$ billion

29. Graphing the function:

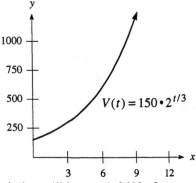

31. The painting will be worth $600 after approximately 6 years.

33.   **a.**   Substitute $t = 3.5$: $V(5) = 450,000(1-0.30)^5 \approx \$129,138.48$

 b.   The domain is $\{t \mid 0 \le t \le 6\}$.

 c.   Sketching the graph:

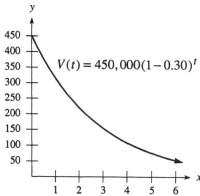

 d.   The range is $\{V(t) \mid 52,942.05 \le V(t) \le 450,000\}$.

 e.   From the graph, the crane will be worth \$85,000 after approximately 4.7 years, or 4 years 8 months.

35.   The domain is $\{1, 3, 4\}$ and the range is $\{1, 2, 4\}$. This is a function.

37.   Find where the quantity inside the radical is non-negative:
$$3x + 1 \ge 0$$
$$3x \ge -1$$
$$x \ge -\tfrac{1}{3}$$
The domain is $\left\{ x \mid x \ge -\tfrac{1}{3} \right\}$.

39.   Evaluating the function: $f(0) = 2(0)^2 - 18 = 0 - 18 = -18$

41.   Simplifying the function: $\dfrac{g(x+h) - g(x)}{h} = \dfrac{2(x+h) - 6 - (2x-6)}{h} = \dfrac{2x + 2h - 6 - 2x + 6}{h} = \dfrac{2h}{h} = 2$

43.   Graphing the function:

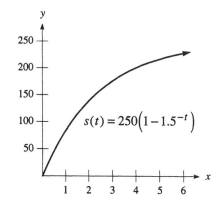

11.2   The Inverse of a Function

1.   Let $y = f(x)$. Switch x and y and solve for y:
$$3y - 1 = x$$
$$3y = x + 1$$
$$y = \frac{x+1}{3}$$
The inverse is $f^{-1}(x) = \frac{x+1}{3}$.

3.   Let $y = f(x)$. Switch x and y and solve for y:
$$y^3 = x$$
$$y = \sqrt[3]{x}$$

The inverse is $f^{-1}(x) = \sqrt[3]{x}$.

5.   Let $y = f(x)$. Switch x and y and solve for y:
$$\frac{y-3}{y-1} = x$$
$$y - 3 = xy - x$$
$$y - xy = 3 - x$$
$$y(1-x) = 3 - x$$
$$y = \frac{3-x}{1-x} = \frac{x-3}{x-1}$$
The inverse is $f^{-1}(x) = \frac{x-3}{x-1}$.

7.   Let $y = f(x)$. Switch x and y and solve for y:

$$\frac{y-3}{4} = x$$
$$y - 3 = 4x$$
$$y = 4x + 3$$

The inverse is $f^{-1}(x) = 4x + 3$.

9.   Let $y = f(x)$. Switch x and y and solve for y:

$$\tfrac{1}{2}y - 3 = x$$
$$y - 6 = 2x$$
$$y = 2x + 6$$

The inverse is $f^{-1}(x) = 2x + 6$.

11.   Let $y = f(x)$. Switch x and y and solve for y:
$$\frac{2y+1}{3y+1} = x$$
$$2y + 1 = 3xy + x$$
$$2y - 3xy = x - 1$$
$$y(2 - 3x) = x - 1$$
$$y = \frac{x-1}{2-3x} = \frac{1-x}{3x-2}$$
The inverse is $f^{-1}(x) = \frac{1-x}{3x-2}$.

13.   Finding the inverse:
$$2y - 1 = x$$
$$2y = x + 1$$
$$y = \frac{x+1}{2}$$
The inverse is $y^{-1} = \frac{x+1}{2}$. Graphing each curve:

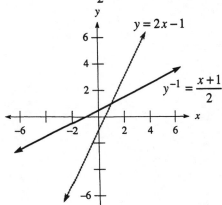

15.   Finding the inverse:
$$y^2 - 3 = x$$
$$y^2 = x + 3$$
$$y = \pm\sqrt{x+3}$$
The inverse is $y^{-1} = \pm\sqrt{x+3}$. Graphing each curve:

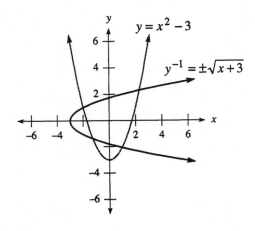

17. Finding the inverse:

$$y^2 - 2y - 3 = x$$
$$y^2 - 2y + 1 = x + 3 + 1$$
$$(y-1)^2 = x + 4$$
$$y - 1 = \pm\sqrt{x+4}$$
$$y = 1 \pm \sqrt{x+4}$$

The inverse is $y^{-1} = 1 \pm \sqrt{x+4}$. Graphing each curve:

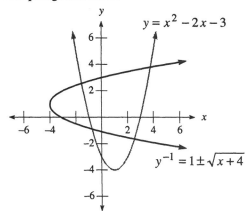

19. The inverse is $x = 3^y$. Graphing each curve:

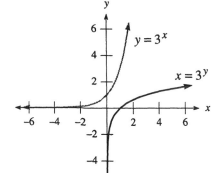

21. The inverse is $x = 4$. Graphing each curve:

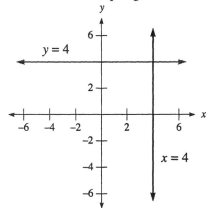

23. Finding the inverse:
$$\frac{1}{2}y^3 = x$$
$$y^3 = 2x$$
$$y = \sqrt[3]{2x}$$
The inverse is $y^{-1} = \sqrt[3]{2x}$. Graphing each curve:

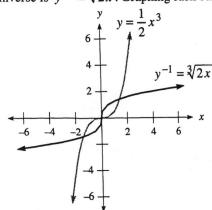

25. Finding the inverse:
$$\frac{1}{2}y + 2 = x$$
$$y + 4 = 2x$$
$$y = 2x - 4$$
The inverse is $y^{-1} = 2x - 4$. Graphing each curve:

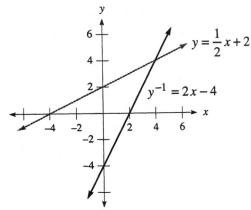

27. Finding the inverse:
$$\sqrt{y+2} = x$$
$$y + 2 = x^2$$
$$y = x^2 - 2$$
The inverse is $y^{-1} = x^2 - 2, x \geq 0$. Graphing each curve:

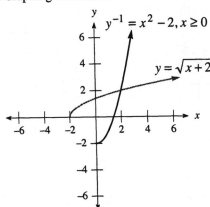

29. **a.** Yes, this function is one-to-one.   **b.** No, this function is not one-to-one.
      c. Yes, this function is one-to-one.

31. **a.** Evaluating the function: $f(2) = 3(2) - 2 = 6 - 2 = 4$

      b. Evaluating the function: $f^{-1}(2) = \dfrac{2+2}{3} = \dfrac{4}{3}$

      c. Evaluating the function: $f\left[f^{-1}(2)\right] = f\left(\frac{4}{3}\right) = 3\left(\frac{4}{3}\right) - 2 = 4 - 2 = 2$

      d. Evaluating the function: $f^{-1}\left[f(2)\right] = f^{-1}(4) = \dfrac{4+2}{3} = \dfrac{6}{3} = 2$

33. Let $y = f(x)$. Switch x and y and solve for y:
$$\frac{1}{y} = x$$
$$y = \frac{1}{x}$$
The inverse is $f^{-1}(x) = \frac{1}{x}$.

35. The inverse is $f^{-1}(x) = 7(x + 2)$.

37. **a.** The value is –3.   **b.** The value is –6.
    c. The value is 2.   **d.** The value is 3.
    e. The value is –2.   **f.** The value is 3.
    g. Each is an inverse of the other.

39. Solving the equation:
$$(2x - 1)^2 = 25$$
$$2x - 1 = \pm\sqrt{25}$$
$$2x - 1 = -5, 5$$
$$2x = -4, 6$$
$$x = -2, 3$$

41. The number is 25, since $x^2 - 10x + 25 = (x - 5)^2$.

43. Solving the equation:
$$x^2 - 10x + 8 = 0$$
$$x^2 - 10x + 25 = -8 + 25$$
$$(x - 5)^2 = 17$$
$$x - 5 = \pm\sqrt{17}$$
$$x = 5 \pm \sqrt{17}$$

45. Solving the equation:
$$3x^2 - 6x + 6 = 0$$
$$x^2 - 2x + 2 = 0$$
$$x^2 - 2x + 1 = -2 + 1$$
$$(x - 1)^2 = -1$$
$$x - 1 = \pm i$$
$$x = 1 \pm i$$

47. Finding the inverse:
$$3y + 5 = x$$
$$3y = x - 5$$
$$y = \frac{x - 5}{3}$$
So $f^{-1}(x) = \frac{x - 5}{3}$. Now verifying the inverse: $f\left[f^{-1}(x)\right] = f\left(\frac{x - 5}{3}\right) = 3\left(\frac{x - 5}{3}\right) + 5 = x - 5 + 5 = x$

49. Finding the inverse:
$$y^3 + 1 = x$$
$$y^3 = x - 1$$
$$y = \sqrt[3]{x - 1}$$
So $f^{-1}(x) = \sqrt[3]{x - 1}$. Now verifying the inverse: $f\left[f^{-1}(x)\right] = f\left(\sqrt[3]{x - 1}\right) = \left(\sqrt[3]{x - 1}\right)^3 + 1 = x - 1 + 1 = x$

51.   Finding the inverse:

$$\frac{y-4}{y-2} = x$$

$$y - 4 = xy - 2x$$

$$y - xy = 4 - 2x$$

$$y(1 - x) = 4 - 2x$$

$$y = \frac{4 - 2x}{1 - x} = \frac{2x - 4}{x - 1}$$

So $f^{-1}(x) = \dfrac{2x-4}{x-1}$. Now verifying the inverse:

$$f\left[f^{-1}(x)\right] = f\left(\frac{2x-4}{x-1}\right) = \frac{\dfrac{2x-4}{x-1} - 4}{\dfrac{2x-4}{x-1} - 2} = \frac{2x-4-4(x-1)}{2x-4-2(x-1)} = \frac{2x-4-4x+4}{2x-4-2x+2} = \frac{-2x}{-2} = x$$

53.   **a.**   From the graph: $f(0) = 1$ **b.**   From the graph: $f(1) = 2$

   c.   From the graph: $f(2) = 5$ **d.**   From the graph: $f^{-1}(1) = 0$

   e.   From the graph: $f^{-1}(2) = 1$ **f.**   From the graph: $f^{-1}(5) = 2$

   g.   From the graph: $f^{-1}[f(2)] = 2$ **h.**   From the graph: $f[f^{-1}(5)] = 5$

11.3   Logarithms Are Exponents

1.   Writing in logarithmic form: $\log_2 16 = 4$ 3.   Writing in logarithmic form: $\log_5 125 = 3$

5.   Writing in logarithmic form: $\log_{10} 0.01 = -2$ 7.   Writing in logarithmic form: $\log_2 \frac{1}{32} = -5$

9.   Writing in logarithmic form: $\log_{1/2} 8 = -3$ 11.   Writing in logarithmic form: $\log_3 27 = 3$

13.   Writing in exponential form: $10^2 = 100$ 15.   Writing in exponential form: $2^6 = 64$

17.   Writing in exponential form: $8^0 = 1$ 19.   Writing in exponential form: $10^{-3} = 0.001$

21.   Writing in exponential form: $6^2 = 36$ 23.   Writing in exponential form: $5^{-2} = \frac{1}{25}$

25.   Solving the equation:

$$\log_3 x = 2$$

$$x = 3^2 = 9$$

27.   Solving the equation:

$$\log_5 x = -3$$

$$x = 5^{-3} = \frac{1}{125}$$

29.   Solving the equation:

$$\log_2 16 = x$$

$$2^x = 16$$

$$x = 4$$

31.   Solving the equation:

$$\log_8 2 = x$$

$$8^x = 2$$

$$x = \frac{1}{3}$$

33.   Solving the equation:

$$\log_x 4 = 2$$

$$x^2 = 4$$

$$x = 2$$

35.   Solving the equation:

$$\log_x 5 = 3$$

$$x^3 = 5$$

$$x = \sqrt[3]{5}$$

37.   Sketching the graph:

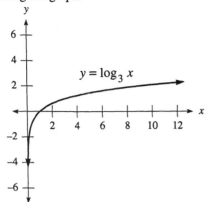

39.   Sketching the graph:

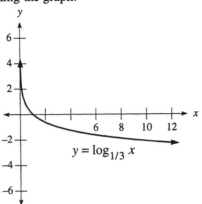

41.   Sketching the graph:

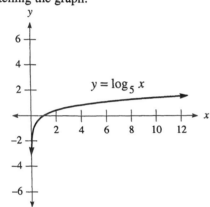

43.   Sketching the graph:

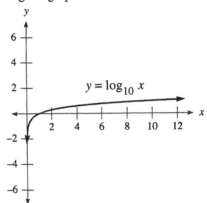

45.   Simplifying the logarithm:

$$x = \log_2 16$$
$$2^x = 16$$
$$x = 4$$

47.   Simplifying the logarithm:
$$x = \log_{25} 125$$
$$25^x = 125$$
$$5^{2x} = 5^3$$
$$2x = 3$$
$$x = \tfrac{3}{2}$$

49.   Simplifying the logarithm:
$$x = \log_{10} 1000$$
$$10^x = 1000$$
$$x = 3$$

51.   Simplifying the logarithm:
$$x = \log_3 3$$
$$3^x = 3$$
$$x = 1$$

53.   Simplifying the logarithm:
$$x = \log_5 1$$
$$5^x = 1$$
$$x = 0$$

55.   First find $\log_6 6$:
$$x = \log_6 6$$
$$6^x = 6$$
$$x = 1$$

Now find $\log_3 1$:
$$x = \log_3 1$$
$$3^x = 1$$
$$x = 0$$

57. First find $\log_2 16$:

$$x = \log_2 16$$
$$2^x = 16$$
$$x = 4$$

Now find $\log_4 2$:

$$x = \log_4 2$$
$$4^x = 2$$
$$2^{2x} = 2$$
$$2x = 1$$
$$x = \tfrac{1}{2}$$

Now find $\log_2 4$:

$$x = \log_2 4$$
$$2^x = 4$$
$$x = 2$$

59. The pH is given by: $\text{pH} = -\log_{10}\left(10^{-7}\right) = -(-7) = 7$

61. The $\left[H^+\right]$ is given by:

$$-\log_{10}\left[H^+\right] = 6$$
$$\log_{10}\left[H^+\right] = -6$$
$$\left[H^+\right] = 10^{-6}$$

63. Using the relationship $M = \log_{10} T$:

$$M = \log_{10} 100$$
$$10^M = 100$$
$$M = 2$$

65. It is 10^8 times as large.

67. Solving the equation:

$$2x^2 + 4x - 3 = 0$$
$$x = \frac{-4 \pm \sqrt{16 + 24}}{4} = \frac{-4 \pm \sqrt{40}}{4} = \frac{-4 \pm 2\sqrt{10}}{4} = \frac{-2 \pm \sqrt{10}}{2}$$

69. Solving the equation:

$$(2y - 3)(2y - 1) = -4$$
$$4y^2 - 8y + 3 = -4$$
$$4y^2 - 8y + 7 = 0$$
$$y = \frac{8 \pm \sqrt{64 - 112}}{8} = \frac{8 \pm \sqrt{-48}}{8} = \frac{8 \pm 4i\sqrt{3}}{8} = \frac{2 \pm i\sqrt{3}}{2}$$

71. Solving the equation:

$$t^3 - 125 = 0$$
$$(t - 5)\left(t^2 + 5t + 25\right) = 0$$
$$t = 5, \frac{-5 \pm \sqrt{25 - 100}}{2} = \frac{-5 \pm \sqrt{-75}}{2} = \frac{-5 \pm 5i\sqrt{3}}{2}$$

73. Solving the equation:

$$4x^5 - 16x^4 = 20x^3$$
$$4x^5 - 16x^4 - 20x^3 = 0$$
$$4x^3\left(x^2 - 4x - 5\right) = 0$$
$$4x^3(x - 5)(x + 1) = 0$$
$$x = -1, 0, 5$$

75. Solving the equation:

$$\frac{1}{x-3}+\frac{1}{x+2}=1$$
$$x+2+x-3=(x-3)(x+2)$$
$$2x-1=x^2-x-6$$
$$x^2-3x-5=0$$
$$x=\frac{3\pm\sqrt{9+20}}{2}=\frac{3\pm\sqrt{29}}{2}$$

77. **a.** Completing the table:

x	-1	0	1	2
$f(x)$	$\frac{1}{8}$	1	8	64

b. Completing the table:

x	$\frac{1}{8}$	1	8	64
$f^{-1}(x)$	-1	0	1	2

c. The equation is $f(x)=8^x$.

d. The equation is $f^{-1}(x)=\log_8 x$.

11.4 Properties of Logarithms

1. Using properties of logarithms: $\log_3 4x=\log_3 4+\log_3 x$

3. Using properties of logarithms: $\log_6 \dfrac{5}{x}=\log_6 5-\log_6 x$

5. Using properties of logarithms: $\log_2 y^5=5\log_2 y$

7. Using properties of logarithms: $\log_9 \sqrt[3]{z}=\log_9 z^{1/3}=\frac{1}{3}\log_9 z$

9. Using properties of logarithms: $\log_6 x^2 y^4=\log_6 x^2+\log_6 y^4=2\log_6 x+4\log_6 y$

11. Using properties of logarithms: $\log_5 \sqrt{x}\cdot y^4=\log_5 x^{1/2}+\log_5 y^4=\frac{1}{2}\log_5 x+4\log_5 y$

13. Using properties of logarithms: $\log_b \dfrac{xy}{z}=\log_b xy-\log_b z=\log_b x+\log_b y-\log_b z$

15. Using properties of logarithms: $\log_{10} \dfrac{4}{xy}=\log_{10} 4-\log_{10} xy=\log_{10} 4-\log_{10} x-\log_{10} y$

17. Using properties of logarithms: $\log_{10} \dfrac{x^2 y}{\sqrt{z}}=\log_{10} x^2+\log_{10} y-\log_{10} z^{1/2}=2\log_{10} x+\log_{10} y-\frac{1}{2}\log_{10} z$

19. Using properties of logarithms: $\log_{10} \dfrac{x^3 \sqrt{y}}{z^4}=\log_{10} x^3+\log_{10} y^{1/2}-\log_{10} z^4=3\log_{10} x+\frac{1}{2}\log_{10} y-4\log_{10} z$

21. Using properties of logarithms:

$$\log_b \sqrt[3]{\dfrac{x^2 y}{z^4}}=\log_b \dfrac{x^{2/3} y^{1/3}}{z^{4/3}}=\log_b x^{2/3}+\log_b y^{1/3}-\log_b z^{4/3}=\frac{2}{3}\log_b x+\frac{1}{3}\log_b y-\frac{4}{3}\log_b z$$

23. Writing as a single logarithm: $\log_b x+\log_b z=\log_b xz$

25. Writing as a single logarithm: $2\log_3 x-3\log_3 y=\log_3 x^2-\log_3 y^3=\log_3 \dfrac{x^2}{y^3}$

27. Writing as a single logarithm: $\frac{1}{2}\log_{10} x+\frac{1}{3}\log_{10} y=\log_{10} x^{1/2}+\log_{10} y^{1/3}=\log_{10} \sqrt{x}\sqrt[3]{y}$

29. Writing as a single logarithm: $3\log_2 x+\frac{1}{2}\log_2 y-\log_2 z=\log_2 x^3+\log_2 y^{1/2}-\log_2 z=\log_2 \dfrac{x^3 \sqrt{y}}{z}$

31. Writing as a single logarithm: $\frac{1}{2}\log_2 x-3\log_2 y-4\log_2 z=\log_2 x^{1/2}-\log_2 y^3-\log_2 z^4=\log_2 \dfrac{\sqrt{x}}{y^3 z^4}$

33. Writing as a single logarithm:

$$\frac{3}{2}\log_{10} x-\frac{3}{4}\log_{10} y-\frac{4}{5}\log_{10} z=\log_{10} x^{3/2}-\log_{10} y^{3/4}-\log_{10} z^{4/5}=\log_{10} \dfrac{x^{3/2}}{y^{3/4} z^{4/5}}$$

35. Solving the equation:

$$\log_2 x + \log_2 3 = 1$$
$$\log_2 3x = 1$$
$$3x = 2^1$$
$$3x = 2$$
$$x = \frac{2}{3}$$

37. Solving the equation:

$$\log_3 x - \log_3 2 = 2$$
$$\log_3 \frac{x}{2} = 2$$
$$\frac{x}{2} = 3^2$$
$$\frac{x}{2} = 9$$
$$x = 18$$

39. Solving the equation:

$$\log_3 x + \log_3 (x-2) = 1$$
$$\log_3 \left(x^2 - 2x\right) = 1$$
$$x^2 - 2x = 3^1$$
$$x^2 - 2x - 3 = 0$$
$$(x-3)(x+1) = 0$$
$$x = 3, -1$$

The solution is 3 (−1 does not check).

41. Solving the equation:

$$\log_3 (x+3) - \log_3 (x-1) = 1$$
$$\log_3 \frac{x+3}{x-1} = 1$$
$$\frac{x+3}{x-1} = 3^1$$
$$x+3 = 3x-3$$
$$-2x = -6$$
$$x = 3$$

43. Solving the equation:

$$\log_2 x + \log_2 (x-2) = 3$$
$$\log_2 \left(x^2 - 2x\right) = 3$$
$$x^2 - 2x = 2^3$$
$$x^2 - 2x - 8 = 0$$
$$(x-4)(x+2) = 0$$
$$x = 4, -2$$

The solution is 4 (−2 does not check).

45. Solving the equation:

$$\log_8 x + \log_8 (x-3) = \frac{2}{3}$$
$$\log_8 \left(x^2 - 3x\right) = \frac{2}{3}$$
$$x^2 - 3x = 8^{2/3}$$
$$x^2 - 3x - 4 = 0$$
$$(x-4)(x+1) = 0$$
$$x = 4, -1$$

The solution is 4 (−1 does not check).

47. Solving the equation:

$$\log_5 \sqrt{x} + \log_5 \sqrt{6x+5} = 1$$
$$\log_5 \sqrt{6x^2 + 5x} = 1$$
$$\frac{1}{2} \log_5 \left(6x^2 + 5x\right) = 1$$
$$\log_5 \left(6x^2 + 5x\right) = 2$$
$$6x^2 + 5x = 5^2$$
$$6x^2 + 5x - 25 = 0$$
$$(3x-5)(2x+5) = 0$$
$$x = \frac{5}{3}, -\frac{5}{2}$$

The solution is $\frac{5}{3}$ $\left(-\frac{5}{2}$ does not check$\right)$.

49. Solving for M: $M = 0.21 \log_{10} \dfrac{1}{10^{-12}} = 0.21 \log_{10} 10^{12} = 0.21(12) = 2.52$

51. Rewriting the expression: $\text{pH} = 6.1 + \log_{10} \left(\dfrac{x}{y}\right) = 6.1 + \log_{10} x - \log_{10} y$

53. Rewriting the formula:

$$D = 10 \log_{10} \left(\frac{I}{I_0}\right)$$
$$D = 10 \left(\log_{10} I - \log_{10} I_0\right)$$

55. Computing the discriminant: $D = (-5)^2 - 4(2)(4) = 25 - 32 = -7$. There are two complex solutions.

57. Writing the equation:
$$(x+3)(x-5) = 0$$
$$x^2 - 2x - 15 = 0$$

59. Writing the equation:
$$(3y-2)(y-3) = 0$$
$$3y^2 - 11y + 6 = 0$$

11.5 Common Logarithms and Natural Logarithms

1. Evaluating the logarithm: $\log 378 \approx 2.5775$

3. Evaluating the logarithm: $\log 37.8 \approx 1.5775$

5. Evaluating the logarithm: $\log 3,780 \approx 3.5775$

7. Evaluating the logarithm: $\log 0.0378 \approx -1.4225$

9. Evaluating the logarithm: $\log 37,800 \approx 4.5775$

11. Evaluating the logarithm: $\log 600 \approx 2.7782$

13. Evaluating the logarithm: $\log 2,010 \approx 3.3032$

15. Evaluating the logarithm: $\log 0.00971 \approx -2.0128$

17. Evaluating the logarithm: $\log 0.0314 \approx -1.5031$

19. Evaluating the logarithm: $\log 0.399 \approx -0.3990$

21. Solving for x:
$$\log x = 2.8802$$
$$x = 10^{2.8802} \approx 759$$

23. Solving for x:
$$\log x = -2.1198$$
$$x = 10^{-2.1198} \approx 0.00759$$

25. Solving for x:
$$\log x = 3.1553$$
$$x = 10^{3.1553} \approx 1,430$$

27. Solving for x:
$$\log x = -5.3497$$
$$x = 10^{-5.3497} \approx 0.00000447$$

29. Solving for x:
$$\log x = -7.0372$$
$$x = 10^{-7.0372} \approx 0.0000000918$$

31. Solving for x:
$$\log x = 10$$
$$x = 10^{10}$$

33. Solving for x:
$$\log x = -10$$
$$x = 10^{-10}$$

35. Solving for x:
$$\log x = 20$$
$$x = 10^{20}$$

37. Solving for x:
$$\log x = -2$$
$$x = 10^{-2} = \tfrac{1}{100}$$

39. Solving for x:
$$\log x = \log_2 8$$
$$\log x = 3$$
$$x = 10^3 = 1,000$$

41. Simplifying the logarithm: $\ln e = \ln e^1 = 1$

43. Simplifying the logarithm: $\ln e^5 = 5$

45. Simplifying the logarithm: $\ln e^x = x$

47. Using properties of logarithms: $\ln 10e^{3t} = \ln 10 + \ln e^{3t} = \ln 10 + 3t$

49. Using properties of logarithms: $\ln Ae^{-2t} = \ln A + \ln e^{-2t} = \ln A - 2t$

51. Evaluating the logarithm: $\ln 15 = \ln(3 \cdot 5) = \ln 3 + \ln 5 = 1.0986 + 1.6094 = 2.7080$

53. Evaluating the logarithm: $\ln \tfrac{1}{3} = \ln 3^{-1} = -\ln 3 = -1.0986$

55. Evaluating the logarithm: $\ln 9 = \ln 3^2 = 2 \ln 3 = 2(1.0986) = 2.1972$

57. Evaluating the logarithm: $\ln 16 = \ln 2^4 = 4 \ln 2 = 4(0.6931) = 2.7724$

59. Computing the pH: $\text{pH} = -\log(6.50 \times 10^{-4}) \approx 3.19$

61. Finding the concentration:
$$4.75 = -\log[\text{H}^+]$$
$$-4.75 = \log[\text{H}^+]$$
$$[\text{H}^+] = 10^{-4.75} \approx 1.78 \times 10^{-5}$$

63. Finding the magnitude:
$$5.5 = \log T$$
$$T = 10^{5.5} \approx 3.16 \times 10^5$$

65. Finding the magnitude:
$$8.3 = \log T$$
$$T = 10^{8.3} \approx 2.00 \times 10^8$$

66. Finding the magnitude:
$$8.7 = \log T$$
$$T = 10^{8.7} \approx 5.01 \times 10^8$$

67.   For the first earthquake:
$$\log T_1 = 6.5$$
$$T_1 = 10^{6.5}$$

For the second earthquake:
$$\log T_2 = 5.5$$
$$T_2 = 10^{5.5}$$

The ratio is $\dfrac{T_1}{T_2} = \dfrac{10^{6.5}}{10^{5.5}} = 10$ times stronger.

69.   Completing the table:

Location	Date	Magnitude (M)	Shockwave (T)
Moresby Island	January 23	4.0	1.00×10^4
Vancouver Island	April 30	5.3	1.99×10^5
Quebec City	June 29	3.2	1.58×10^3
Mould Bay	November 13	5.2	1.58×10^5
St. Lawrence	December 14	3.7	5.01×10^3

71.   Finding the rate of depreciation:
$$\log(1-r) = \tfrac{1}{5}\log\frac{4500}{9000}$$
$$\log(1-r) \approx -0.0602$$
$$1 - r \approx 10^{-0.0602}$$
$$r = 1 - 10^{-0.0602}$$
$$r \approx 0.129 = 12.9\%$$

73.   Finding the rate of depreciation:
$$\log(1-r) = \tfrac{1}{5}\log\frac{5750}{7550}$$
$$\log(1-r) \approx -0.0237$$
$$1 - r \approx 10^{-0.0237}$$
$$r = 1 - 10^{-0.0237}$$
$$r \approx 0.053 = 5.3\%$$

75.   It appears to approach e. Completing the table:

x	$(1+x)^{1/x}$
1	2
0.5	2.25
0.1	2.5937
0.01	2.7048
0.001	2.7169
0.0001	2.7181
0.00001	2.7183

77.   Solving the equation:
$$x^4 - 2x^2 - 8 = 0$$
$$\left(x^2 - 4\right)\left(x^2 + 2\right) = 0$$
$$x^2 = 4, -2$$
$$x = \pm 2, \pm i\sqrt{2}$$

79.   Solving the equation:
$$2x - 5\sqrt{x} + 3 = 0$$
$$\left(2\sqrt{x} - 3\right)\left(\sqrt{x} - 1\right) = 0$$
$$\sqrt{x} = 1, \tfrac{3}{2}$$
$$x = 1, \tfrac{9}{4}$$

11.6   Exponential Equations and Change of Base

1. Solving the equation:
$$3^x = 5$$
$$\ln 3^x = \ln 5$$
$$x \ln 3 = \ln 5$$
$$x = \frac{\ln 5}{\ln 3} \approx 1.4650$$

3. Solving the equation:
$$5^x = 3$$
$$\ln 5^x = \ln 3$$
$$x \ln 5 = \ln 3$$
$$x = \frac{\ln 3}{\ln 5} \approx 0.6826$$

5. Solving the equation:
$$5^{-x} = 12$$
$$\ln 5^{-x} = \ln 12$$
$$-x \ln 5 = \ln 12$$
$$x = -\frac{\ln 12}{\ln 5} \approx -1.5440$$

7. Solving the equation:
$$12^{-x} = 5$$
$$\ln 12^{-x} = \ln 5$$
$$-x \ln 12 = \ln 5$$
$$x = -\frac{\ln 5}{\ln 12} \approx -0.6477$$

9. Solving the equation:
$$8^{x+1} = 4$$
$$2^{3x+3} = 2^2$$
$$3x + 3 = 2$$
$$3x = -1$$
$$x = -\tfrac{1}{3}$$

11. Solving the equation:
$$4^{x-1} = 4$$
$$4^{x-1} = 4^1$$
$$x - 1 = 1$$
$$x = 2$$

13. Solving the equation:
$$3^{2x+1} = 2$$
$$\ln 3^{2x+1} = \ln 2$$
$$(2x+1)\ln 3 = \ln 2$$
$$2x + 1 = \frac{\ln 2}{\ln 3}$$
$$2x = \frac{\ln 2}{\ln 3} - 1$$
$$x = \tfrac{1}{2}\left(\frac{\ln 2}{\ln 3} - 1\right) \approx -0.1845$$

15. Solving the equation:
$$3^{1-2x} = 2$$
$$\ln 3^{1-2x} = \ln 2$$
$$(1-2x)\ln 3 = \ln 2$$
$$1 - 2x = \frac{\ln 2}{\ln 3}$$
$$-2x = \frac{\ln 2}{\ln 3} - 1$$
$$x = \tfrac{1}{2}\left(1 - \frac{\ln 2}{\ln 3}\right) \approx 0.1845$$

17. Solving the equation:
$$15^{3x-4} = 10$$
$$\ln 15^{3x-4} = \ln 10$$
$$(3x-4)\ln 15 = \ln 10$$
$$3x - 4 = \frac{\ln 10}{\ln 15}$$
$$3x = \frac{\ln 10}{\ln 15} + 4$$
$$x = \tfrac{1}{3}\left(\frac{\ln 10}{\ln 15} + 4\right) \approx 1.6168$$

19. Solving the equation:
$$6^{5-2x} = 4$$
$$\ln 6^{5-2x} = \ln 4$$
$$(5-2x)\ln 6 = \ln 4$$
$$5 - 2x = \frac{\ln 4}{\ln 6}$$
$$-2x = \frac{\ln 4}{\ln 6} - 5$$
$$x = \tfrac{1}{2}\left(5 - \frac{\ln 4}{\ln 6}\right) \approx 2.1131$$

21. Evaluating the logarithm: $\log_8 16 = \dfrac{\log 16}{\log 8} \approx 1.3333$

23. Evaluating the logarithm: $\log_{16} 8 = \dfrac{\log 8}{\log 16} = 0.7500$

25. Evaluating the logarithm: $\log_7 15 = \dfrac{\log 15}{\log 7} \approx 1.3917$

27. Evaluating the logarithm: $\log_{15} 7 = \dfrac{\log 7}{\log 15} \approx 0.7186$

29. Evaluating the logarithm: $\log_8 240 = \dfrac{\log 240}{\log 8} \approx 2.6356$

31. Evaluating the logarithm: $\log_4 321 = \dfrac{\log 321}{\log 4} \approx 4.1632$

33. Evaluating the logarithm: $\ln 345 \approx 5.8435$

35. Evaluating the logarithm: $\ln 0.345 \approx -1.0642$

37. Evaluating the logarithm: $\ln 10 \approx 2.3026$

39. Evaluating the logarithm: $\ln 45{,}000 \approx 10.7144$

41.   Using the compound interest formula:

$$500\left(1+\frac{0.06}{2}\right)^{2t}=1000$$

$$\left(1+\frac{0.06}{2}\right)^{2t}=2$$

$$\ln\left(1+\frac{0.06}{2}\right)^{2t}=\ln 2$$

$$2t\ln\left(1+\frac{0.06}{2}\right)=\ln 2$$

$$t=\frac{\ln 2}{2\ln\left(1+\frac{0.06}{2}\right)}\approx 11.72$$

It will take 11.72 years.

45.   Using the compound interest formula:

$$P\left(1+\frac{0.08}{4}\right)^{4t}=2P$$

$$\left(1+\frac{0.08}{4}\right)^{4t}=2$$

$$\ln\left(1+\frac{0.08}{4}\right)^{4t}=\ln 2$$

$$4t\ln\left(1+\frac{0.08}{4}\right)=\ln 2$$

$$t=\frac{\ln 2}{4\ln\left(1+\frac{0.08}{4}\right)}\approx 8.75$$

It will take 8.75 years.

49.   Using the continuous interest formula:

$$500e^{0.06t}=1000$$

$$e^{0.06t}=2$$

$$0.06t=\ln 2$$

$$t=\frac{\ln 2}{0.06}\approx 11.55$$

It will take 11.55 years.

43.   Using the compound interest formula:

$$1000\left(1+\frac{0.12}{6}\right)^{6t}=3000$$

$$\left(1+\frac{0.12}{6}\right)^{6t}=3$$

$$\ln\left(1+\frac{0.12}{6}\right)^{6t}=\ln 3$$

$$6t\ln\left(1+\frac{0.12}{6}\right)=\ln 3$$

$$t=\frac{\ln 3}{6\ln\left(1+\frac{0.12}{6}\right)}\approx 9.25$$

It will take 9.25 years.

47.   Using the compound interest formula:

$$25\left(1+\frac{0.06}{2}\right)^{2t}=75$$

$$\left(1+\frac{0.06}{2}\right)^{2t}=3$$

$$\ln\left(1+\frac{0.06}{2}\right)^{2t}=\ln 3$$

$$2t\ln\left(1+\frac{0.06}{2}\right)=\ln 3$$

$$t=\frac{\ln 3}{2\ln\left(1+\frac{0.06}{2}\right)}\approx 18.58$$

It was invested 18.58 years ago.

51.   Using the continuous interest formula:

$$500e^{0.06t}=1500$$

$$e^{0.06t}=3$$

$$0.06t=\ln 3$$

$$t=\frac{\ln 3}{0.06}\approx 18.31$$

It will take 18.31 years.

53.   Completing the square: $y=2x^2+8x-15=2\left(x^2+4x+4\right)-8-15=2(x+2)^2-23$. The lowest point is $(-2,-23)$.

55.   Completing the square: $y=12x-4x^2=-4\left(x^2-3x+\frac{9}{4}\right)+9=-4\left(x-\frac{3}{2}\right)^2+9$. The highest point is $\left(\frac{3}{2},9\right)$.

57.   Completing the square: $y=64t-16t^2=-16\left(t^2-4t+4\right)+64=-16(t-2)^2+64$

The object reaches a maximum height after 2 seconds, and the maximum height is 64 feet.

59.   Using the population model:

$$32000e^{0.05t}=64000$$

$$e^{0.05t}=2$$

$$0.05t=\ln 2$$

$$t=\frac{\ln 2}{0.05}\approx 13.9$$

The city will reach 64,000 toward the end of the year 2007.

61. Finding when $P(t) = 45,000$:

$$15,000e^{0.04t} = 45,000$$
$$e^{0.04t} = 3$$
$$0.04t = \ln 3$$
$$t = \frac{\ln 3}{0.04} \approx 27.5$$

It will take approximately 27.5 years.

65. Solving for t:

$$A = P2^{-kt}$$
$$2^{-kt} = \frac{A}{P}$$
$$\ln 2^{-kt} = \ln \frac{A}{P}$$
$$-kt \ln 2 = \ln A - \ln P$$
$$t = \frac{\ln A - \ln P}{-k \ln 2} = \frac{\ln P - \ln A}{k \ln 2}$$

63. Solving for t:

$$A = Pe^{rt}$$
$$e^{rt} = \frac{A}{P}$$
$$rt = \ln \frac{A}{P}$$
$$t = \frac{\ln A - \ln P}{r} = \frac{1}{r} \ln \frac{A}{P}$$

67. Solving for t:

$$A = P(1-r)^t$$
$$(1-r)^t = \frac{A}{P}$$
$$\ln(1-r)^t = \ln \frac{A}{P}$$
$$t \ln(1-r) = \ln A - \ln P$$
$$t = \frac{\ln A - \ln P}{\ln(1-r)}$$

Chapter 11 Review

1. Evaluating the function: $f(4) = 2^4 = 16$

3. Evaluating the function: $g(2) = \left(\frac{1}{3}\right)^2 = \frac{1}{9}$

5. Evaluating the function: $f(-1) + g(1) = 2^{-1} + \left(\frac{1}{3}\right)^1 = \frac{1}{2} + \frac{1}{3} = \frac{5}{6}$

7. Graphing the function:

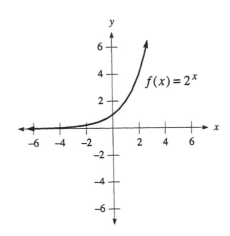

9.   Finding the inverse:
$$2y + 1 = x$$
$$2y = x - 1$$
$$y = \frac{x-1}{2}$$

The inverse is $y^{-1} = \dfrac{x-1}{2}$. Sketching the graph:

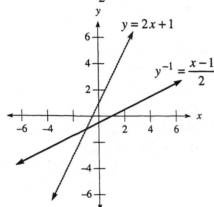

11.   Finding the inverse:
$$2y + 3 = x$$
$$2y = x - 3$$
$$y = \frac{x-3}{2}$$

The inverse is $f^{-1}(x) = \dfrac{x-3}{2}$.

13.   Finding the inverse:
$$\tfrac{1}{2}y + 2 = x$$
$$y + 4 = 2x$$
$$y = 2x - 4$$

The inverse is $f^{-1}(x) = 2x - 4$.

15.   Writing in logarithmic form: $\log_3 81 = 4$

17.   Writing in logarithmic form: $\log_{10} 0.01 = -2$

19.   Writing in exponential form: $2^3 = 8$

21.   Writing in exponential form: $4^{1/2} = 2$

23.   Solving for x:
$$\log_5 x = 2$$
$$x = 5^2 = 25$$

25.   Solving for x:
$$\log_x 0.01 = -2$$
$$x^{-2} = 0.01$$
$$x^2 = 100$$
$$x = 10$$

27.   Graphing the equation:

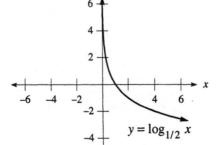

29. Simplifying the logarithm:
$$\log_{27} 9 = x$$
$$27^x = 9$$
$$3^{3x} = 3^2$$
$$3x = 2$$
$$x = \tfrac{2}{3}$$

31. Expanding the logarithm: $\log_2 5x = \log_2 5 + \log_2 x$

33. Expanding the logarithm: $\log_a \dfrac{y^3 \sqrt{x}}{z} = \log_a y^3 + \log_a x^{1/2} - \log_a z = 3\log_a y + \tfrac{1}{2}\log_a x - \log_a z$

35. Writing as a single logarithm: $\log_2 x + \log_2 y = \log_2 xy$

37. Writing as a single logarithm: $2\log_a 5 - \tfrac{1}{2}\log_a 9 = \log_a 5^2 - \log_a 9^{1/2} = \log_a 25 - \log_a 3 = \log_a \tfrac{25}{3}$

39. Solving the equation:
$$\log_2 x + \log_2 4 = 3$$
$$\log_2 4x = 3$$
$$4x = 2^3$$
$$4x = 8$$
$$x = 2$$

41. Solving the equation:
$$\log_3 x + \log_3 (x-2) = 1$$
$$\log_3 \left(x^2 - 2x\right) = 1$$
$$x^2 - 2x = 3^1$$
$$x^2 - 2x - 3 = 0$$
$$(x-3)(x+1) = 0$$
$$x = 3, -1$$
The solution is 3 (–1 does not check).

43. Solving the equation:
$$\log_6 (x-1) + \log_6 x = 1$$
$$\log_6 \left(x^2 - x\right) = 1$$
$$x^2 - x = 6^1$$
$$x^2 - x - 6 = 0$$
$$(x-3)(x+2) = 0$$
$$x = 3, -2$$
The solution is 3 (–2 does not check).

45. Evaluating: $\log 346 \approx 2.5391$

47. Solving for x:
$$\log x = 3.9652$$
$$x = 10^{3.9652} \approx 9,230$$

49. Simplifying: $\ln e = \ln e^1 = 1$

51. Simplifying: $\ln e^2 = 2$

53. Finding the pH: $\text{pH} = -\log\left(7.9 \times 10^{-3}\right) \approx 2.1$

55. Finding $\left[\text{H}^+\right]$:
$$2.7 = -\log\left[\text{H}^+\right]$$
$$-2.7 = \log\left[\text{H}^+\right]$$
$$\left[\text{H}^+\right] = 10^{-2.7} \approx 2.0 \times 10^{-3}$$

57. Solving the equation:
$$4^x = 8$$
$$2^{2x} = 2^3$$
$$2x = 3$$
$$x = \tfrac{3}{2}$$

59. Using a calculator: $\log_{16} 8 = \dfrac{\ln 8}{\ln 16} = 0.75$

61. Using the compound interest formula:

$$5000(1+0.16)^t = 10000$$

$$1.16^t = 2$$

$$\ln 1.16^t = \ln 2$$

$$t\ln 1.16 = \ln 2$$

$$t = \frac{\ln 2}{\ln 1.16} \approx 4.67$$

It will take approximately 4.67 years for the amount to double.

Chapters 1-11 Cumulative Review

1. Simplifying: $-8+2[5-3(-2-3)] = -8+2[5-3(-5)] = -8+2(5+15) = -8+2(20) = -8+40 = 32$

3. Simplifying: $\left(\frac{3}{5}\right)^{-2} - \left(\frac{3}{13}\right)^{-2} = \left(\frac{5}{3}\right)^2 - \left(\frac{13}{3}\right)^2 = \frac{25}{9} - \frac{169}{9} = -\frac{144}{9} = -16$

5. Simplifying: $\sqrt[3]{27x^4y^3} = \sqrt[3]{27x^3y^3 \cdot x} = 3xy\sqrt[3]{x}$

7. Simplifying: $[(6+2i)-(3-4i)]-(5-i) = (6+2i-3+4i)-(5-i) = 3+6i-5+i = -2+7i$

9. Simplifying: $1 + \dfrac{x}{1+\dfrac{1}{x}} = 1 + \dfrac{x}{1+\dfrac{1}{x}} \cdot \dfrac{x}{x} = 1 + \dfrac{x^2}{x+1} = \dfrac{x+1}{x+1} + \dfrac{x^2}{x+1} = \dfrac{x^2+x+1}{x+1}$

11. Multiplying: $\left(3t^2 + \frac{1}{4}\right)\left(4t^2 - \frac{1}{3}\right) = 12t^4 + t^2 - t^2 - \frac{1}{12} = 12t^4 - \frac{1}{12}$

13. Using long division:

$$
\begin{array}{r}
3x+7 \\
3x-4 \overline{\big)\,9x^2 + 9x - 18} \\
\underline{9x^2 - 12x} \\
21x - 18 \\
\underline{21x - 28} \\
10
\end{array}
$$

The quotient is $3x+7+\dfrac{10}{3x-4}$.

15. Subtracting:

$$\frac{7}{4x^2 - x - 3} - \frac{1}{4x^2 - 7x + 3} = \frac{7}{(4x+3)(x-1)} \cdot \frac{4x-3}{4x-3} - \frac{1}{(4x-3)(x-1)} \cdot \frac{4x+3}{4x+3}$$

$$= \frac{28x-21}{(4x+3)(4x-3)(x-1)} - \frac{4x+3}{(4x+3)(4x-3)(x-1)}$$

$$= \frac{24x-24}{(4x+3)(4x-3)(x-1)}$$

$$= \frac{24(x-1)}{(4x+3)(4x-3)(x-1)}$$

$$= \frac{24}{(4x+3)(4x-3)}$$

17. Solving the equation:

$$\frac{2}{3}(6x-5) + \frac{1}{3} = 13$$

$$2(6x-5) + 1 = 39$$

$$12x - 10 + 1 = 39$$

$$12x - 9 = 39$$

$$12x = 48$$

$$x = 4$$

19. Solving the equation:
$$|3x - 5| + 6 = 2$$
$$|3x - 5| = -4$$
Since this statement is impossible, there is no solution.

21. Solving the equation:
$$\frac{1}{x+3} + \frac{1}{x-2} = 1$$
$$(x+3)(x-2)\left(\frac{1}{x+3} + \frac{1}{x-2}\right) = (x+3)(x-2) \cdot 1$$
$$x - 2 + x + 3 = x^2 + x - 6$$
$$2x + 1 = x^2 + x - 6$$
$$x^2 - x - 7 = 0$$
$$x = \frac{1 \pm \sqrt{1+28}}{2} = \frac{1 \pm \sqrt{29}}{2}$$

23. Solving the equation:
$$2x - 1 = x^2$$
$$x^2 - 2x + 1 = 0$$
$$(x-1)^2 = 0$$
$$x = 1$$

25. Solving the equation:
$$\sqrt{7x - 4} = -2$$
$$7x - 4 = 4$$
$$7x = 8$$
$$x = \frac{8}{7}$$
Since $x = \frac{8}{7}$ does not check in the original equation, there is no solution.

27. Solving the equation:
$$x - 3\sqrt{x} + 2 = 0$$
$$\left(\sqrt{x} - 2\right)\left(\sqrt{x} - 1\right) = 0$$
$$\sqrt{x} = 1, 2$$
$$x = 1, 4$$

29. Solving the equation:
$$\log_3 x = 3$$
$$x = 3^3 = 27$$

31. Solving the equation:
$$\log_3(x-3) - \log_3(x+2) = 1$$
$$\log_3 \frac{x-3}{x+2} = 1$$
$$\frac{x-3}{x+2} = 3^1$$
$$x - 3 = 3x + 6$$
$$-2x = 9$$
$$x = -\frac{9}{2} \quad \text{(impossible)}$$
There is no solution.

33. Substituting into the first equation:
$$4(-2y - 2) + 7y = -3$$
$$-8y - 8 + 7y = -3$$
$$-y = 5$$
$$y = -5$$
$$x = -2(-5) - 2 = 8$$
The solution is $(8, -5)$.

35. Solving the inequality:
$$3y - 6 \geq 3 \qquad \text{or} \qquad 3y - 6 \leq -3$$
$$3y \geq 9 \qquad\qquad\qquad 3y \leq 3$$
$$y \geq 3 \qquad\qquad\qquad\quad y \leq 1$$
The solution set is $y \leq 1$ or $y \geq 3$. Graphing the solution set:

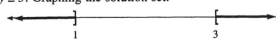

37.   Factoring the inequality:
$$x^3 + 2x^2 - 9x - 18 < 0$$
$$x^2(x+2) - 9(x+2) < 0$$
$$(x+2)(x^2 - 9) < 0$$
$$(x+2)(x+3)(x-3) < 0$$
Forming the sign chart:

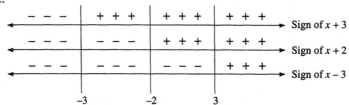

The solution set is $x < -3$ or $-2 < x < 3$. Graphing the solution set:

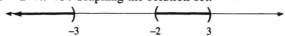

39.   Using the point-slope formula:
$$y - (-3) = -\tfrac{5}{3}(x - 3)$$
$$y + 3 = -\tfrac{5}{3}x + 5$$
$$y = -\tfrac{5}{3}x + 2$$

41.   Writing in scientific notation: $0.0000972 = 9.72 \times 10^{-5}$

43.   Finding the inverse:
$$\tfrac{1}{2}y + 3 = x$$
$$y + 6 = 2x$$
$$y = 2x - 6$$

The inverse function is $f^{-1}(x) = 2x - 6$.

45.   The domain is $\{-3, 2\}$ and the range is $\{-3, -1, 3\}$. This is not a function.

47.   The pattern is to multiply by –3, so the next number is $18(-3) = -54$. This is a geometric sequence.

Chapter 11 Test

1.   Graphing the function:

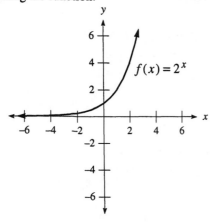

2.   Graphing the function:

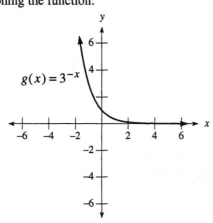

3.   Finding the inverse:
$$2y - 3 = x$$
$$2y = x + 3$$
$$y = \frac{x+3}{2}$$

The inverse is $f^{-1}(x) = \frac{x+3}{2}$. Sketching the graph:

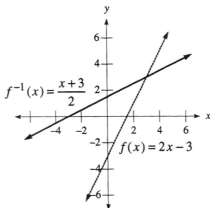

4.   Graphing the function and its inverse:

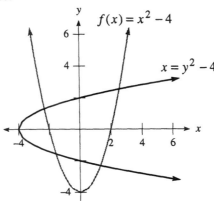

5.   Solving for x:

$$\log_4 x = 3$$
$$x = 4^3 = 64$$

6.   Solving for x:

$$\log_x 5 = 2$$
$$x^2 = 5$$
$$x = \sqrt{5}$$

7.   Graphing the function:

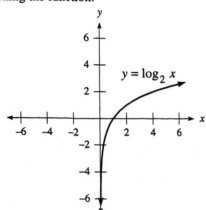

8.   Graphing the function:

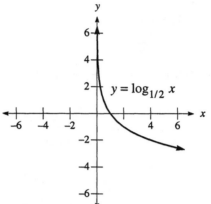

9.   Evaluating the logarithm:
$$x = \log_8 4$$
$$8^x = 4$$
$$2^{3x} = 2^2$$
$$3x = 2$$
$$x = \tfrac{2}{3}$$

10.   Evaluating the logarithm: $\log_7 21 = \dfrac{\ln 21}{\ln 7} \approx 1.5646$

11.   Evaluating the logarithm: $\log 23,400 \approx 4.3692$

12.   Evaluating the logarithm: $\log 0.0123 \approx -1.9101$

13.   Evaluating the logarithm: $\ln 46.2 \approx 3.8330$

14.   Evaluating the logarithm: $\ln 0.0462 \approx -3.0748$

15.   Expanding the logarithm: $\log_2 \dfrac{8x^2}{y} = \log_2 2^3 + \log_2 x^2 - \log_2 y = 3 + 2\log_2 x - \log_2 y$

16.   Expanding the logarithm: $\log \dfrac{\sqrt{x}}{y^4 \sqrt[5]{z}} = \log \dfrac{x^{1/2}}{y^4 z^{1/5}} = \log x^{1/2} - \log y^4 - \log z^{1/5} = \tfrac{1}{2}\log x - 4\log y - \tfrac{1}{5}\log z$

17.   Writing as a single logarithm: $2\log_3 x - \dfrac{1}{2}\log_3 y = \log_3 x^2 - \log_3 y^{1/2} = \log_3 \dfrac{x^2}{\sqrt{y}}$

18.   Writing as a single logarithm: $\tfrac{1}{3}\log x - \log y - 2\log z = \log x^{1/3} - \log y - \log z^2 = \log \dfrac{\sqrt[3]{x}}{yz^2}$

19.   Solving for x:
$$\log x = 4.8476$$
$$x = 10^{4.8476} \approx 70,404$$

20.   Solving for x:
$$\log x = -2.6478$$
$$x = 10^{-2.6478} \approx 0.00225$$

21.   Solving for x:
$$3^x = 5$$
$$\ln 3^x = \ln 5$$
$$x\ln 3 = \ln 5$$
$$x = \dfrac{\ln 5}{\ln 3} \approx 1.4650$$

22.   Solving for x:
$$4^{2x-1} = 8$$
$$2^{4x-2} = 2^3$$
$$4x - 2 = 3$$
$$4x = 5$$
$$x = \tfrac{5}{4}$$

23. Solving for x:

$$\log_5 x - \log_5 3 = 1$$
$$\log_5 \frac{x}{3} = 1$$
$$\frac{x}{3} = 5^1$$
$$x = 15$$

24. Solving for x:

$$\log_2 x + \log_2 (x - 7) = 3$$
$$\log_2 (x^2 - 7x) = 3$$
$$x^2 - 7x = 2^3$$
$$x^2 - 7x - 8 = 0$$
$$(x - 8)(x + 1) = 0$$
$$x = 8, -1$$

The solution is 8 (–1 does not check).

25. Finding the pH: $\text{pH} = -\log\left(6.6 \times 10^{-7}\right) \approx 6.18$

26. Using the compound interest formula: $A = 400\left(1 + \dfrac{0.10}{2}\right)^{2 \cdot 5} = 400(1.05)^{10} \approx 651.56$

There will be \$651.56 in the account after 5 years.

27. Using the compound interest formula:

$$600\left(1 + \frac{0.08}{4}\right)^{4t} = 1800$$
$$\left(1 + \frac{0.08}{4}\right)^{4t} = 3$$
$$\ln(1.02)^{4t} = \ln 3$$
$$4t \ln 1.02 = \ln 3$$
$$t = \frac{\ln 3}{4 \ln 1.02} \approx 13.87$$

It will take 13.87 years for the account to reach \$1,800.

Chapter 12
Conic Sections

12.1   The Circle

1. Using the distance formula: $d = \sqrt{(6-3)^2 + (3-7)^2} = \sqrt{9+16} = \sqrt{25} = 5$

3. Using the distance formula: $d = \sqrt{(5-0)^2 + (0-9)^2} = \sqrt{25+81} = \sqrt{106}$

5. Using the distance formula: $d = \sqrt{(-2-3)^2 + (1+5)^2} = \sqrt{25+36} = \sqrt{61}$

7. Using the distance formula: $d = \sqrt{(-10+1)^2 + (5+2)^2} = \sqrt{81+49} = \sqrt{130}$

9. Solving the equation:
$$\sqrt{(x-1)^2 + (2-5)^2} = \sqrt{13}$$
$$(x-1)^2 + 9 = 13$$
$$(x-1)^2 = 4$$
$$x - 1 = \pm 2$$
$$x - 1 = -2, 2$$
$$x = -1, 3$$

11. Solving the equation:
$$\sqrt{(7-8)^2 + (y-3)^2} = 1$$
$$(y-3)^2 + 1 = 1$$
$$(y-3)^2 = 0$$
$$y - 3 = 0$$
$$y = 3$$

13. The equation is $(x-2)^2 + (y-3)^2 = 16$.

15. The equation is $(x-3)^2 + (y+2)^2 = 9$.

17. The equation is $(x+5)^2 + (y+1)^2 = 5$.

19. The equation is $x^2 + (y+5)^2 = 1$.

21. The equation is $x^2 + y^2 = 4$.

23. The center is (0,0) and the radius is 2.

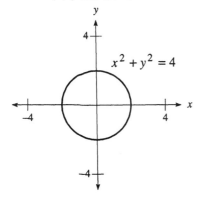

25. The center is (1,3) and the radius is 5.

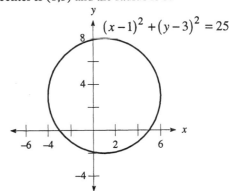

301

27.   The center is $(-2,4)$ and the radius is $2\sqrt{2}$.

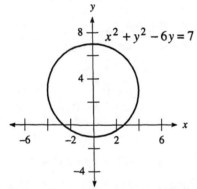

$(x+2)^2 +(y-4)^2 =8$

29.   The center is $(-1,-1)$ and the radius is 1.

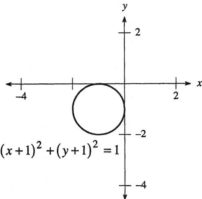

$(x+1)^2 +(y+1)^2 =1$

31.   Completing the square:
$$x^2 +y^2 -6y = 7$$
$$x^2 +\left(y^2 -6y+9\right) = 7+9$$
$$x^2 +(y-3)^2 = 16$$
The center is $(0,3)$ and the radius is 4.

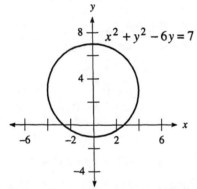

$x^2 +y^2 -6y = 7$

33.   Completing the square:
$$x^2 +y^2 -4x -6y = -4$$
$$\left(x^2 -4x+4\right)+\left(y^2 -6y+9\right) = -4+4+9$$
$$(x-2)^2 +(y-3)^2 = 9$$
The center is $(2,3)$ and the radius is 3.

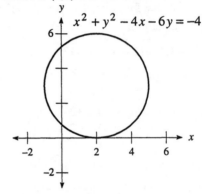

$x^2 +y^2 -4x -6y = -4$

35.   Completing the square:
$$x^2 +y^2 +2x+y = \tfrac{11}{4}$$
$$\left(x^2 +2x+1\right)+\left(y^2 +y+\tfrac{1}{4}\right) = \tfrac{11}{4}+1+\tfrac{1}{4}$$
$$(x+1)^2 +\left(y+\tfrac{1}{2}\right)^2 = 4$$
The center is $\left(-1,-\tfrac{1}{2}\right)$ and the radius is 2.

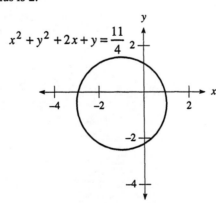

$x^2 +y^2 +2x+y = \dfrac{11}{4}$

37.   The equation is $(x-3)^2+(y-4)^2=25$.

39.   The equations are:

A: $\left(x-\frac{1}{2}\right)^2+(y-1)^2=\frac{1}{4}$

B: $(x-1)^2+(y-1)^2=1$

C: $(x-2)^2+(y-1)^2=4$

41.   The equations are:

A: $(x+8)^2+y^2=64$

B: $x^2+y^2=64$

C: $(x-8)^2+y^2=64$

43.   The x-coordinate of the center is $x=500$, the y-coordinate of the center is $12+120=132$, and the radius is 120. Thus the equation of the circle is $(x-500)^2+(y-132)^2=120^2=14,400$.

45.   Finding the radius of the circular path:

$$2\pi r=1005$$
$$r=\frac{1005}{2\pi}\approx 160$$

The equation of the circular path is $(x-160)^2+(y-160)^2=160^2=25,600$.

47.   Since the radius is 5 meters, the circumference is: $C=2\pi(5)=10\pi$ meters

49.   **a.**   Evaluating: $f(-2)=3^{-2}=\frac{1}{9}$ **b.**   Evaluating: $f(-1)=3^{-1}=\frac{1}{3}$

c.   Evaluating: $f(0)=3^0=1$ **d.**   Evaluating: $f(1)=3^1=3$

e.   Evaluating: $f(2)=3^2=9$

51.   Let $y=f(x)$. Switch x and y and solve for y:

$$2y+3=x$$
$$2y=x-3$$
$$y=\frac{x-3}{2}$$

The inverse is $f^{-1}(x)=\frac{x-3}{2}$.

53.   Let $y=f(x)$. Switch x and y and solve for y:

$$y^2-4=x$$
$$y^2=x+4$$
$$y=\pm\sqrt{x+4}$$

55.   The radius is 2, so the equation is $(x-2)^2+(y-3)^2=4$.

57.   The radius is 2, so the equation is $(x-2)^2+(y-3)^2=4$.

59.   Completing the square:

$$x^2+y^2-6x+8y=144$$
$$\left(x^2-6x+9\right)+\left(y^2+8y+16\right)=144+9+16$$
$$(x-3)^2+(y+4)^2=169$$

The center is $(3,-4)$, so the distance is: $d=\sqrt{(-3)^2+4^2}=\sqrt{9+16}=5$

61.   Completing the square:

$$x^2+y^2-6x-8y=144$$
$$\left(x^2-6x+9\right)+\left(y^2-8y+16\right)=144+9+16$$
$$(x-3)^2+(y-4)^2=169$$

The center is $(3,4)$, so the distance is: $d=\sqrt{3^2+4^2}=\sqrt{9+16}=5$

63.   The equation $y=\sqrt{9-x^2}$ corresponds to the top half of the circle, and the equation $y=-\sqrt{9-x^2}$ corresponds to the bottom half of the circle.

12.2   Ellipses and Hyperbolas

1.   Graphing the ellipse:

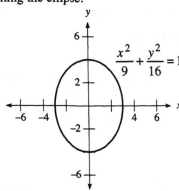

$$\frac{x^2}{9} + \frac{y^2}{16} = 1$$

3.   Graphing the ellipse:

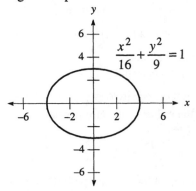

$$\frac{x^2}{16} + \frac{y^2}{9} = 1$$

5.   Graphing the ellipse:

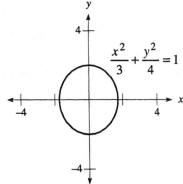

$$\frac{x^2}{3} + \frac{y^2}{4} = 1$$

7.   The standard form is $\frac{x^2}{25} + \frac{y^2}{4} = 1$. Graphing the ellipse:

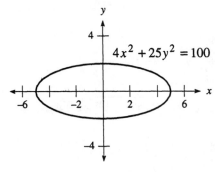

$$4x^2 + 25y^2 = 100$$

9.   The standard form is $\dfrac{x^2}{16} + \dfrac{y^2}{2} = 1$. Graphing the ellipse:

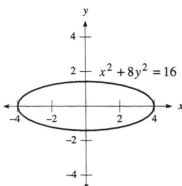

$x^2 + 8y^2 = 16$

11.   Graphing the hyperbola:

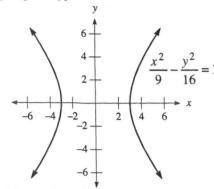

$\dfrac{x^2}{9} - \dfrac{y^2}{16} = 1$

13.   Graphing the hyperbola:

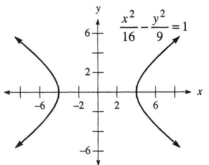

$\dfrac{x^2}{16} - \dfrac{y^2}{9} = 1$

15.   Graphing the hyperbola:

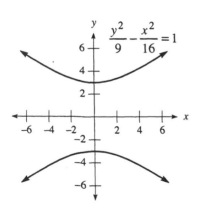

$\dfrac{y^2}{9} - \dfrac{x^2}{16} = 1$

17.   Graphing the hyperbola:

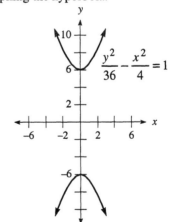

$\dfrac{y^2}{36} - \dfrac{x^2}{4} = 1$

19. The standard form is $\dfrac{x^2}{4} - \dfrac{y^2}{1} = 1$. Graphing the hyperbola:

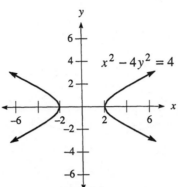

21. The standard form is $\dfrac{y^2}{9} - \dfrac{x^2}{16} = 1$. Graphing the hyperbola:

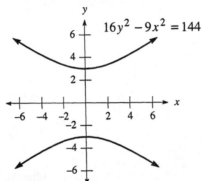

23. For the x-intercepts, set $y = 0$:
$$0.4x^2 = 3.6$$
$$x^2 = 9$$
$$x = \pm 3$$

For the y-intercepts, set $x = 0$:
$$0.9y^2 = 3.6$$
$$y^2 = 4$$
$$y = \pm 2$$

25. For the x-intercepts, set $y = 0$:
$$\frac{x^2}{0.04} = 1$$
$$x^2 = 0.04$$
$$x = \pm 0.2$$

For the y-intercepts, set $x = 0$:
$$-\frac{y^2}{0.09} = 1$$
$$y^2 = -0.09$$
There are no y-intercepts.

27. For the x-intercepts, set $y = 0$:
$$\frac{25x^2}{9} = 1$$
$$x^2 = \frac{9}{25}$$
$$x = \pm\frac{3}{5}$$

For the y-intercepts, set $x = 0$:
$$\frac{25y^2}{4} = 1$$
$$y^2 = \frac{4}{25}$$
$$y = \pm\frac{2}{5}$$

29. Substituting $x = 4$:

$$\frac{4^2}{25} + \frac{y^2}{9} = 1$$

$$\frac{y^2}{9} + \frac{16}{25} = 1$$

$$\frac{y^2}{9} = \frac{9}{25}$$

$$\frac{y}{3} = \pm\frac{3}{5}$$

$$y = \pm\frac{9}{5}$$

31. Substituting $x = 1.8$:

$$16(1.8)^2 + 9y^2 = 144$$

$$51.84 + 9y^2 = 144$$

$$9y^2 = 92.16$$

$$3y = \pm 9.6$$

$$y = \pm 3.2$$

33. The asymptotes are $y = \frac{3}{4}x$ and $y = -\frac{3}{4}x$.

35. The equation has the form $\frac{x^2}{16} + \frac{y^2}{b^2} = 1$. Substituting the point $\left(2, \sqrt{3}\right)$:

$$\frac{(2)^2}{16} + \frac{\left(\sqrt{3}\right)^2}{b^2} = 1$$

$$\frac{1}{4} + \frac{3}{b^2} = 1$$

$$\frac{3}{b^2} = \frac{3}{4}$$

$$b^2 = 4$$

The equation of the ellipse is $\frac{x^2}{16} + \frac{y^2}{4} = 1$.

37. The equation of the ellipse is $\frac{x^2}{20^2} + \frac{y^2}{10^2} = 1$. Substituting $y = 6$:

$$\frac{x^2}{20^2} + \frac{6^2}{10^2} = 1$$

$$\frac{x^2}{400} + \frac{9}{25} = 1$$

$$\frac{x^2}{400} = \frac{16}{25}$$

$$x^2 = 256$$

$$x = 16$$

The man can walk within 16 feet of the center.

39. Substituting $a = 4$ and $c = 3$:

$$4^2 = b^2 + 3^2$$

$$16 = b^2 + 9$$

$$b^2 = 7$$

$$b = \sqrt{7} \approx 2.65$$

The width should be approximately $2(2.65) = 5.3$ feet wide.

41. Writing in logarithmic form: $\log_{10} 100 = 2$

43. Writing in exponential form: $3^4 = 81$

45. Solving for x:

$$\log_9 x = \frac{3}{2}$$

$$x = 9^{3/2} = 27$$

47. Simplifying: $\log_2 32 = \log_2 2^5 = 5$

49. Simplifying: $\log_3 \left[\log_2 8\right] = \log_3 \left[\log_2 2^3\right] = \log_3 3 = 1$

51.   Finding the logarithm: $\log 576 \approx 2.7604$

53.   Finding the logarithm: $\log 0.0576 \approx -1.2396$

55.   Solving for x:
$$\log x = 2.6484$$
$$x = 10^{2.6484} \approx 445$$

57.   Solving for x:
$$\log x = -7.3516$$
$$x = 10^{-7.3516} \approx 4.45 \times 10^{-8}$$

12.3   Second-Degree Inequalities and Nonlinear Systems

1.   Graphing the inequality:

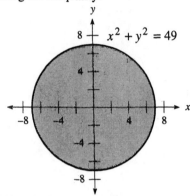

3.   Graphing the inequality:

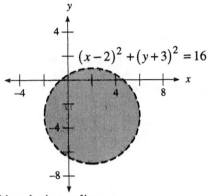

5.   Graphing the inequality:

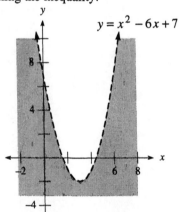

7.   Graphing the inequality:

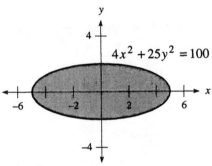

9.   Solving the second equation for y yields $y = 3 - 2x$. Substituting into the first equation:
$$x^2 + (3 - 2x)^2 = 9$$
$$x^2 + 9 - 12x + 4x^2 = 9$$
$$5x^2 - 12x = 0$$
$$x(5x - 12) = 0$$
$$x = 0, \tfrac{12}{5}$$
$$y = 3, -\tfrac{9}{5}$$
The solutions are $\left(0, 3\right), \left(\tfrac{12}{5}, -\tfrac{9}{5}\right)$.

11. Solving the second equation for x yields $x = 8 - 2y$. Substituting into the first equation:

$$(8 - 2y)^2 + y^2 = 16$$
$$64 - 32y + 4y^2 + y^2 = 16$$
$$5y^2 - 32y + 48 = 0$$
$$(y - 4)(5y - 12) = 0$$
$$y = 4, \tfrac{12}{5}$$
$$x = 0, \tfrac{16}{5}$$

The solutions are $(0, 4), \left(\tfrac{16}{5}, \tfrac{12}{5}\right)$.

13. Adding the two equations yields:

$$2x^2 = 50$$
$$x^2 = 25$$
$$x = -5, 5$$
$$y = 0$$

The solutions are $(-5, 0), (5, 0)$.

15. Substituting into the first equation:

$$x^2 + \left(x^2 - 3\right)^2 = 9$$
$$x^2 + x^4 - 6x^2 + 9 = 9$$
$$x^4 - 5x^2 = 0$$
$$x^2\left(x^2 - 5\right) = 0$$
$$x = 0, -\sqrt{5}, \sqrt{5}$$
$$y = -3, 2, 2$$

The solutions are $(0, -3), \left(-\sqrt{5}, 2\right), \left(\sqrt{5}, 2\right)$.

17. Substituting into the first equation:

$$x^2 + \left(x^2 - 4\right)^2 = 16$$
$$x^2 + x^4 - 8x^2 + 16 = 16$$
$$x^4 - 7x^2 = 0$$
$$x^2\left(x^2 - 7\right) = 0$$
$$x = 0, -\sqrt{7}, \sqrt{7}$$
$$y = -4, 3, 3$$

The solutions are $(0, -4), \left(-\sqrt{7}, 3\right), \left(\sqrt{7}, 3\right)$.

19. Substituting into the first equation:

$$3x + 2\left(x^2 - 5\right) = 10$$
$$3x + 2x^2 - 10 = 10$$
$$2x^2 + 3x - 20 = 0$$
$$(x + 4)(2x - 5) = 0$$
$$x = -4, \tfrac{5}{2}$$
$$y = 11, \tfrac{5}{4}$$

The solutions are $(-4, 11), \left(\tfrac{5}{2}, \tfrac{5}{4}\right)$.

21. Substituting into the first equation:

$$-x + 1 = x^2 + 2x - 3$$
$$x^2 + 3x - 4 = 0$$
$$(x + 4)(x - 1) = 0$$
$$x = -4, 1$$
$$y = 5, 0$$

The solutions are $(-4, 5), (1, 0)$.

23. Substituting into the first equation:

$$x - 5 = x^2 - 6x + 5$$
$$x^2 - 7x + 10 = 0$$
$$(x - 2)(x - 5) = 0$$
$$x = 2, 5$$
$$y = -3, 0$$

The solutions are $(2, -3), (5, 0)$.

25. Adding the two equations yields:

$$8x^2 = 72$$
$$x^2 = 9$$
$$x = \pm 3$$
$$y = 0$$

The solutions are $(-3, 0), (3, 0)$.

27.   Solving the first equation for x yields $x = y + 4$. Substituting into the second equation:

$$(y+4)^2 + y^2 = 16$$
$$y^2 + 8y + 16 + y^2 = 16$$
$$2y^2 + 8y = 0$$
$$2y(y+4) = 0$$
$$y = 0, -4$$
$$x = 4, 0$$

The solutions are $(0, -4)$, $(4, 0)$.

29.   **a.**   Subtracting the two equations yields:

$$(x+8)^2 - x^2 = 0$$
$$x^2 + 16x + 64 - x^2 = 0$$
$$16x = -64$$
$$x = -4$$

Substituting to find y:

$$(-4)^2 + y^2 = 64$$
$$y^2 + 16 = 64$$
$$y^2 = 48$$
$$y = \pm\sqrt{48} = \pm 4\sqrt{3}$$

The intersection points are $\left(-4, -4\sqrt{3}\right)$ and $\left(-4, 4\sqrt{3}\right)$.

    b.   Subtracting the two equations yields:

$$x^2 - (x-8)^2 = 0$$
$$x^2 - x^2 + 16x - 64 = 0$$
$$16x = 64$$
$$x = 4$$

Substituting to find y:

$$4^2 + y^2 = 64$$
$$y^2 + 16 = 64$$
$$y^2 = 48$$
$$y = \pm\sqrt{48} = \pm 4\sqrt{3}$$

The intersection points are $\left(4, -4\sqrt{3}\right)$ and $\left(4, 4\sqrt{3}\right)$.

31. Graphing the inequality:

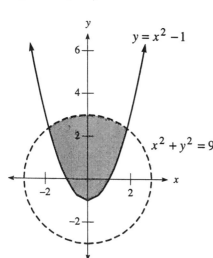

33. Graphing the inequality:

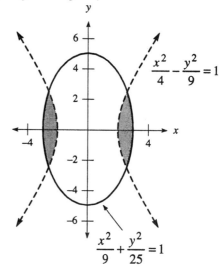

35. The shaded regions have no points in common.

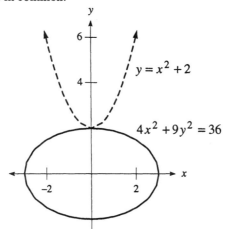

37. The system of inequalities is:
$$x^2 + y^2 < 16$$
$$y > 4 - \tfrac{1}{4}x^2$$

39. The system of equations is:
$$x^2 + y^2 = 89$$
$$x^2 - y^2 = 39$$
Adding the two equations yields:
$$2x^2 = 128$$
$$x^2 = 64$$
$$x = \pm 8$$
$$y = \pm 5$$
The numbers are either 8 and 5, 8 and −5, −8 and 5, or −8 and −5.

41. The system of equations is:
$$y = x^2 - 3$$
$$x + y = 9$$
Substituting into the second equation:
$$x + x^2 - 3 = 9$$
$$x^2 + x - 12 = 0$$
$$(x + 4)(x - 3) = 0$$
$$x = -4, 3$$
$$y = 13, 6$$
The numbers are either –4 and 13, or 3 and 6.

43. Expanding the logarithm: $\log_2 x^3 y = \log_2 x^3 + \log_2 y = 3\log_2 x + \log_2 y$

45. Writing as a single logarithm: $\log_{10} x - \log_{10} y^2 = \log_{10} \dfrac{x}{y^2}$

47. Solving the equation:
$$\log_4 x - \log_4 5 = 2$$
$$\log_4 \frac{x}{5} = 2$$
$$\frac{x}{5} = 4^2$$
$$\frac{x}{5} = 16$$
$$x = 80$$
The solution is 80.

49. Solving the equation:
$$5^x = 7$$
$$\log 5^x = \log 7$$
$$x \log 5 = \log 7$$
$$x = \frac{\log 7}{\log 5} \approx 1.21$$

51. Using the compound interest formula:
$$400\left(1 + \frac{0.10}{4}\right)^{4t} = 800$$
$$(1.025)^{4t} = 2$$
$$\ln(1.025)^{4t} = \ln 2$$
$$4t \ln 1.025 = \ln 2$$
$$t = \frac{\ln 2}{4 \ln 1.025} \approx 7.02$$
It will take 7.02 years.

53. Evaluating the logarithm: $\log_4 20 = \dfrac{\log 20}{\log 4} \approx 2.16$

55. Evaluating the logarithm: $\ln 576 \approx 6.36$

Chapter 12 Review

1. Using the distance formula: $d = \sqrt{(-1-2)^2 + (5-6)^2} = \sqrt{9+1} = \sqrt{10}$

3. Using the distance formula: $d = \sqrt{(-4-0)^2 + (0-3)^2} = \sqrt{16+9} = \sqrt{25} = 5$

5. Solving the equation:
$$\sqrt{(x-2)^2 + (-1+4)^2} = 5$$
$$(x-2)^2 + 9 = 25$$
$$(x-2)^2 = 16$$
$$x-2 = \pm\sqrt{16}$$
$$x-2 = -4, 4$$
$$x = -2, 6$$

7. The equation is $(x-3)^2 + (y-1)^2 = 4$.   9.   The equation is $(x+5)^2 + y^2 = 9$.

11. The equation is $x^2 + y^2 = 25$.

13. Finding the radius: $r = \sqrt{(-2-2)^2 + (3-0)^2} = \sqrt{16+9} = \sqrt{25} = 5$. The equation is $(x+2)^2 + (y-3)^2 = 25$.

15. The center is $(0,0)$ and the radius is 2.

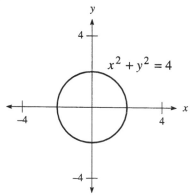

17. Completing the square:
$$x^2 + y^2 - 6x + 4y = -4$$
$$\left(x^2 - 6x + 9\right) + \left(y^2 + 4y + 4\right) = -4 + 9 + 4$$
$$(x-3)^2 + (y+2)^2 = 9$$
The center is $(3,-2)$ and the radius is 3.

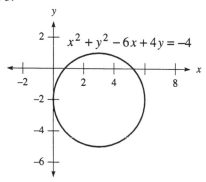

19.   Graphing the ellipse:

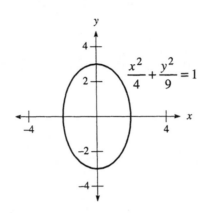

$$\frac{x^2}{4} + \frac{y^2}{9} = 1$$

21.   Graphing the hyperbola:

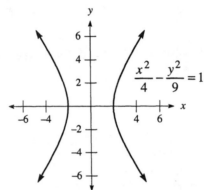

$$\frac{x^2}{4} - \frac{y^2}{9} = 1$$

23.   Graphing the inequality:

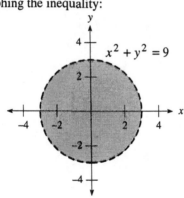

$$x^2 + y^2 = 9$$

25.   Graphing the inequality:

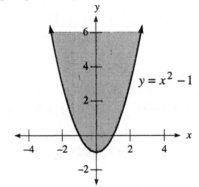

$$y = x^2 - 1$$

27. Graphing the solution set:

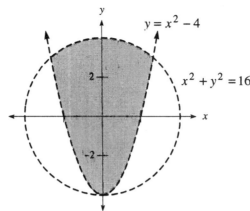

$y = x^2 - 4$

$x^2 + y^2 = 16$

29. Solving the second equation for y yields $y = 4 - 2x$. Substituting into the first equation:

$$x^2 + (4 - 2x)^2 = 16$$
$$x^2 + 16 - 16x + 4x^2 = 16$$
$$5x^2 - 16x = 0$$
$$x(5x - 16) = 0$$
$$x = 0, \frac{16}{5}$$
$$y = 4, -\frac{12}{5}$$

The solutions are $(0, 4), \left(\frac{16}{5}, -\frac{12}{5}\right)$.

31. Adding the two equations yields:

$$18x^2 = 72$$
$$x^2 = 4$$
$$x = \pm 2$$
$$y = 0$$

The solutions are $(-2, 0), (2, 0)$.

Chapters 1-12 Cumulative Review

1. Simplifying: $2^3 + 3(2 + 20 \div 4) = 8 + 3(2 + 5) = 8 + 3(7) = 8 + 21 = 29$

3. Simplifying: $-5(2x + 3) + 8x = -10x - 15 + 8x = -2x - 15$

5. Simplifying: $(3y + 2)^2 - (3y - 2)^2 = 9y^2 + 12y + 4 - 9y^2 + 12y - 4 = 24y$

7. Simplifying: $x^{2/3} \cdot x^{1/5} = x^{2/3 + 1/5} = x^{10/15 + 3/15} = x^{13/15}$

9. Solving the equation:

$$5y - 2 = -3y + 6$$
$$8y = 8$$
$$y = 1$$

11. Solving the equation:

$$|3x - 1| - 2 = 6$$
$$|3x - 1| = 8$$
$$3x - 1 = -8, 8$$
$$3x = -7, 9$$
$$x = -\frac{7}{3}, 3$$

13. Solving the equation:
$$x^3 - 3x^2 - 4x + 12 = 0$$
$$x^2(x-3) - 4(x-3) = 0$$
$$(x-3)(x^2 - 4) = 0$$
$$(x-3)(x+2)(x-2) = 0$$
$$x = -2, 2, 3$$

15. Solving the equation:
$$x - 2 = \sqrt{3x + 4}$$
$$(x-2)^2 = 3x + 4$$
$$x^2 - 4x + 4 = 3x + 4$$
$$x^2 - 7x = 0$$
$$x(x-7) = 0$$
$$x = 0, 7$$
The solution is 7 (0 does not check).

17. Solving the equation:
$$4x^2 + 6x = -5$$
$$4x^2 + 6x + 5 = 0$$
$$x = \frac{-6 \pm \sqrt{36 - 80}}{8} = \frac{-6 \pm \sqrt{-44}}{8} = \frac{-6 \pm 2i\sqrt{11}}{8} = -\frac{3}{4} \pm \frac{i\sqrt{11}}{4}$$

19. Solving the equation:
$$\log_2 x + \log_2 5 = 1$$
$$\log_2 5x = 1$$
$$5x = 2^1$$
$$x = \frac{2}{5}$$

21. Substituting into the first equation:
$$4x + 2(-3x + 1) = 4$$
$$4x - 6x + 2 = 4$$
$$-2x = 2$$
$$x = -1$$
$$y = 4$$
The solution is $(-1, 4)$.

23. Adding the first and third equations yields the equation $4y + z = 7$.
Multiply the third equation by 2 and add it to the second equation:
$$2x - y - 3z = -1$$
$$-2x + 4y + 4z = 6$$
Adding yields the equation $3y + z = 5$. So the system of equations becomes:
$$4y + z = 7$$
$$3y + z = 5$$
Multiply the second equation by -1:
$$4y + z = 7$$
$$-3y - z = -5$$
Adding yields $y = 2$. Substituting to find z:
$$3(2) + z = 5$$
$$6 + z = 5$$
$$z = -1$$
Substituting into the original first equation:
$$x + 2(2) - (-1) = 4$$
$$x + 5 = 4$$
$$x = -1$$
The solution is $(-1, 2, -1)$.

25. Multiplying: $\left(x^{1/5} + 3\right)\left(x^{1/5} - 3\right) = \left(x^{1/5}\right)^2 - (3)^2 = x^{2/5} - 9$

27. Dividing: $\dfrac{3 - 2i}{1 + 2i} = \dfrac{3 - 2i}{1 + 2i} \cdot \dfrac{1 - 2i}{1 - 2i} = \dfrac{3 - 8i + 4i^2}{1 - 4i^2} = \dfrac{3 - 8i - 4}{1 + 4} = \dfrac{-1 - 8i}{5} = -\dfrac{1}{5} - \dfrac{8}{5}i$

29. Graphing the inequality:

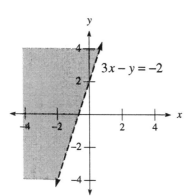

$3x - y = -2$

31. Graphing the curve:

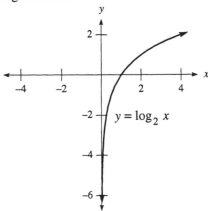

$y = \log_2 x$

33. The standard form is $\dfrac{x^2}{4} - \dfrac{y^2}{9} = 1$. Graphing the hyperbola:

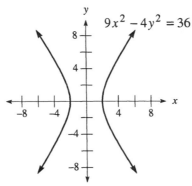

$9x^2 - 4y^2 = 36$

35. Graphing the parabola:

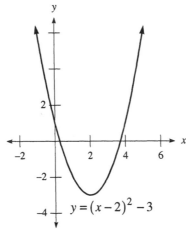

$y = (x-2)^2 - 3$

37. Factoring as a sum of cubes: $x^3 + 8y^3 = (x + 2y)(x^2 - 2xy + 4y^2)$

39. Finding the value: $f(4) = -\frac{3}{2}(4) + 1 = -6 + 1 = -5$

41. First find the slope: $m = \dfrac{-5+1}{-3+6} = -\dfrac{4}{3}$. Using the point-slope formula:

$$y + 5 = -\tfrac{4}{3}(x + 3)$$
$$y + 5 = -\tfrac{4}{3}x - 4$$
$$y = -\tfrac{4}{3}x - 9$$

43. Using the distance formula: $d = \sqrt{(4+3)^2 + (5-1)^2} = \sqrt{49 + 16} = \sqrt{65}$

45. Finding the function values:

$$f(2) = 4 - (2)^2 = 4 - 4 = 0$$
$$g\left(\tfrac{1}{5}\right) = 5\left(\tfrac{1}{5}\right) - 1 = 1 - 1 = 0$$

47. Let $s = 20$ and solve for t:

$$48t - 16t^2 = 20$$
$$16t^2 - 48t + 20 = 0$$
$$4t^2 - 12t + 5 = 0$$
$$(2t - 1)(2t - 5) = 0$$
$$t = \tfrac{1}{2}, \tfrac{5}{2}$$

The object will be 20 feet above the ground after $\tfrac{1}{2}$ second and $\tfrac{5}{2}$ seconds.

49. Factoring the inequality:

$$x^2 + x - 6 > 0$$
$$(x + 3)(x - 2) > 0$$

Forming the sign chart:

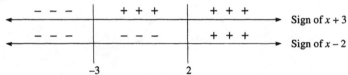

The solution set is $x < -3$ or $x > 2$. Graphing the solution set:

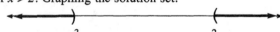

50. The domain is $\{x \mid x \neq 2\}$.

Chapter 12 Test

1. Solving the equation:
$$\sqrt{(x+1)^2 + (2-4)^2} = \left(2\sqrt{5}\right)^2$$
$$(x+1)^2 + 4 = 20$$
$$(x+1)^2 = 16$$
$$x+1 = \pm\sqrt{16}$$
$$x+1 = -4, 4$$
$$x = -5, 3$$

2. The equation is $(x+2)^2 + (y-4)^2 = 9$.

3. Finding the radius: $r = \sqrt{(-3)^2 + (-4)^2} = \sqrt{9+16} = \sqrt{25} = 5$. The equation is $x^2 + y^2 = 25$.

4. Completing the square:
$$x^2 + y^2 - 10x + 6y = 5$$
$$\left(x^2 - 10x + 25\right) + \left(y^2 + 6y + 9\right) = 5 + 25 + 9$$
$$(x-5)^2 + (y+3)^2 = 39$$
The center is $(5,-3)$ and the radius is $\sqrt{39}$.

5. The standard form is $\dfrac{x^2}{4} - \dfrac{y^2}{16} = 1$. Graphing the hyperbola:

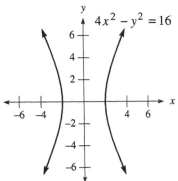

6. Graphing the ellipse:

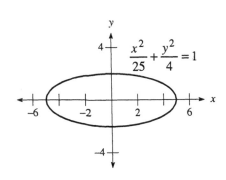

7. Graphing the inequality:

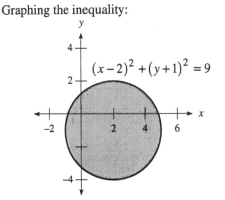

8.   Graphing the inequality:

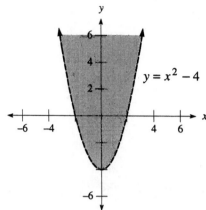

$$y = x^2 - 4$$

9.   Solving the first equation for y yields $y = 5 - 2x$. Substituting into the first equation:

$$x^2 + (5 - 2x)^2 = 25$$
$$x^2 + 25 - 20x + 4x^2 = 25$$
$$5x^2 - 20x = 0$$
$$5x(x - 4) = 0$$
$$x = 0, 4$$
$$y = 5, -3$$

The solutions are $(0,5)$, $(4,-3)$.

10.   Substituting into the first equation:

$$x^2 + \left(x^2 - 4\right)^2 = 16$$
$$x^2 + x^4 - 8x^2 + 16 = 16$$
$$x^4 - 7x^2 = 0$$
$$x^2\left(x^2 - 7\right) = 0$$
$$x = 0, \pm\sqrt{7}$$
$$y = -4, 3$$

The solutions are $(0,-4), \left(\sqrt{7},3\right), \left(-\sqrt{7},3\right)$.

Appendices

Appendix A   Synthetic Division

1. Using synthetic division:

$$\begin{array}{r|rrr} -2 & 1 & -5 & 6 \\ & & -2 & 14 \\ \hline & 1 & -7 & 20 \end{array}$$

The quotient is $x - 7 + \dfrac{20}{x + 2}$.

3. Using synthetic division:

$$\begin{array}{r|rrr} 1 & 3 & -4 & 1 \\ & & 3 & -1 \\ \hline & 3 & -1 & 0 \end{array}$$

The quotient is $3x - 1$.

5. Using synthetic division:

$$\begin{array}{r|rrrr} 2 & 1 & 2 & 3 & 4 \\ & & 2 & 8 & 22 \\ \hline & 1 & 4 & 11 & 26 \end{array}$$

The quotient is $x^2 + 4x + 11 + \dfrac{26}{x - 2}$.

7. Using synthetic division:

$$\begin{array}{r|rrrr} 3 & 3 & -1 & 2 & 5 \\ & & 9 & 24 & 78 \\ \hline & 3 & 8 & 26 & 83 \end{array}$$

The quotient is $3x^2 + 8x + 26 + \dfrac{83}{x - 3}$.

9. Using synthetic division:

$$\begin{array}{r|rrrr} 1 & 2 & 0 & 1 & -3 \\ & & 2 & 2 & 3 \\ \hline & 2 & 2 & 3 & 0 \end{array}$$

The quotient is $2x^2 + 2x + 3$.

11. Using synthetic division:

$$\begin{array}{r|rrrrr} -4 & 1 & 0 & 2 & 0 & 1 \\ & & -4 & 16 & -72 & 288 \\ \hline & 1 & -4 & 18 & -72 & 289 \end{array}$$

The quotient is $x^3 - 4x^2 + 18x - 72 + \dfrac{289}{x + 4}$.

13. Using synthetic division:

$$\begin{array}{r|rrrrrr} 2 & 1 & -2 & 1 & -3 & -1 & 1 \\ & & 2 & 0 & 2 & -2 & -6 \\ \hline & 1 & 0 & 1 & -1 & -3 & -5 \end{array}$$

The quotient is $x^4 + x^2 - x - 3 - \dfrac{5}{x - 2}$.

15. Using synthetic division:

$$\begin{array}{r|rrr} 1 & 1 & 1 & 1 \\ & & 1 & 2 \\ \hline & 1 & 2 & 3 \end{array}$$

The quotient is $x + 2 + \dfrac{3}{x - 1}$.

17. Using synthetic division:

$$\begin{array}{r|rrrrr} -1 & 1 & 0 & 0 & 0 & -1 \\ & & -1 & 1 & -1 & 1 \\ \hline & 1 & -1 & 1 & -1 & 0 \end{array}$$

The quotient is $x^3 - x^2 + x - 1$.

19.   Using synthetic division:

$$\underline{1}\rvert\quad \begin{array}{cccc} 1 & 0 & 0 & -1 \\ & 1 & 1 & 1 \\ \hline 1 & 1 & 1 & 0 \end{array}$$

The quotient is $x^2 + x + 1$.

Appendix B   Introduction to Determinants

1.   Evaluating the determinant: $\begin{vmatrix} 1 & 0 \\ 2 & 3 \end{vmatrix} = 1 \cdot 3 - 0 \cdot 2 = 3 - 0 = 3$

3.   Evaluating the determinant: $\begin{vmatrix} 2 & 1 \\ 3 & 4 \end{vmatrix} = 2 \cdot 4 - 1 \cdot 3 = 8 - 3 = 5$

5.   Evaluating the determinant: $\begin{vmatrix} 0 & 1 \\ 1 & 0 \end{vmatrix} = 0 \cdot 0 - 1 \cdot 1 = 0 - 1 = -1$

7.   Evaluating the determinant: $\begin{vmatrix} -3 & 2 \\ 6 & -4 \end{vmatrix} = (-3) \cdot (-4) - 6 \cdot 2 = 12 - 12 = 0$

9.   Solving the equation:

$$\begin{vmatrix} 2x & 1 \\ x & 3 \end{vmatrix} = 10$$
$$6x - x = 10$$
$$5x = 10$$
$$x = 2$$

11.   Solving the equation:

$$\begin{vmatrix} 1 & 2x \\ 2 & -3x \end{vmatrix} = 21$$
$$-3x - 4x = 21$$
$$-7x = 21$$
$$x = -3$$

13.   Solving the equation:

$$\begin{vmatrix} 2x & -4 \\ 2 & x \end{vmatrix} = -8x$$
$$2x^2 + 8 = -8x$$
$$2x^2 + 8x + 8 = 0$$
$$x^2 + 4x + 4 = 0$$
$$(x+2)^2 = 0$$
$$x = -2$$

15.   Solving the equation:

$$\begin{vmatrix} x^2 & 3 \\ x & 1 \end{vmatrix} = 10$$
$$x^2 - 3x = 10$$
$$x^2 - 3x - 10 = 0$$
$$(x-5)(x+2) = 0$$
$$x = -2, 5$$

17.   Duplicating the first two columns:

$$\begin{vmatrix} 1 & 2 & 0 \\ 0 & 2 & 1 \\ 1 & 1 & 1 \end{vmatrix}\begin{matrix} 1 & 2 \\ 0 & 2 \\ 1 & 1 \end{matrix} = 1 \cdot 2 \cdot 1 + 2 \cdot 1 \cdot 1 + 0 \cdot 0 \cdot 1 - 1 \cdot 2 \cdot 0 - 1 \cdot 1 \cdot 1 - 1 \cdot 0 \cdot 2 = 2 + 2 + 0 - 0 - 1 - 0 = 3$$

19.   Duplicating the first two columns:

$$\begin{vmatrix} 1 & 2 & 3 \\ 3 & 2 & 1 \\ 1 & 1 & 1 \end{vmatrix}\begin{matrix} 1 & 2 \\ 3 & 2 \\ 1 & 1 \end{matrix} = 1 \cdot 2 \cdot 1 + 2 \cdot 1 \cdot 1 + 3 \cdot 3 \cdot 1 - 1 \cdot 2 \cdot 3 - 1 \cdot 1 \cdot 1 - 1 \cdot 3 \cdot 2 = 2 + 2 + 9 - 6 - 1 - 6 = 0$$

21.   Expanding across the first row:

$$\begin{vmatrix} 0 & 1 & 2 \\ 1 & 0 & 1 \\ -1 & 2 & 0 \end{vmatrix} = 0\begin{vmatrix} 0 & 1 \\ 2 & 0 \end{vmatrix} - 1\begin{vmatrix} 1 & 1 \\ -1 & 0 \end{vmatrix} + 2\begin{vmatrix} 1 & 0 \\ -1 & 2 \end{vmatrix} = 0(0-2) - 1(0+1) + 2(2-0) = 0 - 1 + 4 = 3$$

23.   Expanding across the first row:

$$\begin{vmatrix} 3 & 0 & 2 \\ 0 & -1 & -1 \\ 4 & 0 & 0 \end{vmatrix} = 3\begin{vmatrix} -1 & -1 \\ 0 & 0 \end{vmatrix} - 0\begin{vmatrix} 0 & -1 \\ 4 & 0 \end{vmatrix} + 2\begin{vmatrix} 0 & -1 \\ 4 & 0 \end{vmatrix} = 3(0-0) - 0(0+4) + 2(0+4) = 0 - 0 + 8 = 8$$

25. Expanding across the first row: $\begin{vmatrix} 2 & -1 & 0 \\ 1 & 0 & -2 \\ 0 & 1 & 2 \end{vmatrix} = 2\begin{vmatrix} 0 & -2 \\ 1 & 2 \end{vmatrix} + 1\begin{vmatrix} 1 & -2 \\ 0 & 2 \end{vmatrix} + 0\begin{vmatrix} 1 & 0 \\ 0 & 1 \end{vmatrix} = 2(0+2) + 1(2-0) + 0 = 4 + 2 - 6$

27. Expanding across the first row:
$$\begin{vmatrix} 1 & 3 & 7 \\ -2 & 6 & 4 \\ 3 & 7 & -1 \end{vmatrix} = 1\begin{vmatrix} 6 & 4 \\ 7 & -1 \end{vmatrix} - 3\begin{vmatrix} -2 & 4 \\ 3 & -1 \end{vmatrix} + 7\begin{vmatrix} -2 & 6 \\ 3 & 7 \end{vmatrix} = 1(-6-28) - 3(2-12) + 7(-14-18) = -34 + 30 - 224 = -228$$

29. The determinant equation is:
$$\begin{vmatrix} y & x \\ m & 1 \end{vmatrix} = b$$
$$y - mx = b$$
$$y = mx + b$$

31. **a.** Writing the determinant equation:
$$\begin{vmatrix} x & -1.7 \\ 2 & 0.3 \end{vmatrix} = y$$
$$0.3x + 3.4 = y$$
$$y = 0.3x + 3.4$$

b. Substituting $x = 2$: $y = 0.3(2) + 3.4 = 0.6 + 3.4 = 4$ billion dollars

33. Substituting $x = 6$: $y = \begin{vmatrix} 0.1 & 6.9 \\ -2 & 6 \end{vmatrix} = 0.6 + 13.8 = 14.4$ million

Appendix C   Cramer's Rule

1. First find the determinants:
$$D = \begin{vmatrix} 2 & -3 \\ 4 & -2 \end{vmatrix} = -4 + 12 = 8$$
$$D_x = \begin{vmatrix} 3 & -3 \\ 10 & -2 \end{vmatrix} = -6 + 30 = 24$$
$$D_y = \begin{vmatrix} 2 & 3 \\ 4 & 10 \end{vmatrix} = 20 - 12 = 8$$

Now use Cramer's rule:
$$x = \frac{D_x}{D} = \frac{24}{8} = 3 \qquad y = \frac{D_y}{D} = \frac{8}{8} = 1$$

The solution is $(3,1)$.

3. First find the determinants:
$$D = \begin{vmatrix} 5 & -2 \\ -10 & 4 \end{vmatrix} = 20 - 20 = 0$$
$$D_x = \begin{vmatrix} 4 & -2 \\ 1 & 4 \end{vmatrix} = 16 + 2 = 18$$
$$D_y = \begin{vmatrix} 5 & 4 \\ -10 & 1 \end{vmatrix} = 5 + 40 = 45$$

Since $D = 0$ and other determinants are nonzero, there is no solution, or $\varnothing$.

5. First find the determinants:

$$D = \begin{vmatrix} 4 & -7 \\ 5 & 2 \end{vmatrix} = 8 + 35 = 43$$

$$D_x = \begin{vmatrix} 3 & -7 \\ -3 & 2 \end{vmatrix} = 6 - 21 = -15$$

$$D_y = \begin{vmatrix} 4 & 3 \\ 5 & -3 \end{vmatrix} = -12 - 15 = -27$$

Now use Cramer's rule:

$$x = \frac{D_x}{D} = -\frac{15}{43} \qquad y = \frac{D_y}{D} = -\frac{27}{43}$$

The solution is $\left(-\frac{15}{43}, -\frac{27}{43}\right)$.

7. First find the determinants:

$$D = \begin{vmatrix} 9 & -8 \\ 2 & 3 \end{vmatrix} = 27 + 16 = 43$$

$$D_x = \begin{vmatrix} 4 & -8 \\ 6 & 3 \end{vmatrix} = 12 + 48 = 60$$

$$D_y = \begin{vmatrix} 9 & 4 \\ 2 & 6 \end{vmatrix} = 54 - 8 = 46$$

Now use Cramer's rule:

$$x = \frac{D_x}{D} = \frac{60}{43} \qquad y = \frac{D_y}{D} = \frac{46}{43}$$

The solution is $\left(\frac{60}{43}, \frac{46}{43}\right)$.

9. First find the determinants:

$$D = \begin{vmatrix} 1 & 1 & 1 \\ 1 & -1 & -1 \\ 2 & 2 & -1 \end{vmatrix} = 1\begin{vmatrix} -1 & -1 \\ 2 & -1 \end{vmatrix} - 1\begin{vmatrix} 1 & -1 \\ 2 & -1 \end{vmatrix} + 1\begin{vmatrix} 1 & -1 \\ 2 & 2 \end{vmatrix} = 3 - 1 + 4 = 6$$

$$D_x = \begin{vmatrix} 4 & 1 & 1 \\ 2 & -1 & -1 \\ 2 & 2 & -1 \end{vmatrix} = 4\begin{vmatrix} -1 & -1 \\ 2 & -1 \end{vmatrix} - 1\begin{vmatrix} 2 & -1 \\ 2 & -1 \end{vmatrix} + 1\begin{vmatrix} 2 & -1 \\ 2 & 2 \end{vmatrix} = 12 - 0 + 6 = 18$$

$$D_y = \begin{vmatrix} 1 & 4 & 1 \\ 1 & 2 & -1 \\ 2 & 2 & -1 \end{vmatrix} = 1\begin{vmatrix} 2 & -1 \\ 2 & -1 \end{vmatrix} - 4\begin{vmatrix} 1 & -1 \\ 2 & -1 \end{vmatrix} + 1\begin{vmatrix} 1 & 2 \\ 2 & 2 \end{vmatrix} = 0 - 4 - 2 = -6$$

$$D_z = \begin{vmatrix} 1 & 1 & 4 \\ 1 & -1 & 2 \\ 2 & 2 & 2 \end{vmatrix} = 1\begin{vmatrix} -1 & 2 \\ 2 & 2 \end{vmatrix} - 1\begin{vmatrix} 1 & 2 \\ 2 & 2 \end{vmatrix} + 4\begin{vmatrix} 1 & -1 \\ 2 & 2 \end{vmatrix} = -6 + 2 + 16 = 12$$

Now use Cramer's rule:

$$x = \frac{D_x}{D} = \frac{18}{6} = 3 \qquad y = \frac{D_y}{D} = \frac{-6}{6} = -1 \qquad z = \frac{D_z}{D} = \frac{12}{6} = 2$$

The solution is $(3, -1, 2)$.

11. First find the determinants:

$$D = \begin{vmatrix} 1 & 1 & -1 \\ -1 & 1 & 1 \\ 1 & 1 & 1 \end{vmatrix} = 1\begin{vmatrix} 1 & 1 \\ 1 & 1 \end{vmatrix} - 1\begin{vmatrix} -1 & 1 \\ 1 & 1 \end{vmatrix} - 1\begin{vmatrix} -1 & 1 \\ 1 & 1 \end{vmatrix} = 0 + 2 + 2 = 4$$

$$D_x = \begin{vmatrix} 2 & 1 & -1 \\ 3 & 1 & 1 \\ 4 & 1 & 1 \end{vmatrix} = 2\begin{vmatrix} 1 & 1 \\ 1 & 1 \end{vmatrix} - 1\begin{vmatrix} 3 & 1 \\ 4 & 1 \end{vmatrix} - 1\begin{vmatrix} 3 & 1 \\ 4 & 1 \end{vmatrix} = 0 + 1 + 1 = 2$$

$$D_y = \begin{vmatrix} 1 & 2 & -1 \\ -1 & 3 & 1 \\ 1 & 4 & 1 \end{vmatrix} = 1\begin{vmatrix} 3 & 1 \\ 4 & 1 \end{vmatrix} - 2\begin{vmatrix} -1 & 1 \\ 1 & 1 \end{vmatrix} - 1\begin{vmatrix} -1 & 3 \\ 1 & 4 \end{vmatrix} = -1 + 4 + 7 = 10$$

$$D_z = \begin{vmatrix} 1 & 1 & 2 \\ -1 & 1 & 3 \\ 1 & 1 & 4 \end{vmatrix} = 1\begin{vmatrix} 1 & 3 \\ 1 & 4 \end{vmatrix} - 1\begin{vmatrix} -1 & 3 \\ 1 & 4 \end{vmatrix} + 2\begin{vmatrix} -1 & 1 \\ 1 & 1 \end{vmatrix} = 1 + 7 - 4 = 4$$

Now use Cramer's rule:

$$x = \frac{D_x}{D} = \frac{2}{4} = \frac{1}{2} \qquad y = \frac{D_y}{D} = \frac{10}{4} = \frac{5}{2} \qquad z = \frac{D_z}{D} = \frac{4}{4} = 1$$

The solution is $\left(\frac{1}{2}, \frac{5}{2}, 1\right)$.

13. First find the determinants:

$$D = \begin{vmatrix} 3 & -1 & 2 \\ 6 & -2 & 4 \\ 1 & -5 & 2 \end{vmatrix} = 3\begin{vmatrix} -2 & 4 \\ -5 & 2 \end{vmatrix} + 1\begin{vmatrix} 6 & 4 \\ 1 & 2 \end{vmatrix} + 2\begin{vmatrix} 6 & -2 \\ 1 & -5 \end{vmatrix} = 48 + 8 - 56 = 0$$

$$D_x = \begin{vmatrix} 4 & -1 & 2 \\ 8 & -2 & 4 \\ 1 & -5 & 2 \end{vmatrix} = 4\begin{vmatrix} -2 & 4 \\ -5 & 2 \end{vmatrix} + 1\begin{vmatrix} 8 & 4 \\ 1 & 2 \end{vmatrix} + 2\begin{vmatrix} 8 & -2 \\ 1 & -5 \end{vmatrix} = 64 + 12 - 76 = 0$$

$$D_y = \begin{vmatrix} 3 & 4 & 2 \\ 6 & 8 & 4 \\ 1 & 1 & 2 \end{vmatrix} = 3\begin{vmatrix} 8 & 4 \\ 1 & 2 \end{vmatrix} - 4\begin{vmatrix} 6 & 4 \\ 1 & 2 \end{vmatrix} + 2\begin{vmatrix} 6 & 8 \\ 1 & 1 \end{vmatrix} = 36 - 32 - 4 = 0$$

$$D_z = \begin{vmatrix} 3 & -1 & 4 \\ 6 & -2 & 8 \\ 1 & -5 & 1 \end{vmatrix} = 3\begin{vmatrix} -2 & 8 \\ -5 & 1 \end{vmatrix} + 1\begin{vmatrix} 6 & 8 \\ 1 & 1 \end{vmatrix} + 4\begin{vmatrix} 6 & -2 \\ 1 & -5 \end{vmatrix} = 114 - 2 - 112 = 0$$

Since $D = 0$ and the other determinants are also 0, there is no unique solution (dependent).

15. First find the determinants:

$$D = \begin{vmatrix} 2 & -1 & 3 \\ 1 & -5 & -2 \\ -4 & -2 & 1 \end{vmatrix} = 2\begin{vmatrix} -5 & -2 \\ -2 & 1 \end{vmatrix} + 1\begin{vmatrix} 1 & -2 \\ -4 & 1 \end{vmatrix} + 3\begin{vmatrix} 1 & -5 \\ -4 & -2 \end{vmatrix} = -18 - 7 - 66 = -91$$

$$D_x = \begin{vmatrix} 4 & -1 & 3 \\ 1 & -5 & -2 \\ 3 & -2 & 1 \end{vmatrix} = 4\begin{vmatrix} -5 & -2 \\ -2 & 1 \end{vmatrix} + 1\begin{vmatrix} 1 & -2 \\ 3 & 1 \end{vmatrix} + 3\begin{vmatrix} 1 & -5 \\ 3 & -2 \end{vmatrix} = -36 + 7 + 39 = 10$$

$$D_y = \begin{vmatrix} 2 & 4 & 3 \\ 1 & 1 & -2 \\ -4 & 3 & 1 \end{vmatrix} = 2\begin{vmatrix} 1 & -2 \\ 3 & 1 \end{vmatrix} - 4\begin{vmatrix} 1 & -2 \\ -4 & 1 \end{vmatrix} + 3\begin{vmatrix} 1 & 1 \\ -4 & 3 \end{vmatrix} = 14 + 28 + 21 = 63$$

$$D_z = \begin{vmatrix} 2 & -1 & 4 \\ 1 & -5 & 1 \\ -4 & -2 & 3 \end{vmatrix} = 2\begin{vmatrix} -5 & 1 \\ -2 & 3 \end{vmatrix} + 1\begin{vmatrix} 1 & 1 \\ -4 & 3 \end{vmatrix} + 4\begin{vmatrix} 1 & -5 \\ -4 & -2 \end{vmatrix} = -26 + 7 - 88 = -107$$

Now use Cramer's rule:

$$x = \frac{D_x}{D} = -\frac{10}{91} \qquad y = \frac{D_y}{D} = -\frac{63}{91} = -\frac{9}{13} \qquad z = \frac{D_z}{D} = \frac{-107}{-91} = \frac{107}{91}$$

The solution is $\left(-\frac{10}{91}, -\frac{9}{13}, \frac{107}{91}\right)$.

17.   First find the determinants:

$$D = \begin{vmatrix} -1 & -7 & 0 \\ 1 & 0 & 3 \\ 0 & 2 & 1 \end{vmatrix} = -1\begin{vmatrix} 0 & 3 \\ 2 & 1 \end{vmatrix} + 7\begin{vmatrix} 1 & 3 \\ 0 & 1 \end{vmatrix} + 0\begin{vmatrix} 1 & 0 \\ 0 & 2 \end{vmatrix} = 6 + 7 + 0 = 13$$

$$D_x = \begin{vmatrix} 1 & -7 & 0 \\ 11 & 0 & 3 \\ 0 & 2 & 1 \end{vmatrix} = 1\begin{vmatrix} 0 & 3 \\ 2 & 1 \end{vmatrix} + 7\begin{vmatrix} 11 & 3 \\ 0 & 1 \end{vmatrix} + 0\begin{vmatrix} 11 & 0 \\ 0 & 2 \end{vmatrix} = -6 + 77 + 0 = 71$$

$$D_y = \begin{vmatrix} -1 & 1 & 0 \\ 1 & 11 & 3 \\ 0 & 0 & 1 \end{vmatrix} = -1\begin{vmatrix} 11 & 3 \\ 0 & 1 \end{vmatrix} - 1\begin{vmatrix} 1 & 3 \\ 0 & 1 \end{vmatrix} + 0\begin{vmatrix} 1 & 11 \\ 0 & 0 \end{vmatrix} = -11 - 1 + 0 = -12$$

$$D_z = \begin{vmatrix} -1 & -7 & 1 \\ 1 & 0 & 11 \\ 0 & 2 & 0 \end{vmatrix} = -1\begin{vmatrix} 0 & 11 \\ 2 & 0 \end{vmatrix} + 7\begin{vmatrix} 1 & 11 \\ 0 & 0 \end{vmatrix} + 1\begin{vmatrix} 1 & 0 \\ 0 & 2 \end{vmatrix} = 22 + 0 + 2 = 24$$

Now use Cramer's rule:

$$x = \frac{D_x}{D} = \frac{71}{13} \qquad\qquad y = \frac{D_y}{D} = -\frac{12}{13} \qquad\qquad z = \frac{D_z}{D} = \frac{24}{13}$$

The solution is $\left(\frac{71}{13}, -\frac{12}{13}, \frac{24}{13}\right)$.

19.   First find the determinants:

$$D = \begin{vmatrix} 1 & -1 & 0 \\ 3 & 0 & 1 \\ 0 & 1 & -2 \end{vmatrix} = 1\begin{vmatrix} 0 & 1 \\ 1 & -2 \end{vmatrix} + 1\begin{vmatrix} 3 & 1 \\ 0 & -2 \end{vmatrix} + 0\begin{vmatrix} 3 & 0 \\ 0 & 1 \end{vmatrix} = -1 - 6 + 0 = -7$$

$$D_x = \begin{vmatrix} 2 & -1 & 0 \\ 11 & 0 & 1 \\ -3 & 1 & -2 \end{vmatrix} = 2\begin{vmatrix} 0 & 1 \\ 1 & -2 \end{vmatrix} + 1\begin{vmatrix} 11 & 1 \\ -3 & -2 \end{vmatrix} + 0\begin{vmatrix} 11 & 0 \\ -3 & 1 \end{vmatrix} = -2 - 19 + 0 = -21$$

$$D_y = \begin{vmatrix} 1 & 2 & 0 \\ 3 & 11 & 1 \\ 0 & -3 & -2 \end{vmatrix} = 1\begin{vmatrix} 11 & 1 \\ -3 & -2 \end{vmatrix} - 2\begin{vmatrix} 3 & 1 \\ 0 & -2 \end{vmatrix} + 0\begin{vmatrix} 3 & 11 \\ 0 & -3 \end{vmatrix} = -19 + 12 + 0 = -7$$

$$D_z = \begin{vmatrix} 1 & -1 & 2 \\ 3 & 0 & 11 \\ 0 & 1 & -3 \end{vmatrix} = 1\begin{vmatrix} 0 & 11 \\ 1 & -3 \end{vmatrix} + 1\begin{vmatrix} 3 & 11 \\ 0 & -3 \end{vmatrix} + 2\begin{vmatrix} 3 & 0 \\ 0 & 1 \end{vmatrix} = -11 - 9 + 6 = -14$$

Now use Cramer's rule:

$$x = \frac{D_x}{D} = \frac{-21}{-7} = 3 \qquad\qquad y = \frac{D_y}{D} = \frac{-7}{-7} = 1 \qquad\qquad z = \frac{D_z}{D} = \frac{-14}{-7} = 2$$

The solution is $(3, 1, 2)$.

21. First rewrite the system as:
$$-10x + y = 100$$
$$-12x + y = 0$$
Now find the determinants:
$$D = \begin{vmatrix} -10 & 1 \\ -12 & 1 \end{vmatrix} = -10 + 12 = 2$$
$$D_x = \begin{vmatrix} 100 & 1 \\ 0 & 1 \end{vmatrix} = 100 - 0 = 100$$
$$D_y = \begin{vmatrix} -10 & 100 \\ -12 & 0 \end{vmatrix} = 0 + 1200 = 1200$$
Now using Cramer's rule:
$$x = \frac{D_x}{D} = \frac{100}{2} = 50 \qquad\qquad y = \frac{D_y}{D} = \frac{1200}{2} = 600$$
The company must sell 50 items per week to break even.

23. First find the determinants:
$$D = \begin{vmatrix} -164.2 & 1 \\ 1 & 0 \end{vmatrix} = 0 - 1 = -1$$
$$D_x = \begin{vmatrix} 719 & 1 \\ 5 & 0 \end{vmatrix} = 0 - 5 = -5$$
$$D_H = \begin{vmatrix} -164.2 & 719 \\ 1 & 5 \end{vmatrix} = -821 - 719 = -1540$$
Now using Cramer's rule:
$$x = \frac{D_x}{D} = \frac{-5}{-1} = 5 \qquad\qquad H = \frac{D_H}{D} = \frac{-1540}{-1} = 1540$$
There were 1,540 heart transplants in the year 1990.

Appendix D   Systems of Linear Inequalities

1. Graphing the solution set:

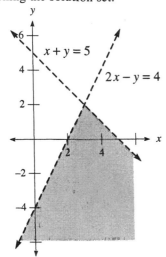

3. Graphing the solution set:

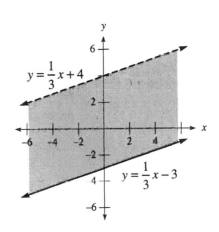

5.   Graphing the solution set:

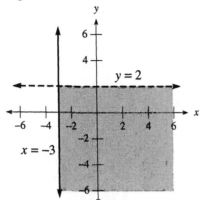

7.   Graphing the solution set:

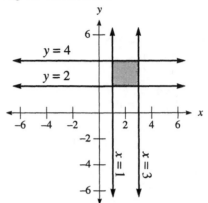

9.   Graphing the solution set:

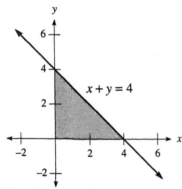

11.   Graphing the solution set:

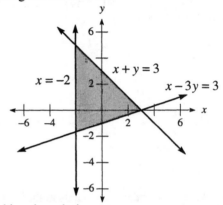

13.   Graphing the solution set:

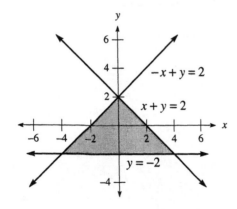

15.   Graphing the solution set:

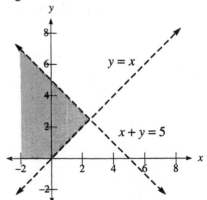

17. Graphing the solution set:

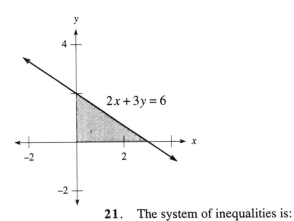

19. The system of inequalities is:
$$x + y \leq 4$$
$$-x + y < 4$$

21. The system of inequalities is:
$$x + y \geq 4$$
$$-x + y < 4$$

23. **a.** The system of inequalities is:
$$0.55x + 0.65y \leq 40$$
$$x \geq 2y$$
$$x > 15$$
$$y \geq 0$$

Graphing the solution set:

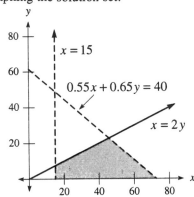

b. Substitute $x = 20$:
$$2y \leq 20$$
$$y \leq 10$$
The most he can purchase is 10 65-cent stamps.